Science and Engineering Ratios

Activation ratios	Displacement ratios	Point-hour r
Air-fuel ratio	Drive ratio	Poisson ratio
Air-liquid ratios	Effector-target ratios	Popped ratios
Air mixing ratios	Efficacy ratio	Pressure ratios
Amplitude ratios	Efficiency ratios	Pulley ratio
Area ratios	Electrooptical ratios	Radius ratio
Bedrock waves ratio	Engineering ratio	Ratio Analysis
Bentonite-sand ratios	Extension/compression ratio	Reinforcement ratios
Brewing ratios	Extraction ratios	Resistance ratios
C/N ratios	Gear ratios	Resolution ratio
Coefficient ratios	Great ratios	Risk ratios
Consolidation ratios	H/V Spectral ratios	Roof Drifts ratios
Chemical ratios	Isotope ratios	Reactivity ratios
Coal-to- air ratios	Likelihood ratio	Reduction ratio
Concentration ratios	Linear ratios	Reactivity ratios
Compression ratio	Male-Female ratios	Reduction ratio
Critical Mix ratios	Mass to charge ratio	Reflux ratio
Damping ratios	Mixing ratios	Settlement ratio
Deformation ratios	Mean ratio	Signal output ratio
Delivery ratios	Mole ratios	Signal-to-noise ratio
Density ratio	Odds ratios	Solubility ratios
Depth-area ratios	Oxygen ratio	Snow-air ratio
Dilution ratio	Power-weight ratios	Step-down ratios
Dimensionless ratios	Phase Distribution ratio	Stiffness-weight ratios
Direction ratios	Producing gas-oil ratio	

Science and Engineering Ratios (continued)	**Biology Ratios**	
Stoichiometric ratios	Birth Sex ratios	Mortality ratios
Surface area/volume ratio	Cell substrate/ adhesion ratio	Nurse-Patient ratios
Taguchi S/N ratios	Extinction ratio	Nutrient ratios
Tensile Strength ratios	Gender ratios	Offspring sex ratios
Torque-weight ratio	Gene expression ratio	Sex-ratio
Trigonometric ratios	Genotype/phenotype ratios	Sons/daughters ratio
Variable aspect ratio	Growth rates	Student/ faculty ratios
Water- cement ratios	Memory ratio	

Mathematics Ratios	
Cauchy's ratio Test	Golden ratio
$\cos\theta$	Inverse ratio
$\cot\theta$	Raabe's ratio Test
Cross ratio	Ratio of Similitude
Derivative	Ratio Test
Divine ratio	$\sin\theta$
Equivalent ratios	Slope

Business Ratios		
Accounting ratios	Debt Service ratio	Management ratios
Actual-Target Cost ratio	Debt/Equity ratio	Operating ratios
Bank Capital ratios	Digit ratios	Optimal ratios
Benefit/cost ratios	Effectiveness ratios	Parity ratio
Bidding ratios	Employment ratios	Payout ratio
Business ratio	Equilibrium ratios	PC ratio
Capital Market ratios	Equity ratios	Performance ratio
Cash Flow ratio	Equivalency ratios	Personnel ratios
Claims ratios	Export ratios	Price index
Company's P/E ratio	Financial ratios	Price ratios
competitor ratios	Growth rates	Price-sales ratio
Consumption ratios	Hedging ratios	Price/Earnings ratio
Contribution ratios	Host–guest ratios	Productivity ratios.
Cost ratios	Income ratios	Profile ratios
Credit ratios	Inflation ratio	Profitability ratios
Cross ratio	Interim ratios	Turnover ratio
Current ratios	Investment ratios.	
Damage ratios	L/B population ratios	
Debt/Credit ratio	Management efficiency ratios	

If you can remember a needed information, you make decisions faster, you learn faster, you work faster, and you are more productive.

Yes, you can memorize them.

Squares of Natural Numbers

1×1=1	26×26=676
2×2=4	27×27=729
3×3=9	28×28=784
4×4=16	29×29=841
5×5=25	30×30=900
6×6=36	31×31=961
7×7=49	32×32=1024
8×8=64	33×33=1089
9×9=81	34×34=1156
10×10=100	35×35=1225
11×11=121	36×36=1296
12×12=144	37×37=1369
13×13=169	38×38=1444
14×14=196	39×39=1521
15×15=225	40×40=1600
16×16=256	41×41=1681
17×17=289	42×42=1764
18×18=324	43×43=1849
19×19=361	44×44=1936
20×20=400	45×45=2025
21×21=441	46×46=2116
22×22=484	47×47=2209
23×23=529	48×48=2304
24×24=576	49×49=2401
25×25=625	50×50=2500

Conversion Factors for Measurements

American System (British System) — **Interc**onversion (Factors) — **Metric System**

Some "**bridges**" for converting from one system to the other

Length

American System:

- 12 inches (in) = 1 foot (ft.)
- 3 feet (ft.) = 1 yard (yd)
- 5280 feet = 1 mile (mi)
- 1760 yards = 1 mile

Interconversion:

- 1 in. = 2.54 cm
- 1 yd = 0.9144 m
- 1 mi = 1.61 km
- 1 km = 0.62 mi

Metric System:

- 1 kilometer (km) = 10^3 m = 1000 m
- 1 hectometer (hm) = 10^2 m = 100 m
- 1 dekameter (dam) = 10^1 m = 10 m
- **1 meter (m)** = 10^0 m = 1m
- 1 decimeter (dm) = 10^{-1} m = 0.1m
- 1 centimeter (cm) = 10^{-2} m = 0.01m
- 1 millimeter (mm) = 10^{-3} m = 0.001m

Some "**bridges**" for converting from one system to the other

Mass

American System:

- 1 *l*b = 16 oz
- 1 ton = 2000 *l*b
- 1 long ton = 2240 *l*b

Interconversion:

- 1 kg = 2.2 *l*b
- 1 *l*b = 454 g
- 1 oz = 28.4 g = 16 drams
- 1 ton = 0.9072 metric ton

Metric System:

- 1 kilogram (kg) = 10^3 g = 1000 g
- 1 hectogram (hg) = 10^2 g = 100 g
- 1 dekagram (dag) = 10^1 g = 10 g
- **1 gram (g)** = 10^0 g = 1 g
- 1 decigram (dg) = 10^{-1} g = 0.1g
- 1 centigram (cg)) = 10^{-2} g = 0.01g
- 1 milligram (mg) = 10^{-3} g = 0.001 g

Some "**bridges**" for converting from one system to the other

Volume

American System:

- 16 fluid oz (fl-oz) = 1 pint (pt)
- 2 pints (pt) = 1 quart (qt)
- 4 quarts = 1 gallon (gal)

Interconversion:

- 1 liter (*l*) = 1.057 qt
- 1 gal = 3.785 *l*
- 1 liter = 2.1 pt
- 1 pt = .473 *l*

Metric System:

- 1 kiloliter (k*l*) = 10^3 *l* = 1000 *l*
- 1 hectoliter (h*l*) = 10^2 *l* = 100 *l*
- 1 dekaliter (da*l*) = 10^1 *l* = 10 *l*
- **1 liter (*l*)** = 10^0 *l* = 1 *l*
- 1 deciliter (d*l*) = 10^{-1} *l* = 0.1*l*
- 1 centiliter (c*l*) = 10^{-2} *l* = 0.01*l*
- 1 milliliter (m*l*) = 10^{-3} *l* = 0.001 *l*

Must remember the following (metric system:)

- 100 cm = 1 m
- 1000 m = 1 km

- 1000 mg = 1 g
- 1000 g = 1 kg

- 1000 m*l* = 1 *l*
- 1 m*l* = 1 cc = 1cm^3
- 1000 cc = 1 *l*

Mnemonic device (metric system)

- k – ilo – 10^3
- h – ecto – 10^2
- d – eka – 10^1
- d – eci – 10^{-1}
- c – enti – 10^{-2}
- m – illi – 10^{-3}

Say the following aloud:

Step 1: First go down vertically as kei-eitch-dii-dii-see-em, then Step 2

Step 2: Kilo-hecto-deka-deci-centi-milli, and then note how the powers decrease vertically downwards.

Examples: 1 Kilometer = 10^3 meter; 1 milligram = 10^{-3} gram;

1 centimeter = 10^{-2} meter = $\frac{1}{100}$ meter --->100 centimeters = 1 meter.

Power of Ratios

Second Edition

A.A.FREMPONG

Power of Ratios

Published by Yellowtextbooks.com. N.Y.

ISBN 978-1-946485-12-0

Printed in the United States of America

Faculty/Student Suggestions for Future Editions

Send Suggestions to the Publisher

Copy and complete the following. This survey will help improve future editions of this book.

1. Which Topics would you like to be added to future Editions? You may include sample problems (and solutions).

....
....
....
....
....
....
....
....

2. Which Topics in this book do you think need more or better coverage?.

....
....
....
....
....
....
....
....
....
....
....

3. Which coverage in this book impressed you most?

....
....
....
....
....
....
....
....
....
....
....

. How useful was this book in preparing for and taking the Final Exam?

Check one: A (Excellent) ; B (Good) ; C (Average) ; D (Fair)

In Memory of My Parents

Mom:
She was a devoted mother, sharing, kind, kinder to strangers and generous to a fault.She never cursed, she never hated; she never cheated, and she never envied. She never lied, and she never got angry. Once, she nursed an almost dying stranger renting a room in her house back to good health to the extent that the relatives of this renter later travelled one hundred miles just to thank mom. She was always peaceloving and forever forgiving. An angel once lived on this earth to serve others.

Dad:
A great dad, kind, generous and forgiving. He emphasized and was an example of both formal education and self-education. A veterinarian, a bacteriologist, an Associate of the Institute of Medical Laboratory Technology (UK), a Fellow of the Royal Society of Health (UK); an incorruptible civil servant; his book on ticks has always inspired me to write whenever the need arises.

NOTE TO THE STUDENT

Applications of Ratio and Proportion

Students in the following areas should master ratios and proportion:
Mathematics, science, engineering, building construction, nursing, respiratory therapy, radiology, nuclear medicine, and similar heath programs; economics and business programs; carpentry; arts and crafts; mural painting and food preparation, and anyone who owns a business.

In engineering design work, if you want to go from a small scale to a larger scale or from a larger scale to a smaller scale, the relationship to apply is that of **proportion.**
In engineering: laws of similarity are very useful. **Model-prototype** relationships have proven to be reliable, and many structures constructed from the tested models have performed well.

Once you have mastered ratios and proportion, you can readily go from one field of application to other fields of applications. Only the names of the terms would be different; but you would be applying the same principles.

Who should have a copy this book?
Every student, every teacher, every house, and every library should have a copy of this book. Why write a book on ratio and proportion only? Am I crazy? No. Just check the contents of all the books on math, science, engineering, economics and business.

The author recommends that every high school should offer a mandatory course entitled "Ratio and Proportion" during the senior year of high school. The content of the course should include applications from various fields.

PS
The manuscript for this book had been ready for publication for at least ten years, but I had been procrastinating its publication, until after working on the **Navier-Stokes equations** of fluid mechanics, I was able to split-up and pair the terms of the equations using ratios. Subsequently, I obtained a solution which checked well in the Navier-Stokes equations. It therefore became appropriate to publish my solutions of the 3-D Navier-Stokes equations in a book with the title "Power of Ratios" since without ratios, I could not split the equations. Splitting-up the equations was critical to solving the equations, and I think the main impediment that has prevented the solutions of these equations in 3-D for over 100 years has been finding a way to split-up the equations. After finding the solutions in 3-D, the author readily wrote the solutions in 4-D. Following the solutions of the Navier-Stokes equations, the hitherto unsolved **Magnetohydrodynamic equation**s were routinely solved. Some months later, I was able to solve another CMI prize problem, P vs NP, for which one was to show either P is equal to NP or P is not equl to P, After solving NP problems and reviewing the solution method, I realized that I had applied a ratio process in solving the NP problems. Another pleasant ratio surprise was in proving Beal Conjecture and Fermat's Last Theorem. The necessary conditions used in proving Beal Conjecture and Fermat's were based on the ratios $(a^x + b^y) : c^z = 1$ and $(a^n + b^n) : c^n = 1$, respectively. Therefore, I decided to include the solutions and proofs in this book. "Power of Ratios".

Books in the series by the author:
Power of Ratios; Integrated Arithmetic; Elementary Algebra; Intermediate Algebra, Elementary Mathematics; Intermediate Mathematics; Elementary & Intermediate Mathematics (combined); College Algebra; College Trigonometry; College Algebra & Trigonometry and Calculus 1 & 2., There are also available Final Exam Review titles for the above titles, but do not buy both.

PREFACE

The original title of this book was to be "Ratios and Proportion", but as I covered the applications of ratios and proportion, it became apparent that "The Power of Ratios" is the more appropriate title, since this title allows me to include topics such as trigonometric ratios, compound interest, fractions, decimals, percent problems, definitions in chemistry, physics, engineering design and business ratios.

A number of concerns have been brewing in my mind over the years regarding "ratios and proportion

One such concern has been that because there are several methods for solving direct proportion problems, at times, I debate as to which method(s) to cover (first) in class. In the various departments of schools and colleges, some methods are preferred: for example, in some departments, "the factor label method" or "the units label method" or the "dimensional analysis" and arithmetic methods are preferred for solving some direct proportion problems (especially, in measurement conversions and stoichiometric calculations); while in some departments, "the ratio = ratio" method is popular". I can recall that a previous student of mine in a nursing program confided to me that on one exam in pharmacology, even though, she had correctly solved a direct proportion problem using a legitimate arithmetic method, the instructor refused to accept the solution, because the instructor had a policy of accepting only the " ratio = ratio" method. This dispute was referred to the mathematics department of the college for resolution. I found her arithmetic method to be very powerful in the hands of anyone who has mastered it, because it can be applied to all the various types of proportion problems, including inverse and compound proportion problems.

Another concern of mine has also been that in most textbooks, "**variation methods**" are usually not covered at the elementary level and when covered at the intermediate level (US), it is covered as a separate chapter apart from proportion, instead of covering these methods amongst the methods for solving proportion problems, since problems stated in terms of proportion can be stated in terms of variation and conversely.

A third concern has been that in recent times, most of the elementary level textbooks around cover **direct proportion** or proportion problems and completely ignore indirect or **inverse proportion** problems; and consequently, the word proportion has more or less been reserved for direct proportion. This should not be the case, since inverse proportion problems are also part of everyday life, and they are around us, whether or not we like them. Even though, it seems that there are more everyday direct proportion problems than inverse proportion problems, a school graduate should also be able to solve the relatively few everyday inverse proportion problems that will surely crop up from time to time. In the technical fields, inverse proportion problems are more or less as common as direct proportion problems. If textbooks continue to ignore everyday inverse problems, students would be graduating from school or college and would not be able to solve simple problems such as:

1. If 2 people can paint a house in 6 days, how many people would paint the house in 3 days?

2. At a speed of 60 miles per hour, a car takes 2 hours to travel from City A to City B. At a speed of 40 miles per hour, how long will it take the car to travel from City A to City B?

A comprehensive coverage of all the various methods (five methods), including variation methods for solving simple direct proportion, inverse proportion, and compound proportion (joint proportion, joint inverse proportion and combined proportion) is presented. In having all these methods in one place, one can compare the relative merits of these methods and choose accordingly. In particular, variation methods are covered as just another method for solving proportion problems, a departure from past practice.

Most of the basic definitions and formulas in mathematics, engineering, science and business are in terms of ratios. Hence, related topics in these areas are presented.

I gratefully acknowledge the help and encouragement of students, colleagues, and friends.

A.A. Frempong
New York

CONTENTS

CHAPTER 1
Ratio

Lesson 1: **Definition; Reduction of Ratios to Lowest Terms**
Lesson 2: **Using Ratios to Compare Quantities**
Lesson 3: **Using Ratios to Divide a Quantity into Parts**

Note: This chapter and the next chapter are very important for everyday life, whether you have your own business or you work for someone. If you master these chapters very well, you would have a good mathematical preparation for a lot of endeavors both at home or at the workplace. We have used both arithmetic and algebraic methods in covering these chapters. If you have no algebraic background, you may skip the algebraic parts and cover only the arithmetic parts. However, you can study the algebraic material in the Appendix, (p.) before covering these chapters.

Lesson 1

Definition and Reduction of Ratios to Lowest Terms

Whenever you go shopping to buy apples, you always want to know the selling price, and the first question you usually ask the seller is "How much are these apples?" A typical answer may be "4 apples for one dollar". As soon as you have agreed to this price, you and the seller have established a ratio between the number of apples the seller is willing to give you and the number of dollars you are willing to pay. Simply, the ratio you have agreed to is 4 apples to 1 dollar, and we can symbolize this as 4**:**1. The seller might also have replied "1 dollar for 4 apples", but in this case, the ratio of dollars to apples is 1 to 4 or 1**:**4. Note the order in which the terms 1 and 4 of the ratio follow "dollar and apples".

If your grandmother left $1000, in her will, to be divided between you and your brother so that you receive $600 and your brother receives $400, then the ratio of your share to your brother's share is $600 to $400 and in simplest terms, this ratio is 3 **to** 2, which means that whenever you receive $3, your brother receives $2.

If the simple interest rate for your savings account at a bank is 5% (5 percent or 5 per hundred), then after one year, the ratio of the interest to the deposit you had in your account is 5 to 100 (That is, your interest is $5 for every $100 in your account.)

We can observe from the above examples that, in everyday life, we are always dealing with ratios, consciously or unconsciously. Uses of ratios include comparison of quantities, division of a quantity into parts, in business decision making and in forming proportions.

Definition

The **ratio** of two quantities a and b or of a to b (written a:b) is a comparison of the two quantities and written as a fraction is $\frac{a}{b}$ (i.e., is the first quantity mentioned divided by the second quantity mentioned), where a and b may or may not be of the same units. The quantities a and b are called the terms of the ratio. However, the ratio of b to a is $\frac{b}{a}$.

Examples

1. The ratio, 3**:**4 is $\frac{3}{4}$ but the ratio 4**:**3 is $\frac{4}{3}$. **Note** that order is important.

2. The ratio, 4 inches to 12 inches is the fraction $\frac{4 \text{ inches}}{12 \text{ inches}} = \frac{1}{3}$

   (Terms have the same units and cancel out)

3. The ratio, 4 inches to 1 foot is the fraction $\frac{4 \text{ inches}}{1 \text{ foot}} = \frac{4 \text{ inches}}{12 \text{ inches}} = \frac{1}{3}$ (1 ft = 12 in.)

(One unit can be converted to the other)

4. The ratio, 4 apples to 1 dollar is the fraction $\frac{4 \text{ apples}}{1 \text{ dollar}}$

(Terms do not have the same units and one unit cannot be converted to the other)

5. The ratio, 50 miles to 2 hours is the fraction $\frac{50 \text{ miles}}{2 \text{ hours}} = \frac{25 \text{ miles}}{1 \text{ hour}}$ (25 mph) <-different units

Note that since ratios can be written as fractions, ratios can be reduced to lowest terms in the same way that we reduce fractions to lowest terms.

Since ratios are used for comparing quantities, a ratio may have three or more terms.

Example The number of pencils, the number of pens, and the number of notebooks are in the ratio of 2:3:4. (i.e., 2 pencils to 3 pens to 4 notebooks)

Rate

If *a* and *b* are **not** of the same units **and** one unit **cannot** be converted to the other unit, some authors call the ratio a **rate.**

1. The rate, 50 miles to 1 hour is the fraction $\frac{50 \text{ miles}}{1 \text{ hour}}$ (50 mph)

Even though **rate** is a common terminology for quantities with different units, in the following example, note that it is actually a ratio, since the terms have the same units or no units.

2. An interest rate of 5% for a savings account (at a bank in the US.) means that the depositor receives \$5 for every \$100 in the account. Note also that 5% is also the fraction $\frac{5}{100}$ or the ratio, 5:100, which is a ratio. Perhaps we should call this interest rate interest ratio.

Example: Reduce the ratio 12:8 to lowest terms

Solution: $12:8 = \frac{12}{8} = \frac{3}{2}$

Answer: In lowest terms 12:8 = 3:2.

Lesson 1 Exercises

Reducing Ratios to Lowest Terms

Reduce to lowest terms (Two terms)

(a). The ratio 2:4

(b). The ratio 6:3

(c). The ratio 8:12

(d) The ratio 12:8

(e). The ratio 5 inches to 10 inches

(f) The ratio 6 apples to 3 dollars

Answers: (a) 1:2; (b) 2:1; (c) 2:3; (d) 3:2; (e) 1:2; (f) 2 apples:1 dollar;

Reduce to lowest terms (Three Terms)

(a) .The ratio 2:4:6 (b) The ratio 6:3:18

(c) The ratio 8:12:6

(d) The ratio 12:8:6

(e) . The ratio 5 inches to 10 inches to 30 inches

(f) The ratio 6 apples to 10 apples to 24 apples

Answers: (a) 1:2:3; (b) 2:1:6; (c) 4:6:3; (d) 6:4:3; (e) 1:2:6; (f) 3:5:12;

Lesson 2
Using Ratios to Compare Quantities

Example 4

At a certain college, there are 2,000 male students and 6,000 female students.

(a) What is the ratio of the number of male students to the number of female students at this college?

(b) What is the ratio of number of female students to the number of male students at this college?

(c) What is the ratio of the number of female students to the total number of students at this college?

Solution

(a) The ratio of males to females is 2000:6000 $= \frac{2000}{6000} = \frac{1}{3}$ or 1 : 3

(b) The ratio of females to males is 6000:2000 = $\frac{6000}{2000} = \frac{3}{1}$ or 3:1

(c) Total number of students = 2000 + 6000 = 8000

The ratio of the number of females to the total number of students is 6000:8000

$= \frac{6000}{8000} = \frac{3}{4}$ or 3 : 4

Example 5

Now let us verify the results of the above problem by doing it backwards.

There are altogether 8,0000 students at a certain college. The ratio of the number of males to the number of females at this college is 1:3.

(a) Find the number of males. **(b)** Find the number of females at this college.

Solution

(a) Fraction of students who are males $\frac{1}{1+3} = \frac{1}{4}$ (Fraction corresponds to first term, 1, of ratio)

The number of males = $\frac{1}{4} \times \frac{8000}{1} = 2{,}000$

(b) Fraction of students who are females $\frac{3}{1+3} = \frac{3}{4}$ (Fraction corresponds to second term, 3)

The number of females = $\frac{3}{4} \times \frac{8000}{1} = 6{,}000$

(c) Let us recalculate the number of females given that the ratio of number of females to the total number of students is 3:4.

Solution: Fraction of students who are females = $\frac{3}{4}$.

The number of females = $\frac{3}{4} \times \frac{8000}{1} = 6{,}000$ <-----This is the same solution as is (b)

Note the difference between how the denominator, 4, of the fraction in (c) was obtained by using only the second term of the ratio, and how the denominator of the fraction in (b) was obtained by adding 1 and 3, the terms of ratio . Note also that in (b) and (a), we added the terms of the ratio to obtain the denominators of the fractions, since the terms, 1 and 3 correspond to the individual genders involved in the problem; while in (c), the second term, 4, of the ratio corresponds to the total number of students. Note particularly that we did **not** add the 3 and the 4 to obtain the denominator of the fraction in (c), but we rather used only the 4 as the denominator. See also p.21. The problem in (c) could have been posed as " 3 out of 4 students at a certain college are females. If there are 8,000 students at this college, how many of the students are females. The solution method is the same as in (b). We could also algebraically, use proportion to obtain the same solution: $\frac{3}{4} = \frac{x}{8000}$, where x is the number of females.

$4x = 3 \times 8000$ and $x = \frac{3}{4} \times \frac{8000}{1} = 6{,}000$.

As we shall soon observe, Example 5 actually belongs to the next lesson.

Lesson 2 Exercises

Comparing two quantities

1. At a certain college, there are 18,000 male students and 32,000 female students.
(a) What is the ratio of the number of male students to the number of female students at this college?
(b) What is the ratio of number of female students to the number of male students at this college?
c) What is the ratio of the number of female students to the total number of students at this college? **Answer** (a) 9:16; (b) 16:9 ; (c) 16:25

2. In a math class, class, there are 15 females and 8 males.
(a) What is the ratio of the number of males to the number of females in this class?
(b) What is the ratio of the number of females to the number of males?
(c) What is the ratio of the number of males to the total number of students?
Answer: (a) 8:15; (b) 15:8; (c) 8:23

3. Rameltha and Albert are to divide $45,000
If Rameltha receives $18,000, what is the ratio of Albert's share to Rameltha's share?
Answer: 3:2

4. A piece of black tape is 4 inches long, and a similar piece of red tape is 2 feet long..
What is the ratio of the length of the black tape to the length of the red tape?
Answer: 1:6

5. Out of 20 games that a team played, the team won 12 games and lost the rest.
(a) What is the ratio of the number of games won to the number of games played?
(b) What is the ratio of the number of games won to the number of games lost?
Answer: (a) 3:5; (b) 3:2

Comparing three quantities

6. In a bag, there are 16 red marbles, 20 black marbles and 12 green marbles.
(a) What is the ratio of red marbles to black marbles to green marbles?
(b) What is the ratio of black marbles to red marbles to green marbles?
(c) What is the ratio of black marbles to the sum of the red and green marbles?
(d) What is the ratio of green marbles to the total number of marbles in the bag?
Answer: (a) red:black:green = 4:5:3 ; (b) black:red:green = 5:4:3; (c) 5:7; (d) 1:4

7. Rameltha, Albert and James are to divide $45,000. If Rameltha receives $18,000, and Albert receives $17,000, what is the ratio of Rameltha's share to James' share to Albert's share? Answer: 18:10:17

8. A piece of black tape is 4 inches long, and similar pieces of red and green tapes are 2 feet , and 3 feet long respectively.
(a) What is the ratio of the length of the black tape to green tape to red tape?
(b) What is the ratio of the length of green tape to red tape to black tape?
Answer: (c) 1:9:6 (b) 9:6:1

9. Out of 20 games that a team played, the team won 12 games, lost 4 and drew the rest
(a) What is the ratio of games won to games lost to games drawn?
(b) What is the ratio of games lost to games won to games played?
(b) What is the ratio of games drawn to games won?
Answer: (a) 3:1:1 ; (b) 1:3:5 ; (c) 1:3

Lesson 3

Using Ratios to Divide a Quantity into Parts

Example 1 Divide 28 in the ratio 3: 4

Method 1: Using Arithmetic

Step 1: Fraction of the number corresponding to the first term: $\frac{3}{3+4} = \frac{3}{7}$

Fraction of the number corresponding to the second term: $\frac{4}{3+4} = \frac{4}{7}$

Step 2: The number corresponding to the first term (of the ratio) = $\frac{3}{7} \times \frac{28}{1} = 12$

The number corresponding to the second term (of the ratio) = $\frac{4}{7} \times \frac{28}{1} = 16$

The numbers are 12 and 16.

Method 2: Using algebra

Let the number corresponding to the first term (of the ratio) = $3k$ (k a constant)
Let the number corresponding to the second term (of the ratio) = $4k$
"The sum of the two numbers is 28" translates to

$$3k + 4k = 28$$
$$7k = 28$$
$$k = 4$$

One of the numbers = $3k = 3(4) = 12$, and
the other number = $4k = 4(4) = 16$
The numbers are 12 and 16.

Method 3: Using Algebra

See Method 3 of Example 2, below.

Example 2 A bag contains red and white marbles in the ratio 3:4. There are altogether 28 marbles in this bag. How many of these marbles are red?

Solution Method 1: Using arithmetic

The number of red marbles = $\frac{3}{7} \times \frac{28}{1} = 12$

Method 2: Using algebra

Let the number of red marbles = $3k$ (k a constant)
Let the number of white marbles = $4k$

$$3k + 4k = 28$$
$$7k = 28 \quad \text{The number of red marbles} = 3k = 3(4) = 12$$
$$k = 4$$

Method 3: Using Algebra

Let the number of red marbles = x
Then the number of white marbles = $(28 - x)$
The proportion "3:4 as x : $(28 - x)$ translates to: (ratios of red marbles to white marbles)

$$\frac{3}{4} = \frac{x}{(28 - x)}$$

Solve for x:

$$3(28 - x) = 4x$$
$$84 - 3x = 4x$$
$$84 = 7x$$
$$12 = x$$

The number of red marbles = 12

Lesson 3 Exercises

Dividing a quantity into two parts

1. The ratio of male students to female students at a certain college is 2:5. if there are 2,800 students at this college, how many of the students are females? Answers: **1.** 2,000;

2. If the ratio (by weight) of hydrogen to oxygen in water is 1:8, find the weight of hydrogen in 27 grams of water. Answer: 3 grams

3. The sum of two numbers is 40. The numbers are in the ratio 2:3. Find these numbers.

Answers: **1.** 16 and 24

4. James and Mary are to divide $72 in the ratio 5:7. How much does each receive?

Answer: James: $30, Mary: $42

5. A grandmother left $45,000 in her will to be divided between two granddaughters, Maria and Diana in the ratio 2 to 3. How much does each receive?

Answer Maria: $18,000; Diana: $27,000

6. John invested $2,000 in a company while James invested $3,000 in the same company. If after 3 years, a profit of $2,000 is to be shared between them in the ratio of their investments. How much does each receive? **Answer** John: $800; James: $1200

7. Mary wants to prepare a solution consisting of alcohol and water in the ratio 1**:**2 by volume. If the total volume of the solution is 12 liters, find the volumes of alcohol and water to be used **Answer** alcohol: 4 liters; water: 8 liters

9. A farmer wants to prepare 24 liters of gasohol which consists of alcohol and gasoline in the ratio 1**:**9. How many liters of alcohol and gasoline are to be mixed?

Answer alcohol: 2.4 liters; gasoline: 21.6 liters

9. A bag contains p fruits, made up of oranges and apples only in the ratio $q{:}r$. Obtain an expression for the number of oranges in terms of p, q, and r. .Answer: $\frac{pq}{q+r}$

10. In a math class, the ratio of males to females is 1**:**3.: If there are 28 students in this class, how many males are in this class? Answer: 7 males

Dividing a quantity into three parts

11. A grandmother left $45,000 in her will to be divided between three granddaughters, Ana, Elizabeth, and Diana in the ratio 1 to 2 to 3. How much does each receive?

Answer: Ana: $7,500, Elizabeth: $15,000, Diana: $22,500

12. John invested $3,000 in a company while James and Alexander invested $4,000 and $5,000 respectively in the same company. If after 4 years, a profit of $4,800 is to be shared between them in the ratio of their investments. How much does each receive?

Answer : John: $1,200, James: 1,600, Alexander: $2,000

13. Mary wants to prepare a solution consisting of alcohol, water and orange juice in the ratio 1**:**2**:3** by volume. If the total volume of the solution is 36 liters, find the volumes of alcohol, water and juice to be used **Answer** Alcohol: 6 liters, water: 12 liters, orange juice: 18 liters

14. A farmer wants to prepare 44 liters of fuel blend which consists of fuel A, fuel B and fuel C in the ratio 1**:**1**:**9. How many liters of each type of fuel are to be mixed?

Answer : fuel A: 4 liters, : fuel B: 4 liters, fuel C: 36 liters

15. A bag contains p fruits, made up of oranges, grapefruits, and apples in the ratio $q{:}r{:}s$. Obtain an expression for the number of oranges in terms of p, q, r, and s. Answer: $\frac{pq}{q+r+s}$

16. On a math test, the number of A's, B's, and C's are in the ratio of 1**:**4**:**2 If there are 28 students in this class, determine the number of (a) A's, (b) B's, and (c) C's. ?

Answer: A's: 4, B's: 16, C''s: 8:

CHAPTER 2

PROPORTION (VARIATION)

Lesson 4

Introduction to Direct and Inverse Proportion

(Definitions, Terminology, Types of Proportion and Principles)

Basically, there are two main **types** of **proportion** (or **variation**):
1. **Direct proportion** or simply, **proportion** (or direct variation).
2. **Inverse proportion** (indirect or inverse variation).

There are also problems which involve both direct and inverse proportion simultaneously.

Direct proportion

Discussion: In the introduction of the lesson on ratios, we discussed an example involving buying apples. There, the seller quoted the price of apples as "4 apples for 1 dollar", and we established the ratio of the number of apples to the number of dollars as 4 to 1. Now, if the buyer wants to buy exactly 4 apples, then the buyer pays exactly 1 dollar. However, if the number of apples is different from 4, then the number of dollars to pay would be different from 1, and the seller has to determine how much the buyer pays for the new number of apples. For example, if the buyer wants 8 apples (instead of 4 apples), then the seller will determine how many dollars the buyer pays (in this case, $2), or if the buyer wants 2 dollars worth of apples, the seller has to determine how many apples to give t he buyer for $2 (in this case, 8 apples). We note above that if the buyer wants to buy exactly 4 apples, then only two quantities are involved in the problem, namely 4 apples and 1 dollar.

However, if the buyer wants 8 apples, then four quantities would be involved in the problem, namely 4 apples, 1 dollar, 8 apples, and the price for 8 apples. We will determine the fourth quantity, namely, the price for 8 apples using these three known quantities and a relationship called a **direct proportion**. In this relationship, as the number of apples increases, the number of dollars also increases; or as the number of apples decreases, the number of dollars also decreases. When the number of apples is zero, the number of dollars is also zero. Uniform or constant changes in the number of apples result in uniform changes in the number of dollars. For example, as the number of apples is doubled, the number of dollars is also doubled; but if he number of apples is halved, the number of dollars is also halved.

Another application of proportion is as follows: Suppose the roof of your house has a triangular shape, and you want to build a much bigger or smaller house such that the corresponding roof has the same shape as your present house, then **direct proportion** can be used to determine the lengths of the sides of the new roof.

Direct proportion occurs very often in the arithmetic of everyday life. It may not be the number of apples and number of dollars, but it may be the number of miles traveled and number of gallons of gasoline used, or the number of miles traveled and number of hours, or the number of grams of hydrogen and the number of grams of oxygen. Direct proportion is usually termed **proportion** in most textbooks, and it is this type that you meet very often.

We should note above that problems stated in terms of proportion can also be stated in terms of variation and conversely. In the examples covered in this chapter, we cover both cases.

Simple Direct Proportion (Direct Variation)

Definition: A **proportion** is a statement that two ratios are equal.

The direct proportion "a is to b as c is to d" (or a:b::c:d) is the equality

$$\frac{a}{b} = \frac{c}{d} \quad \textbf{(quotient = quotient)}$$

In the above proportion, a and d are called the extremes and b and c are called the means of the proportion. We may also define a proportion as a mathematical statement indicating the equality between two **equivalent fractions.** Example: $\frac{2}{3} = \frac{4}{6}$ is a proportion

Proportion Principles

1. We can invert both ratios of a proportion: From (1) above, we obtain after the inversion,

$$\frac{b}{a} = \frac{d}{c}$$

2. We can interchange the extremes of a proportion: From (1) above we would obtain:

$$\frac{d}{b} = \frac{c}{a}$$

3. We can also interchange the means to obtain:

$$\frac{a}{c} = \frac{b}{d}$$

Also $ad = bc$ (That is the product of the extremes = the product of the means) <-- also called cross-multiplication

The proportion principles can be used to obtain desired ratios. For example, if we have $\frac{3}{2}$ as the ratio of one side of a proportion, and we desire the ratio $\frac{2}{3}$ then proportion principle #1 above (inversion of both sides of a proportion) can be used to obtain this ratio, $\frac{2}{3}$, from $\frac{3}{2}$.

More Principles (From $\frac{a}{b} = \frac{c}{d}$, we obtain the following:

4. $\frac{a+b}{b} = \frac{c+d}{d}$ (by addition). Note: $\frac{a+b}{b} = \frac{a}{b} + \frac{b}{b} = \frac{a}{b} + 1$ **and** $\frac{c+d}{d} = \frac{c}{d} + \frac{d}{d} = \frac{c}{d} + 1$

5. $\frac{a-b}{b} = \frac{c-d}{d}$ (by subtraction)

6. $\frac{a-b}{a+b} = \frac{c-d}{c+d}$ (by subtraction and addition)

7. $\frac{a+b}{a-b} = \frac{c+d}{c-d}$ (by addition and subtraction)

Note:: For 6 and 7, divide LHS and RHS of 4 and 5,

8. If $\frac{a}{b} = \frac{c}{d} = \frac{e}{f}$, then $\frac{a+c+e}{b+d+f} = \frac{a}{b} = \frac{c}{d} = \frac{e}{f}$

e.g.,($\frac{2}{10} = \frac{3}{15} = \frac{4}{20} = \frac{2+3+4}{10+15+20} = \frac{9}{45} = \frac{1}{5}$)

Problems Involving Proportion

Example 1 Find x from the proportion x is to 4 as 3 is to 2.

Solution

$$\frac{x}{4} = \frac{3}{2}$$

$2x = 12$ (cross multiplication)

$x = 6$

(You could also solve the above equation by multiplying both sides of the equation by 4)

Example 2 Find x from the proportion: 6 is to x as 4 is to 12.

Solution

$$\frac{6}{x} = \frac{4}{12}$$

$4x = 72$; and $x = 18$

Simple Inverse Proportion (Inverse or indirect Variation)

Discussion: Another type of a relationship called **inverse proportion** (or indirect proportion) also occurs in everyday life. For example, suppose that when you drive your car from City A to City B at a speed of 25 mph, it takes you 4 hours. Now, if you always drive at a speed of 25 mph from City A to City B, and it takes you 4 hours, then only two quantities are involved, namely 25 mph and 4 hours. However, if you want to travel the same distance at a speed of say, 50 mph, then you would like to know the time needed to drive from City A to City B at 50 mph, and in this case, four quantities would be involved in the problem, namely, 25 mph, 4 hours, 50 mph, and the time, *t*, for 50 mph. Since for the same distance, as the speed is increased, the time taken decreases, or if the speed is decreased, the time taken increases, speed is said to be inversely proportional to the time.

Recall the main difference between direct proportion and inverse proportion**:** for direct proportion, both quantities increase or decrease at the same time; whereas for inverse proportion, as one quantity increases, the other quantity decreases and vice versa.
For example, as one quantity is doubled, the other quantity is halved; but if one quantity is halved, the other quantity is doubled.
Example 1 The time taken by a number of people to do a piece of work is **inversely proportional** to the number of people (assuming that each person works at the same rate as everyone else). Thus, as the number of people increases, the time taken to do the work decreases but as the number of people decreases, the time taken increases (i.e., more people, less time; and less people, more time)

Example 2 At constant temperature, the volume of a given mass of a gas is inversely proportional to the pressure on the gas. (This relationship is known as Boyle's law.)

Definition
If *a* is inversely proportional to *b*.

Then a_1 corresponds to b_1, and a_2 corresponds to b_2. and

$$a_1 : \frac{1}{b_1} = a_2 : \frac{1}{b_2}$$

$$a_1 \div \frac{1}{b_1} = a_2 \div \frac{1}{b_2}$$

$$a_1 b_1 = a_2 b_2 \quad \textbf{(product = product)}$$

Or if we use the familiar terms *a, b, c, d* as the terms of the inverse proportion

Then $a : \frac{1}{b} = c : \frac{1}{d}$

$$a \div \frac{1}{b} = c \div \frac{1}{d}$$

$ab = cd$ **(product = product) <------** inverse proportion

We will use the product = product form since this form is easy to remember

Compare:

Direct proportion $\frac{a}{b} = \frac{c}{d}$ <-------- **(quotient = quotient)**

Indirect proportion $ab = cd$ **<------(product = product)**

Note: The terms indirect proportion and inverse proportion can be used interchangeably.

Determining the type of Proportion (Variation) **from the Wording**

Explicit Specification of direct proportion (variation)

There are a number of ways of **explicitly** specifying direct proportion (variation)
If y is a function of x, then the following are equivalent to one another:

(a) y is directly proportional to x, or simply
(b) y is proportional to x. (The word "directly" is omitted.)
(c) y varies directly as x, or simply,
(d) y varies as x, (The word "directly" is omitted.)

The methods we will cover for solving direct proportion problems can be used to solve any of the problems worded as in (a), (b), (c) and (d) above.

Explicit Specification of inverse proportion (variation)

If y is a function of x, then the following are equivalent to one another:

(a) y is inversely or indirectly proportional to x.
(b) y varies inversely as x,

Any of the methods (we will cover) for solving indirect proportion problems can be used to solve any of the problems stated as in (a) and (b) above.

Implicit Specification of proportion (variation)

In most problems, the terms " directly proportional to" , "inversely proportional to", " varies directly as" or "varies inversely as" and similar terms are not used in wording the problem; but rather, we have to deduce from the problem whether it is a direct or an indirect proportion from experience. For example, if 15 oranges cost \$3, what is the cost of 10 oranges?. From life experience, we know that at the same price per an orange, as the **number** of oranges **increases**, the **cost** of oranges **increases** accordingly. Therefore, the relationship in this problem is that of direct **proportion** (variation).

Lesson 4 Exercises A

A. Solve for x: **1.** $\frac{x}{8}=\frac{3}{2}$; **2.** $\frac{5}{8}=\frac{20}{x}$; **3.** $\frac{9}{x}=\frac{3}{5}$; **4.** $\frac{6}{8}=\frac{x}{2}$; **5.** $\frac{x}{9}=\frac{3}{2}$;

6. $\frac{10}{x}=\frac{3}{5}$; **7.** $\frac{x}{8}=\frac{4\frac{1}{2}}{2}$; **8.** $\frac{x}{8\frac{2}{3}}=\frac{3\frac{2}{3}}{2\frac{3}{4}}$; **9.** $\frac{x}{1.8}=\frac{3}{2.1}$; **10.** $\frac{x}{8}=\frac{12\%}{96}$

Answers: **1.** $x=12$; **2.** $x=32$; **3.** $x=15$; **4.** $x=\frac{3}{2}$; **5.** $13\frac{1}{2}$; **6.** $16\frac{2}{3}$; **7.** $x=18$; **8.** $11\frac{5}{9}$;
9. $2\frac{4}{7}$; **10.** 1% or $\frac{1}{100}$

Lesson 4 Exercises B

Determine which of the following are direct proportion or inverse proportion:

1. If 10 gallons of gasoline cost 15 dollars, find the cost of 24 gallons of gasoline.

2. If on 2 gallons of gasoline, a car can travel 24 miles, how many miles can a car travel on 20 gallons of gasoline?

3. If 8 oranges sell for 2 dollars, find the selling price for 24 oranges.

4. If 2 people can paint a house in 6 days, how many people would paint the house in 3 days?

5. At a speed of 60 miles per hour, a car takes 2 hours to travel from City A to City B. At a speed of 40 miles per hour, how long will it take the car to ravel from City A to City B?

6. If 6 mail men can sort a given quantity of mail in 4 hours, how many mail men would be needed to sort the same quantity of mail in 2 hours?

Answers , 1, 2, 3 are direct proportion; 4, 5 and 6 are inverse proportion

Lesson 5

Methods for Solving Direct Proportion Problems

(Direct Variation Problems)

There are a number of methods or approaches for solving **direct proportion** problems. In simple proportion problems, four quantities are involved, and we are (usually) given three quantities and we are required to find the fourth quantity.

We cover **five** methods with examples, and then discuss the relative merits of the methods.

Method 1: Quotient = Quotient or Ratio = Ratio method (Algebraic)

In this method, we word or reword the problem as discussed below.

Consider a direct proportion problem involving four quantities such that a_1 corresponds to b_1, and a_2 corresponds to b_2. We word the proportion as follows:

The proportion a_1 is to b_1 as a_2 is to b_2 translates into the equation

$$\frac{a_1}{b_1} = \frac{a_2}{b_2}$$ **(quotient = quotient or** ratio = ratio) **or**

Using the familiar notation "a is to b as c is to d", we obtain

$$\frac{a}{b} = \frac{c}{d}$$ **(quotient = quotient)**

Example 1

If 2 dollars can buy 8 apples, how many dollars are needed to buy 24 apples?

Method 1: **Quotient = Quotient Method** (Algebraic)

We are required to find the number of dollars needed to buy 24 apples.

Step 1: Let the number of dollars needed to buy 24 apples = x.

(we represent the unknown by a variable)

2 dollars correspond to 8 apples

x dollars correspond to 24 apples.

(Note the order in which the units in the second correspondence is mentioned.)

That is, dollars before apples, and not apples before dollars)

Step 2: 2 dollars are to 8 apples as x dollars are to 24 apples

Step 3: Translating into an equation, $\frac{2 \text{ dollars}}{8 \text{ apples}} = \frac{x \text{ dollars}}{24 \text{ apples}}$

Step 4: Cross-multiplying, $8x = 2(24)$ ($8x$(dollars)(apples) = 2(24)(dollars)(apples))

Step 5 Solve for x: $x = \frac{2(24)}{8}$ (Note above that the units cancel out)

$= 6$

Conclusion: 6 dollars are needed to buy 24 apples.

Note in the above example that the order in which we mention the units, dollars and apples is important.

Note also, however, that in the above example, we could word the proportion as follows:

The proportion, 8 apples are to 2 dollars as 24 apples are to x dollars" translates to $\frac{8}{2} = \frac{24}{x}$

$$8x = 2(24)$$

$$x = \frac{2(24)}{8}$$

$= 6$ (Again, we obtain the same solution as before.)

Method 2 Units Label Method or Dimensions Method (Arithmetic method)

This method is highly recommended in applications such as converting from one unit to another unit in measurements as well as calculations in chemistry. In Chemistry, some authors call this method " The factor label method". In this method, we use the units of the given quantities to guide us in obtaining an expression which on simplification yields the desired result.

We repeat the question: If 2 dollars can buy 8 apples, how many dollars are needed to buy 24 apples?

Solution: We can consider this problem as converting 24 apples to dollars. The justification of this method is in parentheses.

Step 1: $\frac{24 \text{ apples}}{1} \times \frac{?}{?}$

Step 2: Multiply $\frac{24 \text{ apples}}{1} \times \frac{?}{?}$ by a fraction formed by using the quantities 2 dollars and 8 apples as the terms of the fraction such that the denominator has the same units as the 24 apples (the quantity in the numerator), and under such conditions, we can divide out (cancel) the common units in the numerator and the denominator, leaving us the units, dollars.

Then, $\frac{24 \text{ apples}}{1} \times \frac{2 \text{ dollars}}{8 \text{ apples}}$

Step 3: $\frac{24 \cancel{\text{ apples}}}{1} \times \frac{2 \text{ dollars}}{8 \cancel{\text{ apples}}}$ (= Number of apples $\times$ cost per each apple)

$= 6$ dollars

Conclusion: 6 dollars are needed to buy 24 apples.

Let us do the above problem backwards, with the justification of this method in parentheses

If 2 dollars can buy 8 apples, how many apples can one buy for 6 dollars?

Solution Here, we are converting 6 dollars to apples.

$\frac{6 \text{ dollars}}{1} \times \frac{?}{?}$

$= \frac{6 \text{ dollars}}{1} \times \frac{8 \text{ apples}}{2 \text{ dollars}}$ (= total cost $\div$ cost per 1 apple = total cost $\times \frac{1}{\text{cost per 1 apple}}$)

= 24 apples, which is what was given in the original problem.

Method 3 Unitary Method (Arithmetic)

We repeat the question: If 2 dollars can buy 8 apples, how many dollars are needed to buy 24 apples?

Solution: Cost of 8 apples = 2 dollars

Cost of 1 apple = $\frac{2 \text{ dollars}}{8}$ (Divide to obtain the cost for 1 apple))

Cost of 24 apples $= \frac{2 \text{ dollars}}{8} \times \frac{24 \text{ apples}}{1 \text{ apple}}$

$= \frac{2}{8} \times \frac{24}{1}$ dollars (Multiply to obtain the cost for other than 1)

= 6 dollars

Conclusion: 6 dollars are needed to buy 24 apples.

Method 4 "More or Less Method" (Arithmetic method. This method the shorter form of Method 3.

We repeat the question: If 2 dollars can buy 8 apples, how many dollars are needed to buy 24 apple?

If the cost of 8 apples = 2 dollars

$$\text{Then the cost of 24 apples} = \frac{2 \text{ dollars}}{1} \times \frac{24 \text{ \sout{apples}}}{8 \text{ \sout{apples}}}$$

$$= 6 \text{ dollars}$$

Conclusion: 6 dollars are needed to buy 24 apples.

In method 4, we used a very useful principle which states that " If more, the smaller divides, and if less, the larger divides". That is, since we expect 24 apples to cost more than 2 dollars, in forming the fraction involving 8 apples and 24 apples, the smaller of 8 apples and 24 apples is the divisor (smaller divides), and hence we used the fraction,

$\dfrac{24 \text{ apples}}{8 \text{ apples}}$ <----- " smaller divides" (We multiplied by an **improper fraction**)

However, if we had expected the number (answer) to be less than 2 dollars, we would have used the fraction

$\dfrac{8 \text{ apples}}{24 \text{ apples}}$ <----- "larger divides" (i.e., we would have multiplied by a **proper fraction**)

The More or Less Method above is very powerful once you have mastered it, because it can be applied to inverse proportion problems as well as to problems involving both inverse and direct proportion and any number of quantities (variables), especially in science.

Method 5 Variation Method (Algebraic)

We reword the problem: The number of apples a person buys varies directly as the number of dollars paid. If when the number of apples is 8, the number of dollars paid is 2, find the number of dollars paid when the number of apples is 24.

Step 1: Let the number of apples = A

Let the number of dollars paid = D

Then $A = kD$, where k is called the proportionality constant.

Step 2: Substituting $A = 8, D = 2$ *in* $A = kD$,

$8 = 2k$

$4 = k$

Substituting $k = 4$ *in* A = kD, we obtain the formula,

$A = 4D$.

Step 3: When $A = 24$

$24 = 4D$ (substituting 24 for A in the formula, $A = 4D$)

$6 = D$ (solving for D)

$D = 6$

Therefore, 6 dollars would be paid for 24 apples.

Example 2 On 3 gallons of gasoline, one can drive 60 miles. How far can one drive on 8 gallons of gasoline?

Solution

Method 1 Quotient = Quotient Method (Algebraic)

Let the number of miles for 8 gallons of gasoline = x

Proportion statement: 3 gallons are to 60 miles as 8 gallons are to x miles.

$$\text{Translating, } \frac{3 \text{ gallons}}{60 \text{ miles}} = \frac{8 \text{ gallons}}{x \text{ miles}}$$

$$\frac{3}{60} = \frac{8}{x}$$

$$3x = 60(8)$$

$$x = \frac{60(8)}{3}$$

$$x = 160$$

Therefore, on 8 gallons of gasoline, one can drive 160 miles.

Note above that the ratio, gallons to miles = the ratio, gallons to miles; or
the ratio miles to gallons = the ratio miles to gallons
Incorrect: the ratio miles to gallons = the ratio gallons to miles. (violation of the order of the units)

All other proportion problems involving four quantities can be reworded and translated into equations by imitating the examples covered.. Only, the units involved might be different. We may have apples and pounds as the units involved, or hours and dollars, or years and people.

Method 2 Units Label Method. We repeat the problem:
On 3 gallons of gasoline, one can drive 60 miles. How far can one drive on 8 gallons of gasoline?

Solution: We can consider this problem as converting 8 gallons to miles.

Step 1: $\frac{8 \text{ gallons}}{1} \times \frac{?}{?}$

Step 2: Multiply $\frac{8 \text{ gallons}}{1} \times \frac{?}{?}$ by a fraction formed by using the quantities 3 gallons and 60 miles

$$\frac{8 \text{ gallons}}{1} \times \frac{60 \text{ miles}}{3 \text{ gallons}}$$

Step 3: $\frac{8 \,\cancel{\text{gallons}}}{1} \times \frac{60 \text{ miles}}{3 \,\cancel{\text{gallons}}}$

$= 160$ miles

Therefore, on 8 gallons of gasoline, one can drive 160 miles.

Method 3 Unitary Method (Arithmetic)
We repeat the problem: On 3 gallons of gasoline, one can drive 60 miles.
How far can one drive on 8 gallons of gasoline?

Solution

Number of miles driven on 3 gallons of gasoline = 60 miles

Number of miles driven on 1 gallon of gasoline = $\frac{60}{3}$ miles

(Divide to obtain the miles for 1 gallon)

Number of miles driven on 8 gallons of gasoline = $\frac{60}{3} \times 8$ miles (Multiply for other than 1 gallon)
= 160 miles

Therefore, on 8 gallons of gasoline, one can drive 160 miles.

Method 4 More or Less Method We repeat the problem:
On 3 gallons of gasoline, one can drive 60 miles. How far can one drive on 8 gallons of gasoline?

Solution Number of miles driven on 3 gallons of gasoline = 60 miles

Number of miles driven on 8 gallons of gasoline = $\frac{60}{1} \times \frac{8}{3}$ miles
= 160 miles

Therefore, on 8 gallons of gasoline, one can drive 160 miles.

Method 5 Variation Method (Algebraic) We reword the problem:

The number of miles one drives is directly proportional to the number of gallons of gasoline used. If when the number of gallons of gasoline is 3, the number of miles is 60, find the number of miles when the number of gallons of gasoline is 8.

Solution

Step 1: Let the number of gallons of gasoline = n
Let the number of miles driven = d
Then since d is directly proportional to n,
$d = kn$, where k is the proportionality constant

Step 2: Substitute for $d = 60$, $n = 3$, in $d = kn$ and solve for k.
Then $60 = k(3)$ and from which $k = 20$

Step 3: Substitute for $k = 20$ in $d = kn$ to obtain a formula relating d and n.
Then $d = 20n$

Step 4: Substitute for $n = 8$ in $d = 20n$ and determine d

$$d = (20)(8)$$
$$= 160$$

Therefore, on 8 gallons of gasoline, one can drive 160 miles.

Example 3 The ratio of an astronaut's weight on the earth to weight on the moon is 6:1. If an astronaut weighs 180 lb (on earth), what is the astronaut's weight on the moon?

Solution

Method 1 Quotient = Quotient Method (Algebraic)

Let the weight on the moon be x lb, (i.e., x lb corresponds to 180 lb on the earth).

Step 1: The ratio of the weight on earth to weight on the moon is 6:1 or $\frac{6}{1}$.

Step 2: Also, the ratio of the weight on earth to the weight on the moon is 180 lb to x lb or $\frac{180}{x}$.

Step 3 : Equate the two ratios and solve for x.

$$\frac{6}{1} = \frac{180}{x}$$

$$6x = 180(1)$$

$$x = \frac{180(1)}{6}$$

$$x = 30$$

Therefore, the astronaut's weight on the moon is 30 lb.

Method 2 Reword the original problem as a proportion:
6 is to 1 as 180 is to x.

Solution

Translating, $\frac{6}{1} = \frac{180}{x}$

Solving, $x = 30$, and we obtain the same solution as by Methods 1 and 2.

Method 3 Unitary Method (Arithmetic)

Whenever the astronaut weighs 6 lb on the earth, the astronaut weighs 1 lb on the moon.

Whenever the astronaut weighs 1 lb on the earth, the astronaut weighs $\frac{1}{6}$ lb on the moon; and

Whenever the astronaut weighs 180 lb on the earth, the astronaut weighs $\frac{1}{6} \times 180$ lb = 30 lb on the moon.

Method 4: More or Less Method (Arithmetic)

We reword the problem

The weight of an astronaut on earth is directly proportional to the weight on the moon. When the weight on earth is 6 lb, the weight on the moon is 1 lb, find the weight on the moon when the weight on earth is 180 lb..

Solution Weight on the moon when the weight on earth is 6 lb = 1 lb

Weight on moon when weight on earth is 180 lb = $1\ \ell b \times \dfrac{180\ \ell b}{60\ \ell b} = 30$ lb

Therefore, the astronaut 's weight on the moon is 30 lb.

Method 5 Variation Method (Algebraic)

We reword the problem The weight of an astronaut on earth is directly proportional to the weight on the moon. When the weight on earth is 6 lb, the weight on the moon is 1 lb, find the weight on the moon when the weight on earth is 180 lb..

Solution

Step 1: Let the weight on earth = h lb

Let the weight on moon = n lb

Then since h is directly proportional to n,

$h = kn$

Step 2: Determine the value of k by substituting $h = 6, n = 1$ in $h = kn$

Then $6 = k(1)$ and from which $k = 6$

Step 3: Substitute $k = 6$ in $h = kn$ to obtain a formula relating h and n.

Then we obtain $h = 6n$

Step 4: Substitute $h = 180$ in the formula, $h = 6n$ and solve for n.

Then $180 = 6n$

$30 = n$

Therefore, the astronaut 's weight on the moon is 30 lb.

More Examples on the Variation Method

(Note that these examples can also be solved using Methods 1, 2, 3 and 4)

Example A: y varies directly as x. When $y = 48, x = 3$. Find x when $y = 96$

Solution

Step 1: $y = kx$ (1)

Step 2: Substitute $y = 40, x = 2$ in equation (1) and solve for k.

$48 = k(3)$

$48 = 3k$

$16 = k$

Substitute $k = 16$ in equation (1) above to obtain the formula $y = 16x$ (2)

Step 3: Now, to find x when $y = 96$, replace y by 96 in (2)

Then $96 = 16x$

$6 = x$

Therefore when $y = 96, x = 6$.

Example B The distance y an object falls from rest is directly proportional to the square of the time x. When $y = 128, x = 4$. *(a)* Find x when $y = 200$ *(b)* Find y when $x = 6$.

Solution

Step 1: $y = kx^2$ ....(1)

Step 2: Substitute $y = 128, x = 4$ in equation (1) and solve for k.

Then $128 = k(4)^2$

$128 = 16k$

$k = 8$

Substitute $k = 8$ in equation (1) above to obtain the formula $y = 8x^2$ .... (2)

Step 3: (a) To find x when $y = 200$. Replace y by 200 in equation (2) and solve for x.

$200 = 8x^2$

$\frac{200}{8} = x^2$

$25 = x^2$

$5 = x$

$x = 5$ (Since the time taken must be positive in this problem, we reject -5 as a solution)

(b) To find y when $x = 6$, replace x by 6 in equation (2) and evaluate.

$y = 8(6)^2$

$y = 8(36)$

$y = 288$.

Therefore when $x = 6, y = 288$.

EXTRA

Example 4 Assume volume V is directly proportion to temperature T.

Then V_1 corresponds to T_1 as V_2 corresponds to T_2

Since this is direct proportion, we can apply **quotient = quotient** method.

Then $\frac{V_1}{T_1} = \frac{V_2}{T_2}$ (quotient = quotient)

The above equation is a familiar formula for gases in physics.

Relative Merits of the Methods for Solving Proportion Problems

Method 1: Quotient = quotient Method

The quotient = quotient method such as $\frac{a}{b} = \frac{c}{d}$ is very popular and is efficient for a one-step problem involving only four quantities, but is not efficient in a problem such as in Method 2 below, since the method has to be repeated in steps with the introduction of a new variable in each step.

Method 2: Units Label Method (Arithmetic)

The Units Label Method, an arithmetic method, can readily do what Method 1 can do, if the quantities involved have units, and is more efficient than Method 1 in doing repeated calculations. This method is particularly very efficient in calculations involving measurements and in chemistry as in the following example.

Example How many yards are there in 96 inches?

Solution We reword this problem as: Convert 96 in. to yd.

Plan: This plan is based on what is available in the conversion chart or table.
(Using a conversion chart is analogous to using a map to go from one location to another)
The sequence is inches--> feet---> yards.
We use these relationships: 12 in. = 1 ft; 3 ft = 1 yd (We could use 36 in = 1 yd)
(Note: inch = in.)

Step 1: $\frac{96 \text{ in.}}{1} \times \frac{?}{?}$

Step 2: Multiply the $\frac{96 \text{ in.}}{1}$ by a fraction formed by using the quantities 12 in. and 1 ft as the terms of the fraction such that the denominator has the same units as the 96 in.; followed by similar multiplication by a fraction using 3 ft and 1 yd as the terms of the fraction. Then, we obtain

$$\frac{96 \cancel{\text{in}}}{1} \times \frac{1 \cancel{\text{ft}}}{12 \cancel{\text{in}}} \times \frac{1 \text{ yd}}{3 \cancel{\text{ft}}} \quad \left(\frac{96 \cancel{\text{in}}}{1} \times \frac{1 \text{ ft}}{12 \cancel{\text{in}}} = \frac{96 \cancel{\text{in}}}{1} \times \frac{1 \cancel{\text{ft}}}{12 \cancel{\text{in}}} \times \frac{1 \text{ yd}}{3 \cancel{\text{ft}}} = \frac{96 \times 1 \times 1 \text{ yd}}{1 \times 12 \times 3} \right)$$

$$= 2\tfrac{2}{3} \text{ yd}$$

$\therefore$ 96 in. = $2\frac{2}{3}$ yd

Justification of the Units Label Method in the above problem:

From 12 in. = 1 ft, If we divide both sides of this equality by 12 in., we obtain

$$\frac{\cancel{12 \text{ in.}}}{\cancel{12 \text{ in.}}} = \frac{1 \text{ ft}}{12 \text{ in.}}$$

$$1 = \frac{1 \text{ ft.}}{12 \text{ in.}}$$

Similarly, from 3 ft = 1 yd, If we divide both sides by 3 ft., we obtain

$$\frac{\cancel{3\ ft}}{\cancel{3\ ft}} = \frac{1\ yd}{3\ ft}$$

$$1 = \frac{1\ yd}{3\ ft}$$

In step 2, multiplying by $\frac{1 \text{ ft.}}{12 \text{ in.}}$ or by $\frac{1\ yd}{3\ ft}$ is equivalent to multiplying by 1, and therefore, he given value remains equivalent.

Note: In the above example, if we were to use the quotient = quotient method, we would have to apply this method twice, with the introduction of a variable in each step. Note above also that it is easy to generalize the justification of the "Units Label Method" when we have equality such as 12 in. = 1 ft, than when we have equivalence such as 8 apples are equivalent to 2 dollars.

Method 3: Unitary Method (Arithmetic)
This is an arithmetic method and does not require any knowledge of algebra and can be taught as the first arithmetic method for solving proportion problems.

Method 4: More or Less Method (Arithmetic)
This method is a shorter form of Method 3, as one step in Method 3 is skipped. When mastered well, the "More or Less Method" can be extended to solve inverse proportion as well as compound proportion problems, and very efficient in stoichiometric calculations in chemistry as well as conversions in measurements.

Method 5: Variation Method (Algebraic)
This has applications at all levels of mathematics. Some textbooks ignore this method at the elementary level and cover it at the intermediate level.

EXTRA: Two Cases of the Statement of Ratio and the Corresponding Proportion Problems

In solving problems involving ratios and proportion, we must distinguish between the cases in which the terms of a given ratio refer to **disjoint sets** and the case in which the ratios refer to sets which are **not** disjoint.

Case 1

The terms of the ratios refer to disjoint sets (example: the set of females and the set of males in a school)

Example: The ratio of the number of females to the number of males is 3 to 2.

Proportion: The ratio 3 **females** to 2 **males** = the ratio 6 **females** to 4 **males**
or ratio 2 **males** to 3 **females** = ratio 4 **males** to 6 **females**

(Two sets are disjoint if the sets do not have any members common to both sets)

Problems such as " If 8 apples cost 2 dollars, what is the cost of 24 apples" are of case 1

Case 2

The terms do **not r**efer to disjoint sets (example: the set of females and the set of the total number of students at a school)

Example

The ratio of the number of **females** to the **total number** of students is 3 to 5.

Proportion: The ratio 3 **females** to 5 **students** = the ratio 6 **females** to 10 **students.**

Case 2 is sometimes worded as **a number** out of a **total number**

For example, 3 out of 5 students are females or out of every 5 students, 3 are females, which is the same as $\frac{3}{5}$ of the students are females.

Problems such as " if 3 out of 5 students in a school are females, and there are 600 females in this school, how many students are in this school?" can also be set up using

$$\frac{\text{what number out of}}{\text{what number}} = \frac{\text{what number out of}}{\text{what number}}$$ as well as other methods.

Example for Case 2

Example 1 3 out of 5 students at a certain college are females.

(a) If 600 students at this college are females, how many students are there at this college?

(b) If one were to collect a sample of 200 students at this college, how many of these students would be females?

Solution (a):

Method 1 Using ratio = ratio method

Let the number of students at this college = x

Then the proportion 3 females are to 5 students as 600 females are to x students

translates to: $\frac{3\text{ females}}{5\text{ students}} = \frac{600\text{ females}}{x\text{ students}}$

$$\frac{3}{5} = \frac{600}{x}$$

$$3x = 5(600)$$

$$x = \frac{5(600)}{3};\ x = 1000$$

Therefore, there are 1000 students at this college.

Method 2: Using $\frac{\text{what number out of}}{\text{what number}} = \frac{\text{what number out of}}{\text{what number}}$

Let the number of students at this college = x

$$\frac{3 \text{ out of}}{5} = \frac{600 \text{ out of}}{x}$$

$$\frac{3}{5} = \frac{600}{x}$$

$$3x = 5(600)$$

$$x = \frac{5(600)}{3}$$

$$x = 1000$$

Therefore, there are 1000 students at this college

Method 3 Since 3 out of 5 means $\frac{3}{5}$, we can reword this part of the question as: if $\frac{3}{5}$ of a number is 600, what is the number?

Let the number be x

Then $\frac{3x}{5} = 600$

$$3x = 3000$$

$$x = 1000$$

There are 1000 students at this college.

Method 4 Divide 600 by $\frac{3}{5}$ (If $\frac{3}{5}$ of a number is 600, what is the number?)

$$600 \div \frac{3}{5} = \frac{600}{1} \times \frac{5}{3}$$

$$= 1000$$

There are 1000 students at this college.

Method 5: The Units Label Method

Note: 3 out 5 students are females means that for every 5 students, there are 3 females.

Step 1: $\frac{600 \text{ females}}{1} \times \frac{?}{?}$

Step 2: $= \frac{600 \text{ females}}{1} \times \frac{5 \text{ students}}{3 \text{ females}}$

$= 1000$ students

Part (b)

Method 1: We can reword this part of the question as: Find $\frac{3}{5}$ of 200.

$$\frac{3}{5} \text{ of } 200 = \frac{3}{5} \times \frac{200}{1} = 120$$

That is, of a sample of 200 students, 120 students would be females.

Method 2:: $\frac{\text{what number out of}}{\text{what number}} = \frac{\text{what number out of}}{\text{what number}}$

Let the number of females = x

$$\frac{\text{3 out of}}{5} = \frac{x \text{ out of}}{200}$$

$$\frac{3}{5} = \frac{x}{200}$$

$$5x = 3(200)$$

$$x = \frac{3(200)}{5}$$

$$x = 120$$

Therefore, there are 120 females.

Method 3: The Units Label Method

Step 1: $\frac{\text{200 students}}{1} \times \frac{?}{?}$

Step 2: $\frac{\text{200 students}}{1} \times \frac{\text{3 females}}{\text{5 students}}$

= 120 females

Therefore, there are 120 females.

Method 4

Let the number of females. = x

Then the proportion 3 females are to 5 students as x females are to 200 students translates to:

$$\frac{\text{3 females}}{\text{5 students}} = \frac{x \text{ females}}{\text{200 students}}$$

$$\frac{3}{5} = \frac{x}{200}$$

$$5x = 3(200)$$

$$x = \frac{3(200)}{5} = 120$$

Therefore, there are 120 females..

Extra Method

We repeat the question in **Example 1** 3 out of 5 students at a certain college are females.

(a) If 600 students at this college are females, how many students are there at this college?

(b) If one were to collect a sample of 200 students at this college, how many of these students would be females?

This method follows the algebraic method for dividing a quantity into parts, given the ratio.
Since the ratio of females to the total number of students is 3:5,
let the number of females = $3x$, and
let the total number of students = $5x$.
(a) Then $3x = 600$ (there are 600 females); and solving, $x = 200$
Therefore, total number of students =5x = 5(200) =**1000**
Therefore, there are 1000 students at this college.

(b) 5x = 200 (there are 200 students) , and
x =40
Therefore, the number of females = $3x$ = 3(40) =120.

Lesson 5 Exercises A

Problems on direct proportion in which the terms refer to distinct sets

1. If 1 gallon of gasoline costs 2 dollars, how many gallons of gasoline can one buy for 15 dollars? Answer: $7\frac{1}{2}$ gallons

2. If 10 gallons of gasoline cost 15 dollars, find the cost of 24 gallons of gasoline. Answer: 36 dollars

3. If 10 gallons of gasoline cost 15 dollars, how many gallons of gasoline can one buy for 45 dollars? Answer: 30 gallons

4. If x gallons of gasoline cost y dollars, what is the cost of p gallons of gasoline (Express your answer in terms of x y and p) Answer: $\frac{py}{x}$ dollars

5. If on a gallon of gasoline, a car can travel 24 miles, how many miles can the car travel on 10 gallons of gasoline? Answer: 240 miles

6. If on 2 gallons of gasoline, a car can travel 24 miles, how many miles can a car travel on 20 gallons of gasoline? Answer: 240 miles

7. If on 3 gallons of gasoline, a bus can travel 36 miles, how many gallons of gasoline are needed to travel 60 miles? Answer: 5 gallons

8. If on p gallons of gasoline, a bus can travel r miles, how many gallons of gasoline are needed to travel m miles? (Express your answer in terms of the variables.) Answer: $\frac{mp}{r}$ gallons

9. If 4 oranges cost 1 dollar, how many oranges can one buy for 20 dollars? Answer: 80 oranges

10. If 8 oranges sell for 2 dollars, find the selling price for 24 oranges. Answer: 6 dollars

11. If 12 oranges cost 3 dollars, what is the cost of 60 oranges? Answer: 15 dollars

12. If r oranges cost p dollars, what is the cost of m oranges? Answer: $\frac{mp}{r}$ dollars (Express your answer in terms of the variables.)

13. If a household uses one pound of sugar in 3 weeks, how many pounds of sugar would be used in 6 weeks? Answer: 2 pounds

14. If a household uses 2 pounds of sugar in 3 weeks, how many pounds of sugar would be used in 51 weeks? Answer: 34 pounds

15. If a household uses 2 pounds of sugar in 3 weeks, 12 pounds of sugar would be used in how many weeks? Answer: 18 weeks

16. If a household uses x pounds of sugar in r weeks, how many pounds of sugar would be used in y weeks? (Express your answer in terms of the variables.) Answer $\frac{xy}{r}$ pounds

17. If one teacher is needed for a class of 22 students, how many teachers would be needed to teach 88 students, assuming that each class has 22 students? Answer: 4 teachers

18. If 6 teachers are needed to teach 102 students, how many students would be taught by 3 teachers? Answer: 51 students

19. If t teachers are needed to teach s students, how many students would be taught by b teachers? (Express your answer in terms of the variables.) Answer: $\frac{bs}{t}$ students

20. If 4 grams of hydrogen are required to produce 36 grams of water, how many grams of hydrogen would be required to produce 108 grams of water? Answer:: 12 grams

21, If from 4 grams of hydrogen, 36 grams of water are produced, how many grams of hydrogen are needed to produce 216 grams of water? Answer: 24 grams

22. If from t grams of hydrogen, s grams of water are produced, how many grams of hydrogen are needed to produce b grams of water (Express your answer in terms of the variables Answer: $\frac{bt}{s}$ grams

23. If p gallons of water are needed to produce x pounds of a concrete mixture, how many gallons of water are needed to produce y pounds of this mixture? Answer: $\frac{py}{x}$ gallons

24. 1f each room in a building has 3 windows, how many windows are there in 6 rooms,? Answer: 18 windows

25. If 3 rooms in a building have 12 windows, how many windows are in 36 rooms, assuming that each room has the same number of windows? Answer: 144 windows

26. If p rooms in a building have q windows, how many windows are in r rooms, assuming that each room has the same number of windows? (Express answer in terms of the variables.) Answer: $\frac{qr}{p}$ windows

27. A buyer purchased 20 items for 60 dollars. How many items can be purchased for 30 dollars, assuming that the cost is the same for each of the items. Answer: 10 items

28. A buyer purchased r items for s dollars. How many items can be purchased for t dollars, assuming that the cost is the same for each of the items. (Answer in terms of the variables.) Answer: $\frac{rt}{s}$ items

29. In 2 hours, a car can travel 120 miles. Find the distance the car can travel in 3 hours, assuming the same speed Answer:: 180 miles

30. In x hours, a car can travel y miles. Find the distance the car can travel in k hours, assuming the same speed. Answer: $\frac{ky}{x}$ miles

31. In 2 hours, a car can travel 120 miles, At the same speed, how long will it take the car to travel 180 miles?. Answer: 3 hours

32. In x hours, a car can travel y miles. At the same speed, how long will take the car to travel t miles? (Express your answer in terms of the variables) Answer: $\frac{xt}{y}$ hours

33. If 40 liters of air are needed to ventilate a patient for 2 minutes, how many liters of air are needed to ventilate this patient for 24 minutes? Answer: 480 liters

34. If a liters of air are needed to ventilate a patient for b minutes, how many liters of air are needed to ventilate this patient for c minutes? Answer $\frac{ac}{b}$ liters

35. If $4\frac{1}{2}$ gallons of gasoline cost 6 dollars, how many gallons of gasoline can one buy for 16 dollars? Answer: 12 gallons

36. If there are 12 inches in 1 foot, how many inches are there in 4 feet? Answer: 48 inches

37. If there are 12 inches in I foot, how many inches are there in x feet? (answer in terms of x) answer: $12x$ inches

38. If a person's heart beats 4 times every 3 seconds, how many times will it beat in 21 seconds. Answer: 28 times

39. To prepare an optimum mixture of gasohol which consists of alcohol and gasoline, Alexander used 20 liters of alcohol and 180 liters of gasoline. How many liters of alcohol are needed if 270 liters of gasoline are used and the composition ratio of the mixture remains unchanged? Answer: 30 liters

40. Charles' Law states that at constant pressure, the volume of a gas is directly proportional the temperature of the gas. If when the volume of a given mass of a gas is 4.2 L, the temperature is $309°K$, find the volume when the temperature is $297°K$ Answer : 4.0 L

41. Gay-Lussac's Law states that at constant volume, the pressure on a gas is directly proportional to the temperature of the gas. If when the pressure on a given mass of a gas is 720 mm Hg, the temperature is $303°K$, find the pressure when the temperature is $309°K$? Answer: 734 mm Hg

42 The ratio of a man's weight on the earth to his weight on the moon is 6:1. If the man weighs (on earth) 240 lb, what is his weight on the moon? Answer: 40 lb

Lesson 5 Exercises B

Problems on direct proportion in which some terms refer to non distinct sets

1. If 2 out of 5 students at a college are males and there are 2,800 students at this college, how many of the students are males? Answers: 1,120

2. If 2 out of 5 students at a college are males and there are 2,800 students at this college, how many of the students are females? Answers: 1,680

3. If 2 out of 5 students at a college are males and there are 1,120 males at this college, how many students are at this college? Answers: 2,800

4.. If 40% of students at a college are males and there are 1,120 males at this college, how many students are at this college? Answers: 2,800

5. If 40% of the students at a college are males and there are 2,800 students at this college, how many of the students are males? Answers: 1,120

6. Is $\frac{2}{5}$ of a number is 16, what is the number? Answer: 40

7. James and Mary are to divide some amount of money and James is to receive $\frac{5}{12}$ of this amount. If James receives 30 dollars, what is the amount of money to be divided? Answer: 72 dollars

8. If 2 out 3 dollars in a bank account are to be given to Diana, If Diana was given $30,000 how much money is in this bank account? Answer: $45,000

9. If 1 out of 3 liters of an alcohol-water solution is pure alcohol, and this solution contains 12 liters of pure alcohol, what is the total volume of this alcohol-water solution? **Answer** 36 liters

10. If q out of p fruits are oranges, and there are r oranges, how many fruits are there. Obtain an expression for the number of fruits in terms of p, q, and r. Answer: $\frac{pr}{q}$ fruits

11. In a math class, 2 out of 4 students passed the final exam. If 32 students passed the exam, how many students are in this class? Answer: 64

Lesson 5 Exercises C

1. y varies as x. When $y = 12, x = 4$. Find y when $x = 10$.
2. The distance S an object falls from rest is directly proportional to the square of the time t. If when $S = 64, t = 2$, find S when $t = 4$.
3. W is proportional to V. When $W = 120, V = 3$, find W when $V = 8$.
4.. S is directly proportional to t.. When $S = 64, t = 2$. Find t when $S = 288$.
5. Express by means of an equation: $(a - b)$ is proportional to $(c - d)$.
6. The distance S an object falls from rest varies directly as the square of the time t. If when $S = 32, t = 1$, find S when $t = 5$.

Answers: 1. 30; **2.** 256; **3.** 320; **4.** 9; **5.** $a - b = k(c - d)$, where k is a constant; **6.** 800.

Lesson 5 Exercises D

1. At constant pressure, the volume V of a perfect gas varies as the absolute temperature T. If when the temperature is 2730 Absolute, the volume of one gram of a gas is 1.7 cm^3, find the volume when the absolute temperature is 4830°.
2. According to Hooke's law, the force required to stretch a spring is directly proportional to the elongation of the spring. If a 20 lb force stretches a spring 6 in., what force will be required to stretch it 8 in.?
3. The cost C of gasoline is proportional to the number of gallons N of gasoline. If 10 gallons of gasoline cost $12, what is the cost of 25 gallons of gasoline? How many gallons of gasoline can $18 purchase?

Answers: 1. 3.0 cm^3; **2.** 27 lb; **3.** $30; 15 gallons

Lesson 6
Methods for Solving Inverse Proportion Problems

It is very interesting but not surprising that the operations in the steps for solving inverse proportion problems are the opposite operations (inverse operations) in the steps for solving direct proportion problems. By operations, we mean multiplication and division are opposite (inverse) operations. Thus if we know the operations in the steps for solving a direct proportion problem, we can deduce the operations in the steps for solving an inverse proportion problem (by using the opposite operations). For every method for solving direct proportion problems, there is a corresponding method for solving inverse proportion problems, with the steps in the inverse method obtained from the direct proportion method by using opposite operations.

Method 1: **Product = Product Method** (Algebraic)

Assume a is inversely proportional to b.

(same as a is proportional to the **reciprocal** of b)

Then a_1 corresponds to b_1, and a_2 corresponds to b_2. and

$$a_1 : \frac{1}{b_1} \text{ as } a_2 : \frac{1}{b_2}$$

$$a_1 \div \frac{1}{b_1} = a_2 \div \frac{1}{b_2}$$

$$a_1 b_1 = a_2 b_2 \qquad \textbf{(product = product)}$$

Or if we use the familiar terms a, b, c, d as the terms of the inverse proportion

Then $a : \frac{1}{b}$ as $c : \frac{1}{d}$

$a \div \frac{1}{b}$ as $c \div \frac{1}{d}$

$ab = cd$ **(product = product) <------** inverse proportion

We will use the product = product form since this form is easy to remember.

Compare: **Direct proportion** $\frac{a}{b} = \frac{c}{d}$ <-------**(quotient = quotient)**

Inverse proportion $ab = cd$ **<------(product = product)**

Example 1 2 people can paint a house in 6 days. How many people would be required to paint the house in 3 days? Derive a formula relating the number of people and the number of days; and use this formula to find the number of people required to paint the house in 3 days.

Method 1 Product = Product Method

It is known that as the number of people increases, the number of days required decreases, and vice versa.

Let the number of people = P,

Let the number of days D.

Then P is indirectly proportional to D,

P_1 corresponds indirectly to D_1 as P_2 corresponds indirectly to D_2.

Since this is indirect proportion, $P_1D_1 = P_2D_2$.

From the problem, $P_1 = 2$, $D_1 = 6$, $P_2 = ?$. $D_2 = 3$.

Substituting these values in the equation

$$2(6) = P_2(3)$$

$$\frac{2(6)}{3} = P_2$$

$$4 = P_2$$

Therefore, 4 people are required to paint the house in 3 days.

Method 2: Units Label Method for Inverse Proportion

By reversing the steps in the corresponding direct proportion method (Method 2), we obtain the approach covered below.

We repeat the question: 2 people can paint a house in 6 days. How many people would be required to paint the house in 3 days?

Step 1: Find the product, 2 people•(6 days)

Step 2: Divide : $\frac{2 \text{ people} \bullet (6 \text{ days})}{3 \text{ days}}$

$$\frac{2 \text{ people} \bullet (\overset{2}{\cancel{6 \text{ days}}})}{\cancel{3 \text{ days}}}$$

$= 4$

Therefore, 4 people are required to paint the house in 3 days.

Method 3: Unitary Method (Arithmetic)

We repeat the question: 2 people can paint a house in 6 days. How many people would be required to paint the house in 3 days?

Solution

The number of people required for 6 days = 2

The number of people required for 1 day = 2×6 (Multiply. More people are needed)

The number of people required for 3 days = $\frac{2 \times 6}{3}$ (We divide. Less people are needed)

$= 4$

Therefore, 4 people are required to paint the house in 3 days.

Method 4: More or Less Method (Arithmetic)

We repeat the question: 2 people can paint a house in 6 days. How many people would be required to paint the house in 3 days?

Solution

Step 1: The number of people needed to paint the house in 6 days = 2

Step 2: The number of people needed to paint the house in 3 days = $2 \times \frac{6}{3}$

$= 4$ (more people needed)

Therefore, 4 people are required to paint the house in 3 days.

(Note that in Step 2, in forming a fraction using 3 and 6, we used an improper faction, $\frac{6}{3}$ (fraction greater than or equal to 1), since we expect the number of people to be greater than 2. For the work to be done faster, more people are needed. If we had expected the number of people to be less than 2, we would have used the proper fraction $\frac{3}{6}$). Master this arithmetic method.

Method 5: Variation Method (Algebraic) **We reword the question:**

The number of people needed to paint a house varies inversely as the number of days taken to paint the house. If when the number of people is 2, the number of days is 6, find the number of people when the number of days is 3.

Step 1: Let the number of people = n

Let the number of days = d

Then since n varies inversely as d, (or n is proportional to the **reciprocal** of d

$$n = \frac{k}{d}$$

Step 2: Determine the value of k by substituting $n = 2, d = 6$ in $n = \frac{k}{d}$

Then $2 = \frac{k}{6}$ and from which $k = 12$.

Step 3: Substitute $k = 12$ in $n = \frac{k}{d}$ to obtain a formula relating n and d.

Then we obtain $n = \frac{12}{d}$

Step 4: Substitute for $d = 3$ in the formula . $n = \frac{12}{d}$ and solve for n

Then $n = \frac{12}{3} = 4$

Thus when $d = 3, n = 4$

Therefore, 4 people are required to paint the house in 3 days.

One more example on Method 5

Example 2 The number of people required to sort a quantity of letters varies inversely as the time taken to sort the letters. If 8 people can sort a quantity of letters in 3 hours, how many people would be required to sort the same quantity of letters in 2 hours?

Let the number of people required to sort the letters in x hours be y

Step 1: Then $y = \frac{k}{x}$ (1)

Step 2: When $x = 3, y = 8$. Substitute these values in equation (1) and solve for k.

$8 = \frac{k}{3}$.

Solving, $k = 24$.

Substitute for $k = 24$ in equation (1) above to obtain

the formula $y = \frac{24}{x}$ ....(2)

Step 3. when $x = 2, y = \frac{24}{2}$ (Replacing x by 2 in equation (2))

$y = 12$

Therefore, to sort the letters in 2 hours would require 12 people.

Example 4 Assume pressure P is indirectly proportional to volume V

Then P_1 corresponds to V_1 as P_2 corresponds to V_2

Since P is inversely proportional to V,

$P_1 : \frac{1}{V_1}$ as $P_2 : \frac{1}{V_2}$ and from which

Then $P_1V_1 = P_2V_2$ (product = product)

The above equation is a familiar formula for gases in physics.

Lesson 6 Exercises A (Inverse Proportion)

1. If 2 people can paint a house in 6 days, how many people would paint the house in 3 days?

Answer: 4 people

2. At a speed of 60 miles per hour, a car takes 2 hours to travel from City A to City B. At a speed of 40 miles per hour, how long will it take the car to ravel from City A to City B?

Answer: 3 hours

3. If 6 mail men can sort a given quantity of mail in 4 hours, how many mail men would be needed to sort the same quantity of mail in 2 hours?

Answer: 12 men

4. According to Boyle's law, at constant temperature, the volume of a given mass of a gas is inversely proportional to pressure on the gas. When the volume of a gas is 50 cu. in, the pressure is 10 lb per sq. in. What is the pressure (a) when the volume is 80 cu. in.? ;
(b) when the volume is 15 cu. in.?;
(c) What is the volume when the pressure is 40 lb per sq. in.?

Answer: (a) $6\frac{1}{4}$ lb per sq. in.; (b) $33\frac{1}{3}$ lb per sq. in; (c) $12\frac{1}{2}$ u. in;

5. The speed of rotation of meshed gears is inversely proportional to the number of teeth. A gear with 12 teeth rotates at 840 rpm. What is the rotation speed of a 16-tooth gear?

Answer: 630 rpm

6. The electric current in a circuit is inversely proportional to the resistance of the circuit. If the current is 22 amps when the resistance is 5 ohms, find the current when the resistance is 10 ohms.

Answer: 11 amps

Lesson 6 Exercises B

1. At constant pressure, the volume V of a perfect gas varies as the absolute temperature T. If when the temperature is 2730° Absolute, the volume of one gram of a gas is 1.7 cm^3, find the volume when the Absolute temperature is 4830°.
2. According to Hooke's law, the force required to stretch a spring is directly proportional to the elongation of the spring. If a 20 lb force stretches a spring 6 in., what force will be required to stretch it 8 in.?
3. The cost C of gasoline is proportional to the number of gallons N of gasoline. If 10 gallons of gasoline cost $12, (a) what is the cost of 25 gallons of gasoline? and (b) how many gallons of gasoline can $18 buy?

Answers: 1. 3.0 cm^3; **2.** $26\frac{2}{3}$ lb; **3.** (a) $30, (b) 15 gallons

Lesson 6 Exercises C

1. y varies inversely as x. If when $y = 6, x = 2$, determine y when $x = 8$.

2. F is indirectly proportional to the square of D. When $F = 30, D = 3$. Find F when $D = 2$.

3. W is indirectly proportional to V. When $W = 15, V = 45$ find W when $V = 60$.

Answers: **1**. $\frac{3}{2}$; **2**. $67\frac{1}{2}$; **3**. $11\frac{1}{4}$;

Lesson 6 Exercises D

1 According to Boyle's law, at constant temperature, the volume of a given mass of a gas varies inversely as the pressure on the gas. When the volume of a gas is 50 in.^3, the pressure is 10 lb per $in.^2$. What is the pressure (a) when the volume is 80 in.^3?; (b) when the volume is 15 in.^3 ?; (c) Find the volume when the pressure is 40 lb per in.^2.?

2. If 8 people can complete a piece of work in 40 days, how many people will complete the same piece of work in I0 days? Assume that all the people work at the same rate as one another.

3. At constant temperature, the volume of a given mass of a gas is inversely proportional to the pressure on the gas. When the volume is 150 cu. in,. the pressure is 30 lb per $in.^2$. Find the pressure (a) when the volume is 60 in.^3?; (b) when the volume is 300 in.^3.

Ans:**1**. (a) $6\frac{1}{4}$ lb; (b) $33\frac{1}{3}$ lb per in.^2; (c) $12\frac{1}{2}$ in.^3 **2**. 32 people; **3**. (a) 75 lb per in.^2.; (b) 15 lb per in.^2.

Lesson 7

Compound Proportion (Variation) Problems

Note: If you have no algebraic background, you may skip the algebraic part and cover only the arithmetic part (p.35). However, you can study the algebraic material in the Appendix, (p.159-167) before covering these chapters.

Deriving formulas for any proportion problem, with any number of variables and involving both direct and inverse relationships

Case 1: Joint Proportion (Joint Variation)

If a quantity varies **directly** as **two** or more other quantities (i.e. as the product of two or more other quantities), we call such a proportion **joint proportion or joint variation**.

Example 1: If z varies directly as x and y, then

$$z = kxy \text{ or } \frac{z}{xy} = k \text{ (joint variation)}$$

or using subscripted variables, $\frac{z_1}{x_1y_1} = \frac{z_2}{x_2y_2}$ <-- Faster form since you do not have to find k first.

Case 2: Joint Inverse Proportion (Joint Inverse Variation)

The following type of proportion may not be found in current textbooks. However, the author believes that for completeness, this proportion should be included. See the two application examples below.

If a quantity varies **indirectly** as **two** or more other quantities (i.e. as the inverse of the produc t of two or more other quantities), we will call such a proportion **inverse joint proportion or** inverse **joint variation**.

Example If z varies **indirectly** as x and y, then

$$z = \frac{k}{xy} \text{ or } xyz = k \text{ (inverse joint variation)}$$

or using subscripted variables, $z_1x_1y_1 = z_2x_2y_2$ <-- Faster form

Examples

1. A weight watcher exercising 3 hours per day, 5 times a week lost 10 pounds in 4 months. How many months will it take the weight watcher to lose 10 pounds exercising 2 hours per day, 6 times a week? **Answer:** 5 months

2. A car traveling at 48 mph and for 6 hours per day covered a certain distance in 10 weeks How many weeks will it take the car traveling at 60 mph, and for 4 hours per day to cover the same distance? **Answer:** 12 weeks

Case 3 Combined Proportion (Combined Variation)

If a quantity varies **directly** as one quantity (or as two or more quantities) and **inversely** as another quantity (or other quantities), we call such a variation **combined variation**.

Example 2: If z varies directly as x and inversely as y, then

$$z = k\frac{x}{y} \text{ or } \frac{zy}{x} = k \text{ (combined variation)}$$

or using subscripted variables, $\frac{z_1y_1}{x_1} = \frac{z_2y_2}{x_2}$ <-- Faster form since you do not have to find k first.

Example 3: Assume that pressure P is directly proportional to the temperature T, and inversely proportional to the volume V.

Step 1: Then P_1 corresponds to T_1 directly, and to V_1 inversely as P_2 corresponds to T_2 directly, and to V_2 inversely.

Step 2: Write down P_1 (the first quantity.)

Step 3: For each of the other quantifies, if it is directly proportional to P_1, then it divides P_1 (i.e., write it as a divisor), but if it is inversely proportional to P_1, it multiplies P_1 (that is write it as a factor in the numerator). Similarly, repeat the process beginning with P_2.

Then $\frac{P_1V_1}{T_1} = \frac{P_2V_2}{T_2}$, a familiar gas law in physics.

Example 4: Assume that A is directly proportional to B and directly proportional to C.

Step 1: Then A_1 corresponds to B_1 directly, and to C_1 directly as A_2 corresponds to B_2 directly, and to C_2 directly.

Step 2: Write down A_1 (the first quantity).

Step 3: For each of the other quantities if it is directly proportional to A_1, then it divides A_1 (That is write it as a divisor) but if it is inversely proportional to A_1, it multiplies A_1 (i.e., write it as a factor in the numerator) . Similarly, on the right side of the equality symbol, repeat the process beginning with A_2.

Then $\frac{A_1}{B_1C_1} = \frac{A_2}{B_2C_2}$

Example 5 Assume that the temperature T is directly proportional to the pressure, P, and directly proportional to the volume, V.

Step 1: Then T_1 corresponds to P_1 directly, and to V_1 directly as T_2 corresponds to P_2 directly, and to V_2 directly.

Step 2: Write down T_1 (the first quantity).

Step 3: For each of the other quantifies if it is directly proportional to T_1, then it divides T_1 (i.e., write it as a divisor) but if it is inversely proportional to T_1, it multiplies T_1 (i.e., is write it in the numerator as a factor) Similarly, repeat the process beginning with T_2.

Then $\frac{T_1}{P_1V_1} = \frac{T_2}{P_2V_2}$

By inverting both sides of the above equation, we obtain

$\frac{P_1V_1}{T_1} = \frac{P_2V_2}{T_2}$, the familiar gas law.

Question: Can we obtain the same equation in above example, given the following information?: Assume volume V is directly proportional to temperature T. and inversely proportional to the pressure P. Then V_1 corresponds to T_1 directly, to P_1 inversely as V_2 corresponds to T_2 directly, and to P_2 inversely.

Translating into an equation $\frac{V_1P_1}{T_1} = \frac{V_2P_2}{T_2}$ or equivalently , $\frac{P_1V_1}{T_1} = \frac{P_2V_2}{T_2}$. Answer: Yes.

Example 6: Assume that A is inversely proportional to B and inversely proportional to C.

Then A_1 corresponds to B_1 inversely, and to C_1 inversely as A_2 corresponds to B_2 inversely, and to C_2 inversely.

Then, following the agreement made in the previous examples, we obtain

$A_1B_1C_1 = A_2B_2C_2$ **(product = product)**

Solving Compound Proportion Problems Using Arithmetic

In the previous examples, we covered how to solve compound proportion problems using the variation method. However, we can also solve compound proportion problems using **arithmetic.** We can apply **Method 4** of Lesson 18 (p.66) and Lesson 19 (p.81), (The "More or Less Method")

Example 7. If 8 men can make 20 tables in 6 days, how many men would be required to make 60 tables in 9 days?

Solution:

The number of men required to make 20 tables in 6 days = 8 men

The number of men required to make 60 tables in 9 days = 8 men $\times \frac{60 \text{ tables}}{20 \text{ tables}} \times \frac{6 \text{ days}}{9 \text{ days}}$

(See p.14 and p.29 for setting up the fractions) $= 8 \text{ men} \times \frac{60}{20} \times \frac{6}{9} = 16 \text{ men}$

Lesson 7 Exercises A

(Compound Proportion)

1. If 5 people working 6 hours per day can complete a job in 8 days, how many people working 2 hours per day can complete the same job in 10 days? **Answer:** 12 people

2 If 12 people working 4 hours per day, 3 days per week can complete a job in 4 years, how many people working 9 hours per day, 4 days per week can complete the same job in 2 years? **Answer:** 8 people

3. A weight watcher exercising 3 hours per day, 5 times a week lost 10 pounds in 4 months. How many months will it take the weight watcher to lose 10 pounds exercising 2 hours per day, 6 times a week? **Answer:** 5 months

4. A car traveling at 48 mph and for 6 hours per day covered a certain distance in 10 weeks. How many weeks will it take the car traveling at 60 mph, and for 4 hours per day to cover the same distance? **Answer:** 12 weeks

5 If 10 people working 6 hours per day, 7 days per week earned a certain amount of income each week, how many people working 4 hours per day, 5 days per week would earn the same income in one week? **Answer:** 21 people

6 Six men working 2 days a week can build a fence in 8 weeks. How many men working 4 days a week will build the same fence in 12 weeks. **Answer:** 2 men

7. Six men working 2 days a week can build a fence in 8 weeks. How many weeks will it take 3 men working 4 days a week to build the same fence. **Answer:** 8 weeks

8. If 8 men can make 20 tables in 6 days, how many men would be required to make 60 tables in 9 days? **Answer:** 16 men

Lesson 7 Exercises B

1. Z varies jointly as x and y. If when $x = 2, y = 3, Z = 16$, what is Z when $x = 5$ and $y = 4$?

2. V is indirectly proportional to P and directly proportional to T. When $V = 30, T = 25, P = 15$. Find V when $T = 40$, and $P = 3$.

3. F is directly proportional to G and varies inversely as L. When $F = 256, G = 36$, and $L = 30$. What is G, when $F = 128$ and $L = 15$?

4. The kinetic energy E of a moving object varies as the mass M. of the object and the square of the velocity V. If when $E = 50, M = 4, V = 5$, find E when $M = 16$ and $V = 3$.

5. If 12 people working 2 days a week can complete a piece of work in 4 weeks, how many people working 3 days per week can complete the same piece of work in 2 weeks? (Assume that all the people involved work at the same rate as one another)

Answers: **1.** $53\frac{1}{3}$; **2.** 240; **3.** 9; **4.** 72; **5.** 16 people;

EXTRA

Determining the type of relationship, given an equation

Determine how the first variable is related to each of the other variables in each equation.

(a) $\frac{P_1V_1}{T_1} = \frac{P_2V_2}{T_2}$; (b) $\frac{A_1B_1}{C_1} = \frac{A_2B_2}{C_2}$; **(c)** $\frac{B_1}{D_1} = \frac{B_2}{D_2}$; **(d)** $\frac{A_1B_1C_1}{D_1E_1} = \frac{A_2B_2C_2}{D_2E_2}$

Solution

(a) *P* is inversely proportional to *V* and directly proportional to *T*
(b) *A* is inversely proportional to *B* and directly proportional to *C*
(c) *B* is directly proportional to *D*.
(d) *A* is inversely proportional to *B*, inversely proportional C, directly proportional to *D*, and directly proportional *E*.

Comparison of Sample Formulations of Proportion Problems

Proportion Formulation (subscript notation)	**Variation Formulation**	**Statement of relationship**
1. $\frac{V_1}{T_1} = \frac{V_2}{T_2}$	$\frac{V}{T} = k$ or $V = kT$	*V* is directly proportional to *T*
2. $P_1V_1 = P_2V_2$	$PV = k$ or $P = \frac{k}{V}$	*P* is inversely proportional to *V*
3. $\frac{P_1V_1}{T_1} = \frac{P_2V_2}{T_2}$	$\frac{PV}{T} = k$ or $PV = kT$	*P* is inversely proportional to *V* and directly proportional to *T*.
4. $I_1R_1 = I_2R_2$	$IR = k$ or $I = \frac{k}{R}$	*I* is inversely proportional to *R*
5. $E_1I_1 = E_2I_2$	$EI = k$ or $E = \frac{k}{I}$	*E* is inversely proportional to *I*
6. $\frac{E_1}{N_1} = \frac{E_2}{N_2}$	$\frac{E}{N} = k$ or $E = kN$	*E* is directly proportional to *N*
7. $I_1N_1 = I_2N_2$	$IN = k$ or $I = \frac{k}{N}$	*I* is inversely proportional to *N*
8.		
9.		

Lesson 8
Graphs (Pictures) of Proportional Relationships
Direct Proportion

Given: y is directly proportional to x

Table 1: y = 2x

x =	0	1	2	3	4
y =	0	2	4	6	8

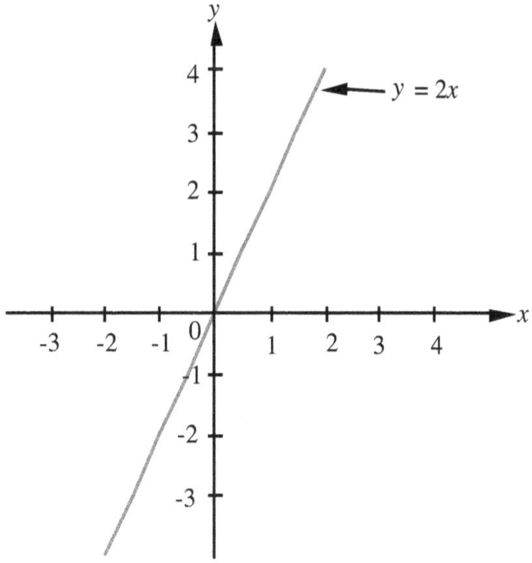

Figure: Graph of $y = 2x$

The graph of $y = kx$ or $\frac{y}{x} = k$

Inverse Proportion

Given: y is inversely proportional to x

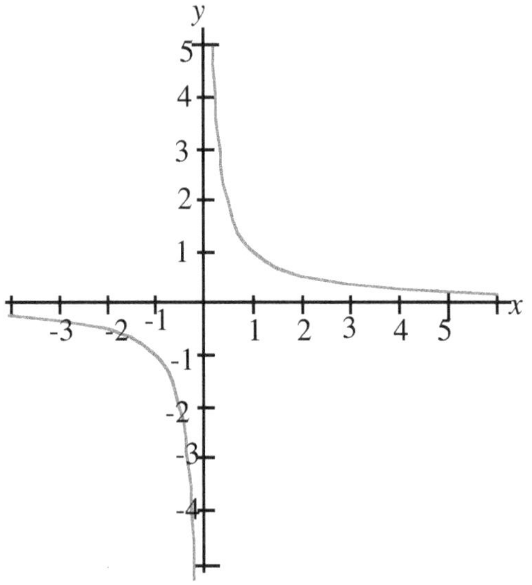

Figure: Graph of $f(x) = \frac{1}{x}$

$y = \frac{k}{x}$ or $xy = k$

CHAPTER 3

Geometric Applications of Ratios
Similar Triangles

Lesson 9 **Properties of Similar Triangles; Theorems and Proofs**
Lesson 10: **Comparison of Congruency and Similarity of Triangles and Applications of Similarity Theorems**

Lesson 9

Properties of Similar Triangles; Theorems and Proofs

Preliminaries

Ratio: The ratio of two quantities a and b (written $a{:}b$) is the fraction $\frac{a}{b}$ (i.e., is the first quantity divided by the second quantity). The quantities a and b are called the terms of the ratio.

But the ratio $b{:}\,a$ is $\frac{b}{a}$

The ratio 3:4 is $\frac{3}{4}$; but the ratio 4:3 is $\frac{4}{3}$

Note above that order is important.

Proportion : A proportion is a statement that two ratios are equal.
The proportion a is to b as c is to d (symbolically $a{:}b{::}c{:}d$) is the equality

$$\frac{a}{b}=\frac{c}{d} \qquad (1)$$

In the above proportion, a and d are called the extremes and b and c are called the means of the proportion.

Proportion Principles

1. We can invert both ratios of the proportion: From (1) above, we would obtain after the inversion:

$$\frac{b}{a}=\frac{d}{c}$$

2. We can interchange the extremes of a proportion: From (1) above we would obtain:

$$\frac{d}{b}=\frac{c}{a}$$

3. We can also interchange the means to obtain:

$$\frac{a}{c}=\frac{b}{d}$$

Proportion principles can be used to obtain desired ratios. For example, if we have $\frac{a}{b}$ as the ratio of one side of a proportion, and we desire the ratio $\frac{b}{a}$, then proportion principle #1 above (inversion of both sides of a proportion) can be used to obtain this ratio, $\frac{b}{a}$, from $\frac{a}{b}$.

Example: If $\frac{a}{b}=\frac{2}{3}$, find $\frac{b}{a}$.

Solution By inverting both sides of the equation, we obtain $\frac{b}{a}=\frac{3}{2}$

Identifying corresponding sides and angles of two similar triangles

1. Corresponding sides are opposite congruent angles.
2. Congruent angles are opposite corresponding sides.

To identify corresponding sides in two similar triangles say, Δ #1 and Δ #2:

Case 1: Given or knowing the lengths of two sides and the included angle in each triangle (incl'd angles being congr.).
The side with the smaller length in Δ #1 corresponds to the side with the smaller length in Δ #2.
The side with the greater length in Δ #1 corresponds to the side with the greater length in Δ #2.
The third side (length known or unknown) in Δ #1 corresponds to the third side of Δ #2.

Case 2: Given the lengths of all three sides of each triangle.
The side with the smallest length in Δ #1 corresponds to the side with smallest length in Δ #2.
The side with the greatest length Δ #1 corresponds to the side with the greatest length in Δ #2.
The side with intermediate length in Δ #1 corresponds to the side with intermediate length in Δ #2,

Case 3: Given the measures of two pairs of congruent angles in the two triangles.
Step 1: Pick a side in Δ #1, and note the angle this side is opposite to.
Step 2: Go to Δ #2 and locate the angle which is congruent to the angle noted in Δ #1. The side opposite to this located angle in Δ #2 is the corresponding side.

To identify corresponding angles in two similar triangles say Δ's #1 and #2:

Given or knowing the corresponding sides
Step 1: Pick an angle in Δ #1, note the side this angle is opposite to.
Step 2: Go to Δ #2 and locate the side which corresponds to the side noted in Δ #1. The angle opposite to this located side in Δ #2 is the corresponding angle. Note of course that if the angles in Δ #1 and Δ #2 have the same measure, these angles correspond and they are also congruent.

Note: The above identification guidelines assume that the two triangles are given to be or have been proved to be similar. Do not therefore apply them until you know that the two triangles are similar. However, a modified form of this guide can be helpful in picking corresponding sides, even if the triangles are not similar.
See also Examples 1 & 2, pages 43 & 45 respectively: the "Plan "parts.

Example Identify the corresponding lengths and corresponding angles in the Δ's given below

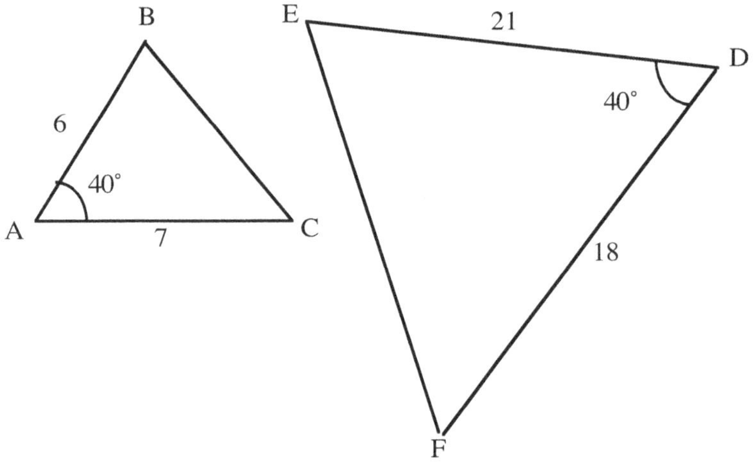

Solution: Smaller length of one triangle corresponds to the smaller length of the other triangle; the larger length of one triangle corresponds to the larger length of the other triangle. Therefore 6 corresponds to 18; 7 corresponds to 21. ∠ D corresponds to ∠ A; ∠ C to ∠ E; and ∠ B to ∠ F.

Definition: Two triangles are similar ($\sim$) if and only if the corresponding angles are congruent and the ratio of corresponding sides is the same for each pair of corresponding sides.

Corresponding angles being congruent implies that corresponding angles have equal measures. The statement that "the ratio of corresponding sides is the same for each pair of corresponding sides" implies that the corresponding sides are in proportion.

Properties of Similar Triangles

1. Each pair of corresponding angles are congruent (i.e. have equal measure).
2. Corresponding sides are in proportion (that is, the ratio of the lengths of any pair of corresponding sides is equal to the ratio of the lengths of any other pair of corresponding sides).

Ordered naming of triangles in stating the similarity of triangles and identification of corresponding parts from the similarity statement

As it is in the case of the congruency of triangles, it is also good communication to follow a certain order in stating that two triangles are similar. The logical order is to match the letters of the vertices in stating that two triangles are similar.

Example Stating that $\Delta ABC \sim \Delta DEF$ (read "triangle *ABC* is similar to triangle *DEF*"), implies that the vertex *A* (of ΔABC) corresponds to the vertex *D* (of ΔDEF); the vertex *B* (of ΔABC) corresponds to the vertex *E* (of ΔDEF); and the vertex *C* (of ΔABC) corresponds to the vertex *F* (of ΔDEF).
By matching the vertices, the sides $\overline{AB}$ and $\overline{DE}$ naturally correspond, $\overline{BC}$ and $\overline{EF}$ correspond; and $\overline{AC}$ and $\overline{DF}$ correspond.

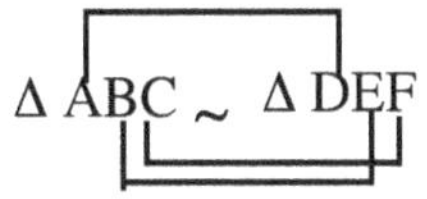

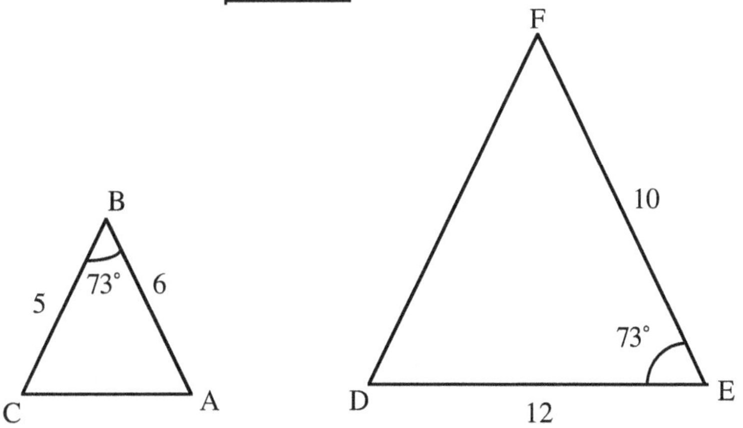

Theorems and Proofs

Conditions or criteria for proving similarity of triangles

* The conditions or criteria for proving that two triangles are similar are embodied in the following theorems:

Theorem 1: If two pairs of corresponding sides of two triangles are in proportion **and** the included angles (the angles between the sides involved) are congruent, then the two triangles are similar. Abbreviated **SAS.**

In the figure below, Δ ABC ~ Δ DEF (read "triangle ABC is similar to triangle DEF"), since the conditions in Theorem 1 have been satisfied. In the next example we will prove the similarity of these two triangles.

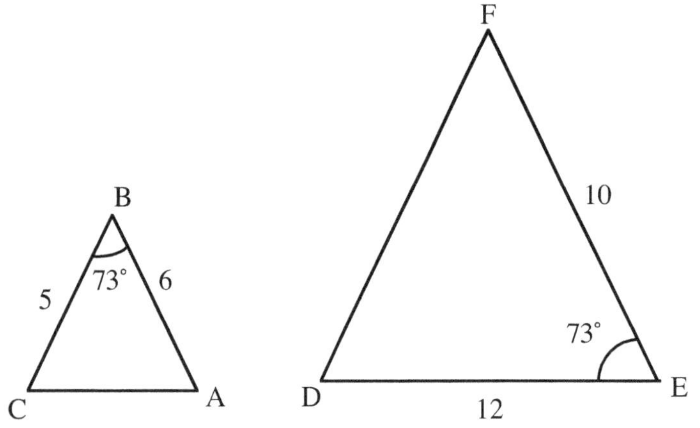

* In the Appendix, all the theorems on similar triangles covered have been stated in terms of the lengths of the sides, and the measures of the angles. These forms of the theorems may be useful when dealing with the lengths of the sides of the triangles.

Example 1 In the triangles shown (**Figure 1**), prove that $\Delta ABC \sim \Delta DEF$

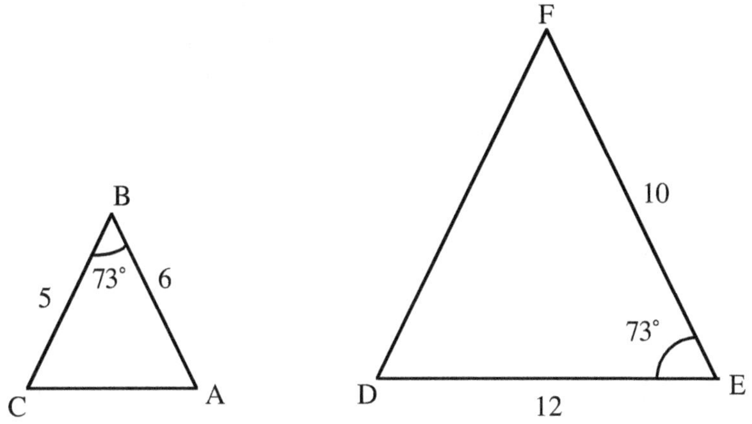

Figure 1

Given: 1. In ΔABC, $m\angle B = 73°$, $AB = 6$, $BC = 5$.

2. In ΔDEF, $m\angle E = 73°$, $DE = 12$, $EF = 10$.

To prove: That $\Delta ABC \sim \Delta DEF$

Plan: Arrange the dimensions of each triangle in increasing order (or decreasing order) ,i.e. smaller length first, etc. For ΔABC, we have 5,6; for ΔDEF, we have 10 ,12.
Divide the smaller length of ΔABC by the smaller length of ΔDEF, and note the line segments involved in the ratio. Next, divide the larger length of ΔABC by the larger length of ΔDEF and again note the sides involved.. If we obtain the same ratio (same quotient), the sides of the triangles are in proportion, otherwise, they are not in proportion. We will also check to see if the given angles have the same measure (or are congruent) **and** are also included (i.e., the angles are also between the sides involved in the ratios).

Proof:

Statements	Reasons
1. The ratio, $5:10 = \frac{5}{10} = \frac{1}{2} = \frac{BC}{EF}$	**1.** The ratio of two quantities is the first quantity divided by the second quantity
2. The ratio, $6:12 = \frac{6}{12} = \frac{1}{2} = \frac{AB}{DE}$	**2.** The ratio of two quantities is the first quantity divided by the second quantity.
3. $\frac{AB}{DE} = \frac{BC}{EF}$	**3.** If quantities are equal to the same quantity, they are equal to each other.
4. AB, DE, BC and EF are in proportion.	**4.** A proportion is the equality of two ratios.
5. $m\angle B = 73°$	**5. Given**
6. $m\angle E = 73°$	**6.** Given
7. $m\angle B = m\angle E$	**7.** If quantities are equal to the same quantity, they are equal to each other.
8. $\angle B \cong \angle E$	**8.** If two angles have the same measure, they are congruent.
9. $\angle$ **B and** $\angle$ **E are included angles.**	
10. $\therefore \Delta ABC \sim \Delta DEF$	**10.** SAS
	Q.E.D.

How to find an unknown side (Knowing that the triangles involved are similar)

Pick the unknown side from one triangle and then pick the corresponding side in the other triangle and form a ratio. Go back to the first triangle and pick a known side whose corresponding side in the second triangle is known, and form another ratio. Equate the two ratios and solve for the unknown.

You may also pick corresponding sides and set up a proportion as "a is to b as c is to d" and translate to $\frac{a}{b} = \frac{c}{d}$ (where a, b, c, and d are the lengths of the corresponding sides).

Example 1b If the two triangles below are similar, find x.

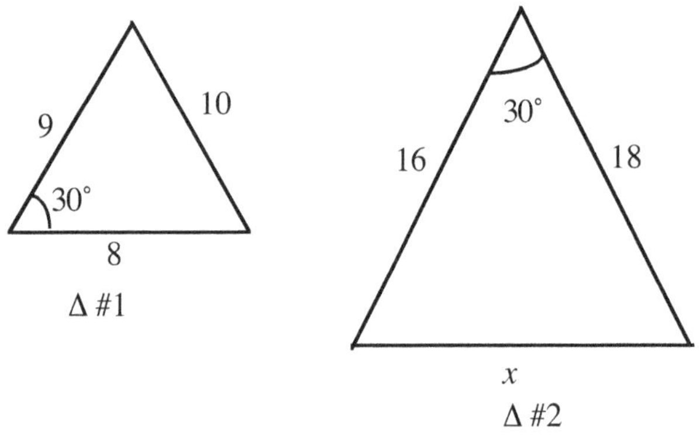

Solution

Step 1: Pick the x in say, Δ #2; note that x is opposite the angle whose measure is 30°

Step 2 : Go to the other triangle say, Δ #1; pick the side opposite the angle whose measure is 30°. The length of this corresponding side is 10.

Step 3: Form a ratio using these corresponding sides x and 10 to obtain $\frac{x}{10}$.

Step 4: Go back to Δ #2, pick a known side whose corresponding side in Δ #1 is also known. In this problem, we can pick either the "18" or the "16". We pick 16; and pick its corresponding side in Δ #1 which is 8.

Step 5: Form another ratio using 16 and 8 to obtain $\frac{16}{8}$.

Step 6 : Equate the ratios from Step 3 and Step 5, and solve for x.

$$\frac{x}{10} = \frac{16}{8}$$
$$8x = 10(16)$$
$$x = 20$$

Theorem 2: If all three pairs of corresponding sides of two triangles are in proportion (i.e., the ratio of corresponding sides is the same for each pair of corresponding sides) then the two triangles are similar. Abbreviated **SSS** (Side-Side-Side: but note implied are the **ratios of sides,** and **not** equality of sides)

In **Figure 2** below, Δ ABC ~ Δ FDE (read "triangle ABC is similar to triangle FDE", since the conditions in Theorem 2 have been satisfied. We will prove this similarity in the following example.

Example 2 In Figure 2 below, prove that Δ ABC ~ Δ FDE

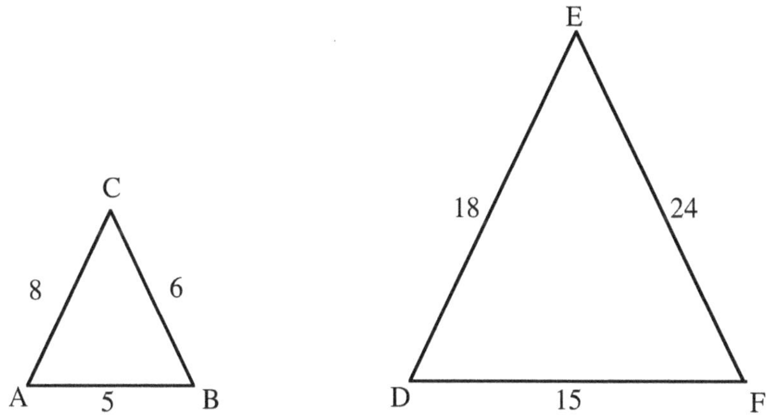

Figure 2

Given: **1**. In Δ ABC, AB = 5, BC = 6, CA = 8.
2. In Δ FDE, DF = 15, FE = 24, ED = 18.

To prove: That Δ ABC ~ Δ FDE

Plan: Arrange the dimensions of each triangle in increasing order (or decreasing order) ,i.e. smallest length first, etc. For Δ ABC, we have 5, 6, 8; for Δ FDE, we have 15 ,18, 24. Divide the smallest length of Δ ABC by the smallest length of Δ FDE, and note the line segments involved in the ratio. Next, divide the larger length of Δ ABC by the larger length of Δ FDE and note the sides; and similarly divide the largest length of Δ ABC by the largest length of Δ FDE, and again note the sides.. If we obtain the same ratio (same quotient) for all three cases, the three pairs of sides of the triangles are in proportion, and the triangles are similar; otherwise, the sides are not in proportion, and the triangles are not similar.

Proof:

Statements	Reasons
1. The ratio, $5{:}15 = \frac{5}{15} = \frac{1}{3} = \frac{AB}{DF}$	**1.** The ratio of two quantities is the first quantity divided by the second quantity.
2. The ratio, $6{:}18 = \frac{6}{18} = \frac{1}{3} = \frac{BC}{DE}$	**2.**The ratio of two quantities is the first quantity divided by the second quantity.
3. The ratio, $8{:}24 = \frac{8}{24} = \frac{1}{3} = \frac{AC}{FE}$	**3.** The ratio of two quantities is the first quantity divided by the second quantity.
4. $\frac{AB}{DF} = \frac{BC}{DE} = \frac{AC}{FE}$	**4.** If quantities are equal to the same quantity, they are equal to each other. Also all the three pairs of corresponding sides are in proportion.
5. Δ ABC~ Δ FDE	**5.** SSS

Q.E.D

Theorem 3: If two angles of one triangle are congruent to two corresponding angles of another triangle, then the two triangles are similar. Abbreviated **AA** (Angle-Angle).

Note that if two pairs of angles are congruent, the third pair are congruent, and therefore, AA implies AAA.

In **Figure 3** below, Δ ABC ~ Δ DEF (read "triangle ABC is similar to triangle DEF"), since the conditions in Theorem 3 have been satisfied. In the following example, will prove that the two triangles given are similar .

Example 3 In Figure 3 below, prove that Δ ABC ~ Δ DEF.

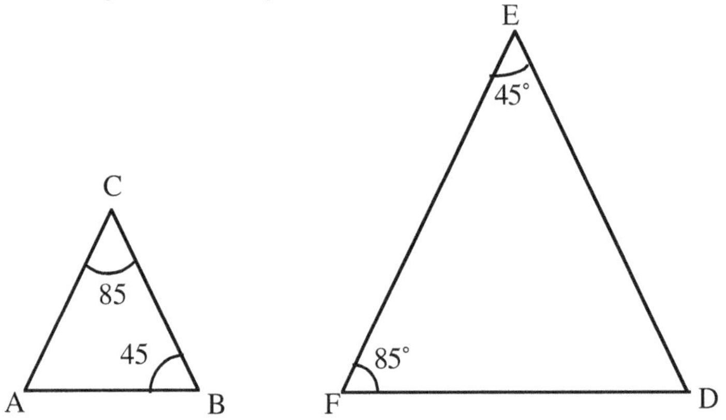

Figure 3

Given: **1**. In Δ ABC, m∠ B = 45°, m∠ C = 85°
2. In Δ DEF, m∠ E = 45° , m∠ F = 85°

To prove: That Δ ABC ~ Δ DEF

Proof:

Statements	Reasons
1. m∠ B = 45°	**1**. Given
2. m∠ E = 45°,	**2**. Given
3. m∠ B = m∠ E	**3**. If quantities are equal to the same quantity, they are equal to each other.
4. ∠ B ≅ ∠ E	**4**. If two angles have the same measure, they are congruent.
5. m∠ C = 85°	**5**. Given
6. m∠ F = 85°	**6**. Given
7. m∠ C = m∠ F	**7**. If quantities are equal to the same quantity, they are equal to each other.
8. ∠ C ≅ ∠ F	**8**. If two angles have the same measure, they are congruent.
9. Δ ABC ~ Δ DEF	**9.** AA

Q.E.D

Lesson 9 Exercises

A Determine which of the triangles in Fig.1, Fig.2 and Fig 3 are similar; and state the reason.

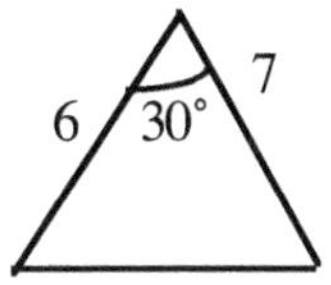

Fig.1

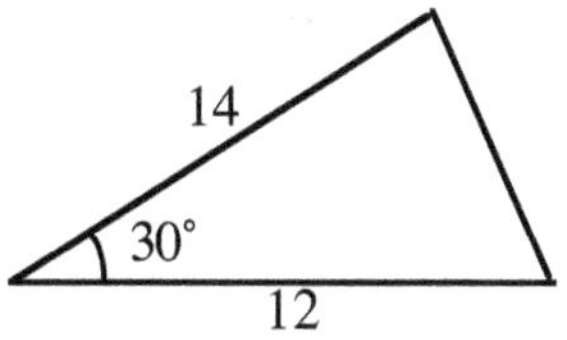

Fig. 2

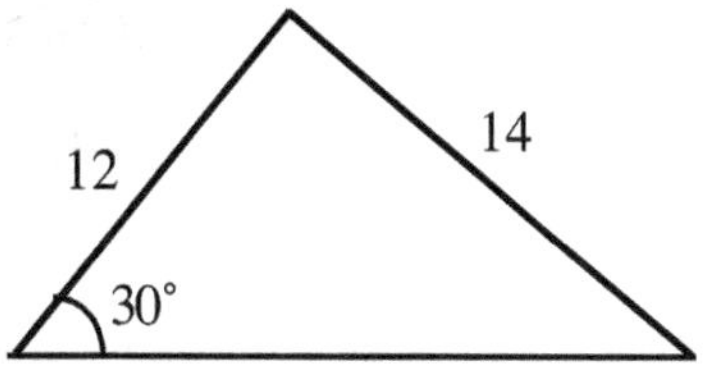

Fig. 3

Answers: **1.** Fig 1 and Fig.2 by SAS

B In the figures below, (a) Show that the Δ's are similar.
(b) Find x.

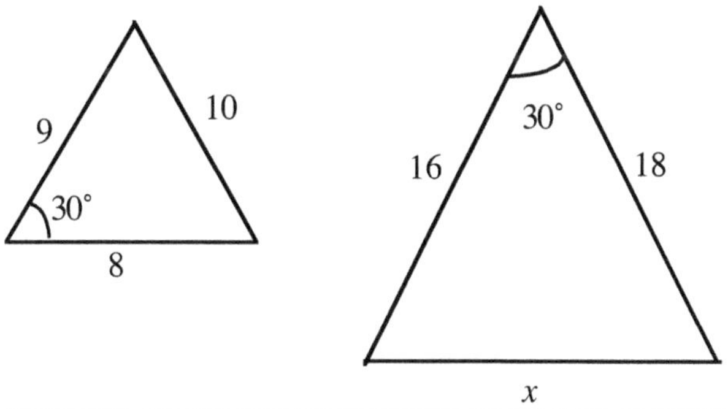

Solution: (a) See lesson and imitate; (b) 20.

C Determine if the two triangles in each problem with the give dimensions are similar:

1. Triangle #1 of sides 12, 16, 24 ; Triangle #2 of sides 8, 6, 12

2. Triangle #1 of sides 4, 6, 8; Triangle #2 of sides 8, 18, 12.

Answers: **1.** Similar; **2.** Not similar

D

Determine if the two triangles in each of the following problems are similar according to the data given.

1. Δ #1 of sides 3 and 4; included angle 75°; Δ #2 of sides 8 and 6, included angle 25°.

2. Δ #1 of sides 3, and 4; included angle 65°; Δ #2 of sides 8, and 6, included angle 65°

Answers: **1.** Not similar; **2.** Similar

Lesson 10

Comparison of Congruency and Similarity of Triangles; Applications of Similarity Theorems

Comparison of congruency and similarity of triangles

1. Corresponding angles: For **both** congruency and similarity, the **corresponding angles** are **congruent.**
2. Corresponding sides: For congruency we have **equality of sides** (equality of lengths)
   For similarity, we have **equality of ratios of sides** (equality of ratio of lengths)
3. For congruency, each ratio of corresponding sides = 1. (Ratio of similitude = 1.)
   For similarity, each ratio of corresponding sides could be 1 or another number.
   (Ratio of similitude = 1, or another number)

Note 1: If two triangles are congruent, they are also similar; but if two triangles are similar, they are not necessarily congruent.

Note 2: There is no need for **ASA** for similar triangles (the "**S**" in **ASA** is redundant), since **ASA** implies **AA** which is sufficient for two triangles to be similar.

Example 4 In Figure 4 below, $\overline{DE}$ is parallel to $\overline{BC}$.
(a) Prove that $\Delta ADE \sim \Delta ABC$. (b) If BC = 12, DE = 4, AD = 6, Find AB.

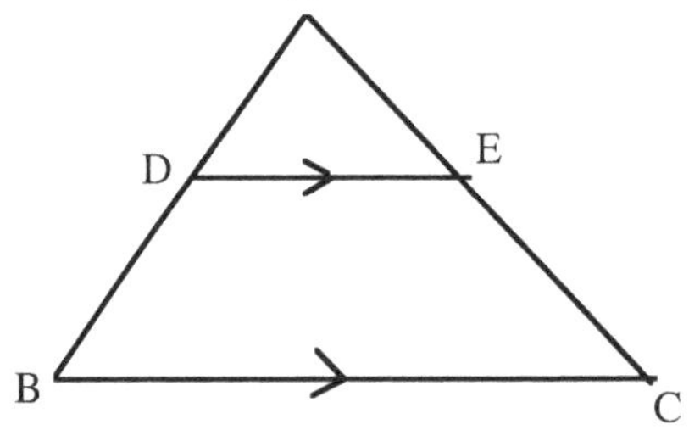

Figure 4

(a) **Given** : ΔABC in which DE || BC.
To prove: That $\Delta ADE \sim \Delta ABC$.

Proof:

Statements	Reason
1. $\angle ADE \cong \angle ABC$	**1.** Corresponding angles, || lines $\overline{DE}$, $\overline{BC}$ cut by transversal $\overline{AB}$.
2. $\angle AED \cong \angle ACB$	**2.** Corresponding angles, || lines $\overline{DE}$, $\overline{BC}$ cut by transversal $\overline{AC}$.
3. $\Delta ADE \sim \Delta ABC$	**3.** AA

(b)
BC = 12 ,DE = 4, AD = 6.
Let AB = x
Since $\Delta ADE \sim \Delta ABC$, the corresponding sides are in proportion:

Therefore, $\frac{AD}{x} = \frac{DE}{BC}$ (AD corresponds to AB. They are opposite congruent angles)

Substituting: AD = 6, DE = 4, BC =12.

$$\frac{6}{x} = \frac{4}{12}$$
$$4x = 72$$
$$x = 18$$
$$\therefore AB = 18$$

Example 5 A tree cast a 18 ft. shadow at noon when a nearby upright pole of length 7 ft. cast a shadow of length 3 ft. Find the height of the tree assuming that both the tree and pole meet the ground at right angles.

Solution

We will assume that the measure of the angle between the sun's ray and the ground is the same for both the tree and the nearby pole.

Step 1: The tree, its shadow and the sun's ray form a right triangle. Similarly, the pole, its shadow together with the sun's ray form another right triangle. The two triangles formed are similar by AA. See Figure 1 below.

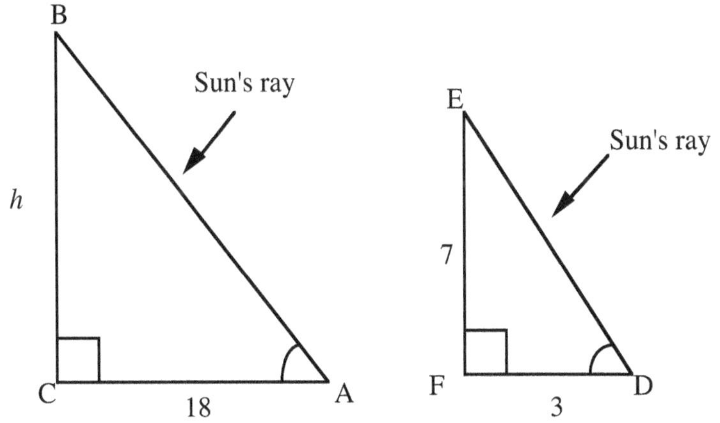

Figure 1

Step 2:

Since $\Delta\, ABC \sim \Delta\, DEF$,

$$\frac{h}{7} = \frac{18}{3}$$ ($\overline{BC}$ and $\overline{EF}$ are corresponding sides; $\overline{CA}$ and $\overline{FD}$ are corresponding sides)

and $h = \dfrac{18(7)}{3}$

$h = 42$

The height of the tree is 42 ft.

Extra If you study calculus in the future, you will apply similarity of triangles when covering problems on related rates and when finding volumes of solids of known cross-sections.

Lesson 10 Exercises

A tree cast a 18 ft. shadow at noon when a nearby upright pole of length 7 ft. cast a shadow of length 3 ft. Find the height of the tree assuming that both the tree and pole meet the ground at right angles.

Answer: 42 ft.

CHAPTER 4

Lesson 11: **Radian-Degree Conversions**
Lesson 12: **Right Triangle Trigonometry and Applications**
Lesson 13: **Special Δ 's: The 30°-60°-90° and 45°-45°-90° triangles**

Angle: In **geometry**, we define an angle as being formed by two rays (or line segments) meeting at a common point called the vertex of the angle. The main unit for measuring angles in geometry is the **degree**.

In **trigonometry**, we define an **angle** as being formed by rotating a line segment (or ray) about an end point, from an initial position to a final or terminal position. We call the initial position , the initial side and the terminal position the terminal side. The angle formed is measured by the amount of rotation. If the rotation is in a counterclockwise direction, then the measure of the angle formed is positive. If the rotation is in a clockwise direction, the measure of the angle formed is negative. In trigonometry, we measure angles in degrees as well as in **radians**.

Lesson 11
Radian-Degree Conversions

Converting from Degrees to Radians

Converting from one unit to another unit is a direct proportion problem, and as such any of the methods of solving direct proportion problems are applicable.

Example 1 Convert 240° to radians.

Solution $360^\circ = 2\pi$ radians, or $\boxed{180^\circ = \pi \text{ radians}}$

Method 1 Let x radians $= 240^\circ$............(1)
π radians $= 180^\circ$ ....(2)

Divide equation (1) by equation (2)

Then $\dfrac{x \text{ radians}}{\pi \text{ radians}} = \dfrac{240^\circ}{180^\circ}$, from which

$$x = \frac{\pi(240)}{180}$$

$$x = \frac{\pi(\cancel{240^\circ})^4}{{}_3\cancel{180^\circ}}$$

$$x = \frac{4\pi}{3}; \qquad \therefore 240^\circ = \frac{4\pi}{3} \text{ radians.}$$

Method 2 Using the units label method

Step 1: $\dfrac{240 \text{ degrees}}{1} \times \dfrac{?}{?}$

Step 1: $\dfrac{240 \text{ degrees}}{1} \times \dfrac{\pi \text{ radians}}{180 \text{ degree}}$

$= \dfrac{240}{1} \times \dfrac{\pi \text{ radians}}{180}$

$= \dfrac{4\pi}{3}$ radians

Method 3 We will use simple proportion to solve the problem. Let x radians correspond to 240°. Then x radians is to 240° as π radians is to 180°.

Step 1: " x radians is to 240° as π radians is to 180°" translates to:

$$\frac{x \text{ radians}}{240°} = \frac{\pi \text{ radians}}{180°}$$

Step 2: Solve for x:

$$x = \frac{\pi(240)}{180}$$

$$x = \frac{\pi\cancel{(240°)}^{4}}{{}_{3}\cancel{180°}} = \frac{4\pi}{3}\text{ ; Therefore } 240° = \frac{4\pi}{3} \text{ radians.}$$

Converting from Radians to Degrees

Example 2 Convert $\frac{7\pi}{5}$ radians to degrees.

Method 1 Let $x° = \frac{7\pi}{5}$ radians..................................(1)

180° = π radians.......................................(2)

Divide equation (1) by equation (2): $\frac{x°}{180°} = \frac{\frac{7\pi}{5} \text{ radians}}{\pi \text{ radians}}$

$$\frac{x°}{180°} = \frac{7\pi}{5}\ \frac{1}{\pi}$$

$$x = \frac{7\pi}{5}\ \frac{1(180)}{\pi}$$

Solving for x,

$$x = \frac{7\cancel{\pi}}{{}_{1}\cancel{5}}\ \frac{1\cancel{(180)}\,36}{\cancel{\pi}}$$

$$x = 252$$

$$\therefore \frac{7\pi}{5} \text{ radians} = 252°.$$

Method 2

Step 1: $\dfrac{\frac{7\pi}{5}\text{ radians}}{1} \times \frac{?}{?}$

Step 1: $\dfrac{\frac{7\pi}{5}\text{ radians}}{1} \times \dfrac{180 \text{ degrees}}{\pi \text{ radians}}$

$= \frac{7}{5} \times \frac{180 \text{ degrees}}{1}$

= 252 degrees

$\frac{7\pi}{5}$ radians = 252°

Method 3

Let $x°$ correspond to $\frac{7\pi}{5}$ radians.

Then, "$x°$ is to $\frac{7\pi}{5}$ radians as 180° is to π radians." translates to:

$$\frac{x°}{\frac{7\pi}{5}\text{ radians}} = \frac{180°}{\pi\text{ radians}}$$

Solving for x,

$$x = \frac{180}{\pi}\frac{7\pi}{5}$$

$$x = 252°$$

$$\therefore \frac{7\pi}{5}\text{ radians} = 252°.$$

We could have solved Examples 1 and 2 using proportion, knowing that π radians = 180°.

In Example 1: If 180° = π

$$\text{Then, } 240° = \frac{\pi\text{ radians}}{1}\cdot\frac{\overset{4}{\cancel{240}}}{\underset{3}{\cancel{180}}}$$

$$\therefore 240° = \frac{4\pi}{3}\text{ radians.}$$

Lesson 11 Exercises

1. Convert 270° to radians

2. Convert $\frac{6\pi}{5}$ radians to degrees

Answers: **1**. $\frac{3\pi}{2}$ radians ; **2.** 216°

Lesson 12

Right Triangle Trigonometry and Applications

Trigonometric Ratios (Trigonometric Functions)

sine (**sin**); cosine (**cos**); tangent (**tan**); cosecant (**csc**); secant (**sec**); and cotangent (**cot**).

Definitions: The following definitions are applicable to **right** triangles only.

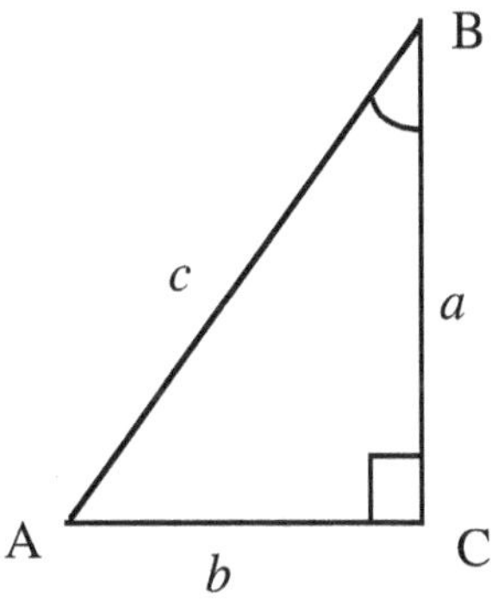

1. **sin A** $= \frac{\text{opposite side}}{\text{hypotenuse}} = \frac{a}{c}$ (sin A is abbreviation for the sine of A)

2. **cos A** $= \frac{\text{adjacent side}}{\text{hypotenuse}} = \frac{b}{c}$ (cos A is abbreviation for the cosine of A)

3. **tan A** $= \frac{\text{opposite side}}{\text{adjacent side}} = \frac{a}{b}$ (tan A is abbreviation for the tangent of A)

4. **csc A** $= \frac{1}{\sin A} = \frac{\text{hypotenuse}}{\text{opposite side}} = \frac{c}{a}$ (csc A is abbreviation for the cosecant of A)

5. **sec A** $= \frac{1}{\cos A} = \frac{\text{hypotenuse}}{\text{adjacent side}} = \frac{c}{b}$ (sec A is abbreviation for the secant of A)

6. **cot A** $= \frac{1}{\tan A} = \frac{\text{adjacent side}}{\text{opposite side}} = \frac{b}{a}$ (cot A is abbreviation for the cotangent of A)

Note above that **sin B** $= \frac{b}{c}$; **cos B** $= \frac{a}{c}$; **tan B** $= \frac{b}{a}$

Note also that for $\angle$ B, the opposite side is $\overline{AC}$, and the adjacent side is $\overline{BC}$. For $\angle$ A, the opposite side is $\overline{BC}$ and the adjacent side is $\overline{AC}$

Some mnemonic devices will be discussed after Example 1

Example 1 Solve the right triangle ABC for the unknown parts, given that $b = 4$, $m\angle B = 30^\circ$

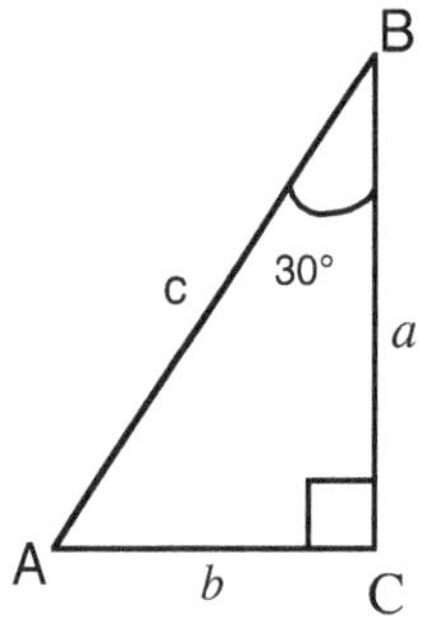

Solution

(a) $m\angle A = 180^o - (30^o + 90^o)$ ($m\angle C = 90^\circ$; sum of the measures of the $\angle$'s of a Δ is 180)

$= 180^o - 120^o$

$m\angle A = 60^o$

(b) To find c:

c is the hypotenuse, $b = 4$ is opposite $\angle B$.

$\sin 30 = \frac{4}{c}$ ($m\angle B = 30^\circ, b = 4$)

$c \sin 30^\circ = 4$

$c = \frac{4}{\sin 30^o} = \frac{4}{\frac{1}{2}} = 8$ ($\sin 30^\circ = \frac{1}{2}$)

now $c = 8$, $b = 4$

Apply the Pythagorean theorem (We may apply this theorem since Δ ABC is a right triangle)

$8^2 = 4^2 + a^2$

$64 = 16 + a^2$

$48 = a^2$

$a = \sqrt{48} = \sqrt{16}\sqrt{3} = 4\sqrt{3}$

Note: In (b), we could have found a first using the tangent relationship

$\tan A = \frac{a}{4}$ ($\frac{\textbf{O}\text{pposite}}{\textbf{A}\text{djacent side}}$ "T**OA**" <-------- mnemonic device)

$a = 4 \tan A$

$a = 4 \tan 60^o$

$a = 4\sqrt{3}$

Also, $\cos 30^\circ = \frac{a}{c}$

$\frac{\sqrt{3}}{2} = \frac{a}{8}$ and $a = \frac{8(\sqrt{3})}{2} = 4\sqrt{3}$

Note: The determination of which trigonometric relationship to use depends on what we are given and what we want to find. A good practice is to scribble down the following three relationships:

$\sin\theta = \frac{\text{opposite side}}{\text{hypotenuse}}$ ($\frac{\textbf{O}\text{pposite}}{\textbf{H}\text{ypotenuse}}$ "**SOH**")

$\cos\theta = \frac{\text{adjacent. side}}{\text{hypotenuse.}}$ ($\frac{\textbf{A}\text{djacent}}{\textbf{H}\text{ypotenuse}}$ "**CAH**")

$\tan\theta = \frac{\text{opposite side}}{\text{adjacent side}}$ ($\frac{\textbf{O}\text{pposite}}{\textbf{A}\text{djacent}}$ "T**OA**")

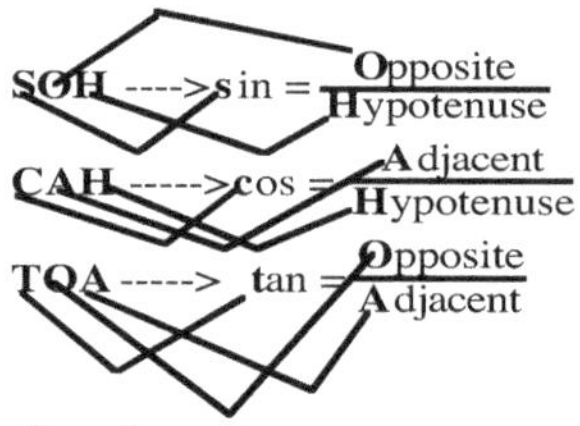

and use the relationship in which you know all but one quantity. Sometimes, you may use two relationships simultaneously to find two unknowns.

Applications of Right Angle Trigonometry

Definition: For an observer sighting (looking) at something above the observer, **the angle of elevation** is the angle between the horizontal line (x-axis) from the observer's eye and the line of sight to the object, the angle being measured from the horizontal to the line of sight.

Example Find the height reached by a kite, if the length of the kite is 400 ft, and the angle of elevation of the kite is 60°.

Step 1: Sketch a diagram for the system under consideration.
Assume that the hand holding on to the kite and the eye coincide at one point.

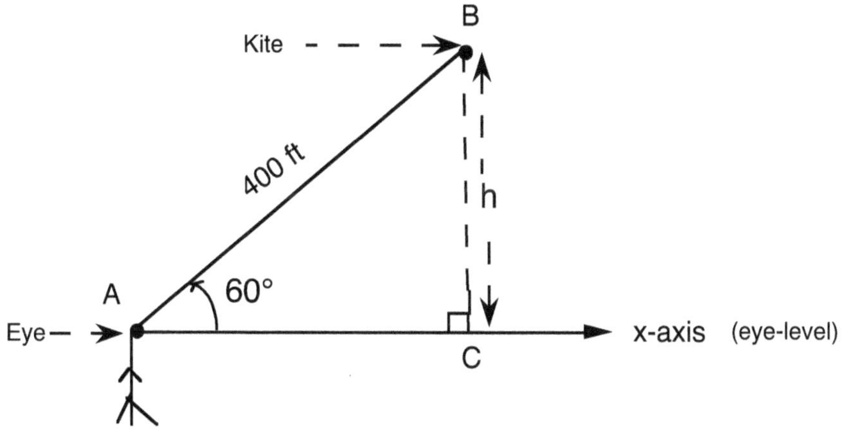

AB = Length of kite = 400 ft.
BC = h = height of kite (vertical distance from eye-level to the kite)

Step 2: Δ ABC is a right triangle, and therefore we may apply the trigonometric relationships.

$\sin 60° = \dfrac{h}{400}$ $\qquad$ (SOH : $\dfrac{\textbf{O}\text{pposite}}{\textbf{H}\text{ypotenuse}}$)

$h = 400 \sin 60°$

$= 400 \left(\dfrac{\sqrt{3}}{2}\right)$ ft.

$= 200\sqrt{3}$ ft. $\qquad$ ($\sqrt{3} \approx 1.73$)

≈ 346.41 ft.

The height reached is $200\sqrt{3}$ ft. ($\approx$346.41 ft.)

Angle of depression

Definition:

For an observer sighting (looking) at something below the observer, the angle of depression is the angle between the horizontal line (x-axis) from the observer's eye and the line of sight to the object, the angle being measured from the horizontal to the line of sight.

In the figure below: $\angle\alpha$ is the angle of elevation of B from A.
$\angle\beta$ is the angle of depression of A from B.
The angle of elevation is congruent to the angle of depression.

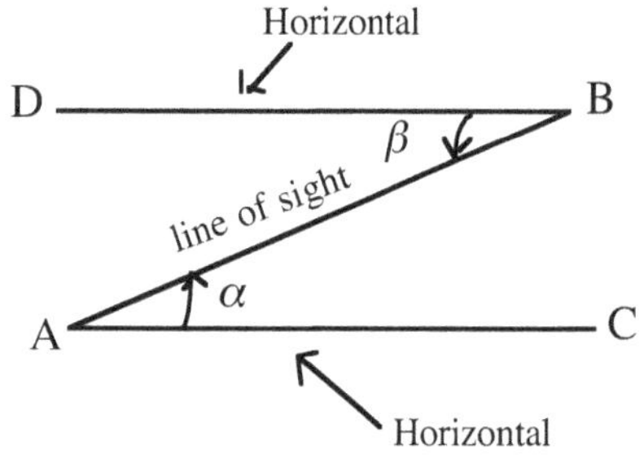

Lesson 12 Exercises

A Solve the right triangle ABC for the unknown parts, given that $b = 6$, and $m\angle A = 60^\circ$

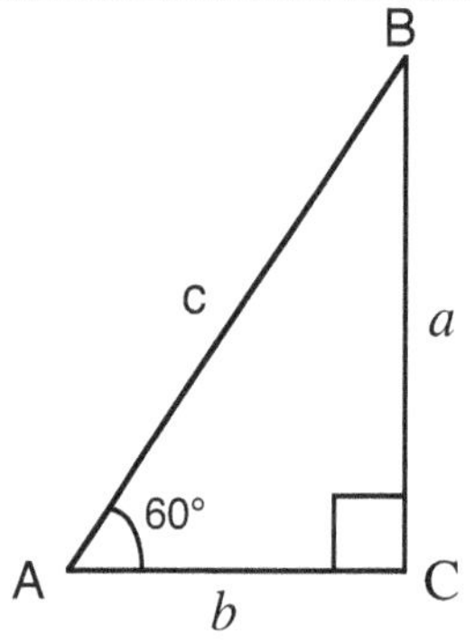

Answers: $m\angle B = 30^\circ$; $a = 6\sqrt{3}$; $c = 12$

B Find the height reached by a kite, if the length of the kite is 600 ft, and the angle of elevation of the kite is 30°.

Answer: 300ft.

Lesson 13

Special triangles: The 30°-60°-90° triangle and the 45°-45°-90° triangle

The trigonometric functions for 30°, 45°, 60°, and 90° occur very frequently in mathematics that students are usually required to memorize them.

The 45°-45°-90° triangle

By sketching an isosceles right triangle with measures of 45°, 45°, and 90° (Figure 1) we will be able, using the basic definitions, write down the trigonometric functional value for 45°.

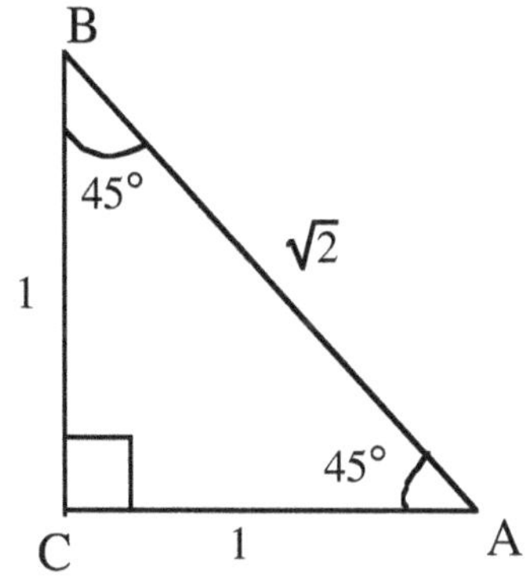

Figure 1

In an isosceles triangle, the sides opposite the congruent angles have the same length. If we let the length of the sides opposite the congruent angles be 1 unit each, we can calculate the length of the hypotenuse using the Pythagorean theorem.

$$AB^2 = AC^2 + BC^2$$
$$AB^2 = 1^2 + 1^2$$
$$AB^2 = 2$$
$$AB = \sqrt{2}.$$

Example $\sin 45^o = \dfrac{\text{opposite side}}{\text{hypotenuse}} = \dfrac{\mathbf{1}}{\mathbf{\sqrt{2}}} = \dfrac{1}{\sqrt{2}} \cdot \dfrac{\sqrt{2}}{\sqrt{2}} = \dfrac{\mathbf{\sqrt{2}}}{\mathbf{2}}$ (Rationalizing the denominator)

$$\cos 45^o = \frac{\text{adjacent side}}{\text{hypotenuse}} = \frac{\mathbf{1}}{\mathbf{\sqrt{2}}} = \frac{1}{\sqrt{2}} \cdot \frac{\sqrt{2}}{\sqrt{2}} = \frac{\mathbf{\sqrt{2}}}{\mathbf{2}}$$

$$\tan 45^o = \frac{\text{opposite side}}{\text{adjacent side}} \frac{1}{1} = 1.$$

The 30°-60°-90° triangle

For the 30°-60°-90° triangle (Figure 2), the dimensions are obtained by considering an equilateral triangle whose sides are each 2 units long. From geometry, the length of the hypotenuse is twice the length of the shortest side. The third side is calculated by the Pythagorean theorem.
Thus if AB = 2, then AC = 1 and $2^2 = 1^2 + BC^2$ and from which BC = $\sqrt{3}$. Usually, after having gone through the derivation, using geometric considerations and the Pythagorean theorem, students become familiar with the dimensions 1, 2, $\sqrt{3}$. The problem of recall that remains then is how to place these dimensions on the 30°-60°-90° triangle, if you do not want to draw the equilateral triangle, but draw only Δ ABC. The following mnemonic device will be helpful:

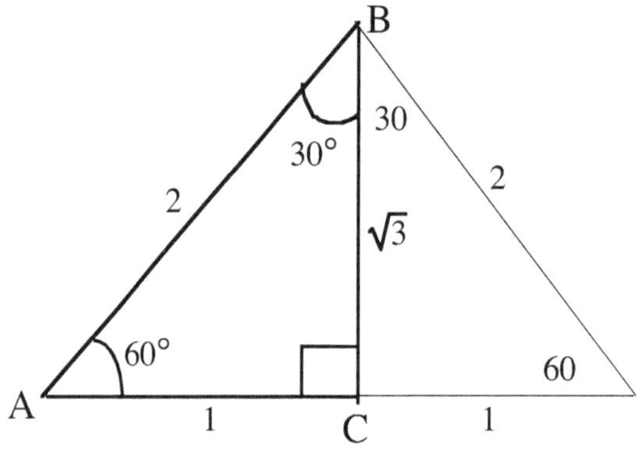

Figure 2

Mnemonic device: If you can remember the numbers 1, 2, $\sqrt{3}$ as being the lengths of the sides of this triangle, and also remember that in any triangle, the longest side is opposite the largest angle, and that the shortest side is opposite the smallest angle, you should be able to place these dimensions accordingly on the triangle. Thus, 1 is opposite the 30° angle, 2 is opposite the 90° angle (the right angle) , and $\sqrt{3}$ is opposite the 60° angle. ($\sqrt{3} \approx 1.73$)

By applying the mnemonic devices, SOH , CAH, TOA (see p.54) we can write down some functional values.

Examples

$$\sin 30^\circ = \frac{\text{opposite side}}{\text{hypotenuse}} = \frac{1}{2}$$

$$\cos 30^\circ = \frac{\text{adjacent side}}{\text{hypotenuse}} = \frac{\sqrt{3}}{2}$$

$$\sin 60^\circ = \frac{\text{opposite side}}{\text{hypotenuse}} = \frac{\sqrt{3}}{2}.$$

Lesson 13 Exercises

In Figure 2, above, find the following:

1. $\tan 30^\circ$; **2.** $\tan 60^\circ$; 3. $\csc 30^\circ$; **4,.** $\csc 60^\circ$; **5.** $\sec 30^\circ$; **6.** $\sec 60^\circ$ •

CHAPTER 5

Lesson 14: **Straight Lines: Introductory Theme: Two** points
Slopes of lines;
Lesson 15: **Equations of Straight Lines**

Lesson 14
Introductory Theme: Two points

1. Why two points? Two points, because given or knowing two points, a straight line can be drawn by connecting the two points, using a straight edge and pencil.

2. Why two points? Two points, because given or knowing two points

the slope, m, of the line segment connecting the two points $P_1(x_1, y_1)$ and $P_2(x_2, y_2)$ can be found

by applying $m = \dfrac{y_2 - y_1}{x_2 - x_1}$

3a. Why two points? Two points, because if we know the **slope, *m***, and the ***y*-intercept, *b*,** of the line, we can obtain two points and draw the graph of the line

3b Note: y–intercept,b implies the point $(0, b)$, By choosing a point (x, y) on a line, we have two points, and the slope (as well as an equation) of the line connecting the two pints $(0, b)$ and (x, y) is given by

$m = \dfrac{y - b}{x - 0}$ <--**slope = slope**

$mx = y - b$ or

$\boxed{y = mx + b}$ <------**slope-intercept form** of the equation of a line

4. Why two points? Two points, because given or knowing two points

an equation of the line segment connecting the two points $P_1(x_1, y_1)$ and $P_2(x_2, y_2)$ can be found

by applying $\boxed{y - y_1 = \left(\dfrac{y_2 - y_1}{x_2 - x_1}\right)(x - x_1)}$ (from $\dfrac{y - y_1}{x - x_1} = \dfrac{y_2 - y_1}{x_2 - x_1}$ <-- **is slope = slope**

or $\boxed{y - y_1 = m(x - x_1)}$ <------**point-slope form,** where $m = \dfrac{y_2 - y_1}{x_2 - x_1}$

5. Why two points? Two points, because given or knowing the two-intercept points $(a, 0)$, $(0, b)$ an equation of the line segment connecting the two points $P_1(a, 0)$ and $P_2(0, b)$ can be found by

applying $y - y_1 = \left(\dfrac{y_2 - y_1}{x_2 - x_1}\right)(x - x_1)$ to obtain

$y - 0 = \dfrac{b - 0}{0 - a}(x - a)$ (or from $\dfrac{y - 0}{x - a} = \dfrac{b - 0}{0 - a}$) <--- **slope = slope**

or $y = \frac{b}{-a}(x - a)$ or $y = \frac{b}{-a}x + (\frac{b}{-a})(-a)$ or $\boxed{y = -\frac{b}{a}x + b}$ also $\frac{x}{a} + \frac{y}{b} = 1$ **(Two intercept form)**

6. Why two points? Two points, because given the graph (picture) of a line , we are given infinitely many points from which we can read the coordinates of any two points on the line and write an equation of a line by applying **3, 4** or **5** above. A picture is worth a thousand words

The above theme summarizes this chapter

Slopes of Lines

Finding the slope of a line given the coordinates of two points on the line

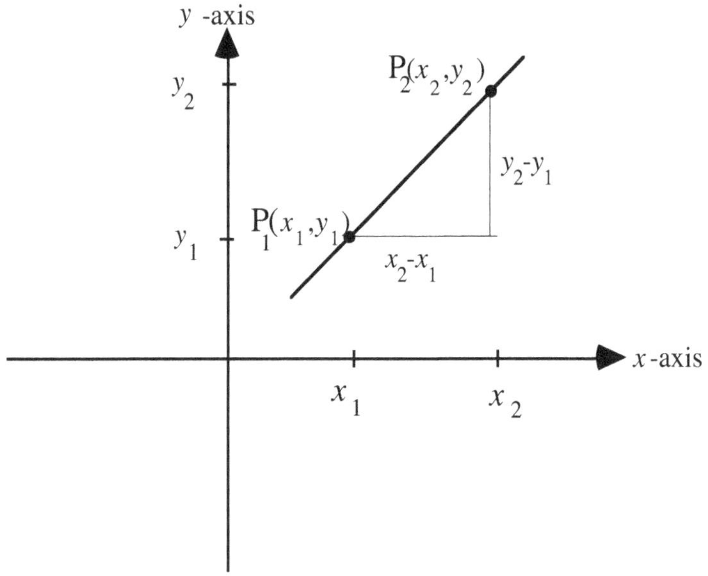

The slope, m, of the line segment connecting the points $P_1(x_1, y_1)$ and $P_2(x_2, y_2)$ is given by

$$m = \frac{y_2 - y_1}{x_2 - x_1} \qquad \text{(a ratio)}$$

Example Find the slope of the line passing through the points (2,3) and (6,8).

Solution : Identify the first point as $P_1(2,3)$ and the second point as $P_2(6,8)$

Then $x_1 = 2, y_1 = 3$; and $x_2 = 6$, $y_2 = 8$

Applying the slope formula, $m = \dfrac{y_2 - y_1}{x_2 - x_1}$

$$m = \frac{8 - 3}{6 - 2}$$

$$m = \frac{5}{4}$$

The slope is $\frac{5}{4}$.

Special Cases of the Slopes of Lines

The special cases are for horizontal and vertical lines

Slope of a horizontal Line

The **slope** of a **horizontal** line is **zero**, since the vertical change is zero. For example, the slope, m, of the horizontal line in Figure **1**, below, by the slope formula is $m = \frac{3-3}{5-2} = \frac{0}{3} = 0$.

(Note that $m = \frac{\text{vertical change}}{\text{horizontal change}} = \frac{\text{change in } y}{\text{change in } x}$)

Slope of a Vertical Line

The **slope** of a **vertical** line is **undefined** since the horizontal change is zero. For example, the slope, m, of the vertical line in Figure **2**, below, by the slope formula is $m = \frac{2-(-3)}{4-4} = \frac{5}{0}$ is undefined.

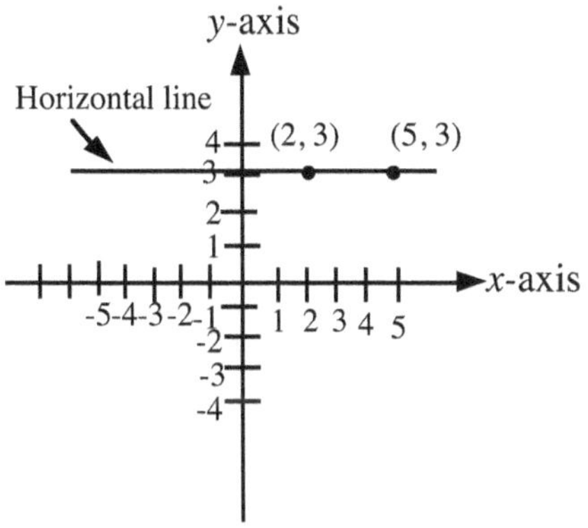

Figure 1

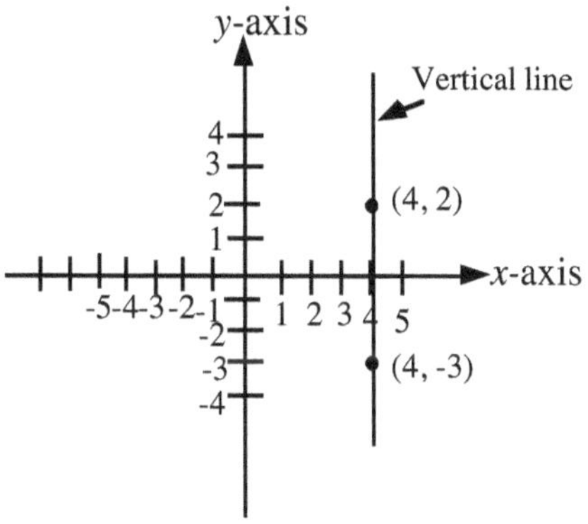

Figure 2

B Find the slope of each line passing through the given points:

1. (2, 3) and (6, 4) **2**. (3, -2) and (2, -4) **3**. (5, 2) and (1, -1)

4. (2, 2) and (-3, 3) **5**. (-4, 2) and (5, 2) **6**. (-3, -5) and (-3, -6)

7. Find c so that the slope of the line passing through the points $(c, 2c)$ and (-6, 8) is 1.

Answers: **1**. $\frac{1}{4}$; **2**. 2; **3**. $\frac{3}{4}$; **4**. $-\frac{1}{5}$; **5**. 0; **6**. undefined; **7**. $c = 14$

Lesson 15
Equations of Straight Lines

There are a number of approaches that we can use in **finding equations** of straight lines. Each approach depends upon what we are given.

Generally, we can easily write down an equation of a line if we know any of the pairs of properties or characteristics of (information about) a line in the formulas presented below.

These formulas are based on finding two different expressions for the slope of a line and equating these expressions to each other.

1. If we know the **slope, *m***, and the ***y*-intercept, *b*,** of the line, we can apply the slope-intercept form, $\boldsymbol{y = mx + b}$.

2. If we know the slope, $\boldsymbol{m}$, and the coordinates $\boldsymbol{(x_1,y_1)}$ of one point on the line, we can apply the point-slope form, $\boldsymbol{(y - y_1) = m(x - x_1)}$.

3. If we know the coordinates $\boldsymbol{(x_1,y_1)}$ and $\boldsymbol{(x_2,y_2)}$ of two points on the line,

we can apply $\boldsymbol{(y - y_1) = \dfrac{y_2 - y_1}{x_2 - x_1}(x - x_1)}$.

4. If we are given the graph (picture), we can determine any of the above pairs of properties (information) from the graph, and then apply the formulas in the above cases; however, if we want to memorize one more formula, we can apply the equation $\boldsymbol{y = -\dfrac{b}{a}x + b}$, where a is the x-intercept and b is the y-intercept. (If $b = 0$ or $a = 0$ use the other methods.)

(This equation is another form of the two-intercept form: $\dfrac{x}{a} + \dfrac{y}{b} = 1$)

We will now cover the above **four** cases in detail with examples.

Case 1: Finding an equation of a line given the slope and the *y*-intercept of the line

Example 1 Find an equation of the line with slope 3 and y-intercept of 4.

Solution We will apply the slope-intercept form of the equation of a straight line, $y = mx + b$.

Substituting the slope, $m = 3$, and the y-intercept, $b = 4$ in this equation,

$$y = 3x + 4$$

Example 2 Find an equation of the line with slope - 2 and y-intercept of - 5.

Solution: Substituting $m = -2$, $b = -5$ in $y = mx + b$, we obtain the equation $y = -2x - 5$

Case 2: Finding an equation of a line given the slope of the line and the coordinates of one point on the line

Example 1 Find an equation of the line passing through the point (3,-2) and having a slope 4.

Solution. We will cover two methods.

Method 1

Step 1: Find the y-intercept, b, by substituting $x_1 = 3$, $y_1 = -2$, $m = 4$ in

$$y = mx + b$$

then, $-2 = 4(3) + b$

$$-2 = 12 + b$$

$$-14 = b$$

Step 2: Now, since we know that $m = 4$, $b = -14$, we can apply $y = mx + b$ (as in Case 1, above) and then, $y = 4x - 14$

Method 2

We will use the point-slope form of the equation of a straight line which is given by

$$y - y_1 = m(x - x_1) \text{ <-------point-slope form.} \qquad (1)$$

Substituting the coordinates, $x_1 = 3$, $y_1 = -2$ and slope, $m = 4$ in equation (1), we obtain

$$y - (-2) = 4(x - 3)$$

$$y + 2 = 4(x - 3) \qquad (2)$$

Equation (2) is the point-slope form of the required equation.

By solving equation (2) for y, we obtain

$$y = 4x - 14 \quad \text{<----------- slope-intercept form} \qquad (3)$$

Equation (3) is the slope-intercept form of the required equation.
In this form, we can, by inspection, determine the slope and the y-intercept.

The author recommends the slope-intercept form (for this course), unless otherwise specified.

Case 3: Finding an equation of a line given the coordinates of two different points on the line

Example Find an equation of the line passing through the points (2,1) and (-3,-4).

Solution

Step 1: Find the slope, m, with $x_1 = 2, y_1 = 1, x_2 = -3$ $y_2 = -4$

$$m = \frac{y_2 - y_1}{x_2 - x_1}$$

$$m = \frac{-4 - 1}{-3 - 2}$$

$$m = 1$$

Scrapwork

$$\frac{-4 - 1}{-3 - 2} = \frac{-5}{-5} = 1$$

Now, $m = 1, x_1 = 2, y_1 = 1, x_2 = -3, y_2 = -4$, and we can apply the procedure in Case 2.

Step 2: Applying Method 2 of Case 2 and
Substituting $m = 1, x_1 = 2, y_1 = 1$ in $y - y_1 = m(x - x_1)$, we obtain

$$y - 1 = 1(x - 2)$$

$$y = x - 1$$

Also, If we substitute $m = 1, x_2 = -3, y_2 = -4$ in $y - y_2 = m(x - x_2)$, we obtain

$$y - (-4) = 1(x - (-3))$$

$$y + 4 = x + 3$$

$$y = x - 1$$

Again, we obtain the same equation. An equation of the line is $y = x - 1$.
We conclude also that any of the two given points can be used in finding an equation of the straight line. From the above solution, we can state a general formula for Case 3 as

$$(y - y_1) = \frac{y_2 - y_1}{x_2 - x_1}(x - x_1)$$

Case 4: Finding an equation of a line given the graph (picture) of the line

If we are given the graph (picture), we can determine any of the above pairs of properties (information) from the graph (see p.62, case 4).

It is also useful to be able to tell immediately from the graph if the line has a positive slope, a negative slope, a zero slope, or an undefined slope.

The signs of the slopes of lines

The lines in Figure 1 have positive slopes.

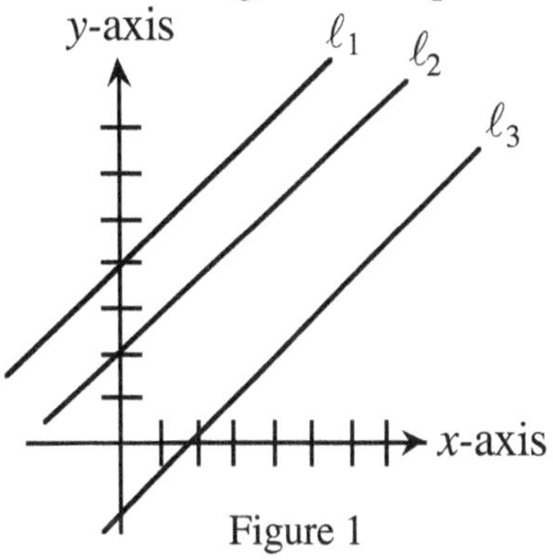

Figure 1

Lines ℓ_1, ℓ_2, and ℓ_3 have positive slopes. (Simply, these lines **lean** to the **right** in the page; or these lines rise as one moves ones head from the left to the right in the page.)

The lines in Figure 2 have negative slopes.

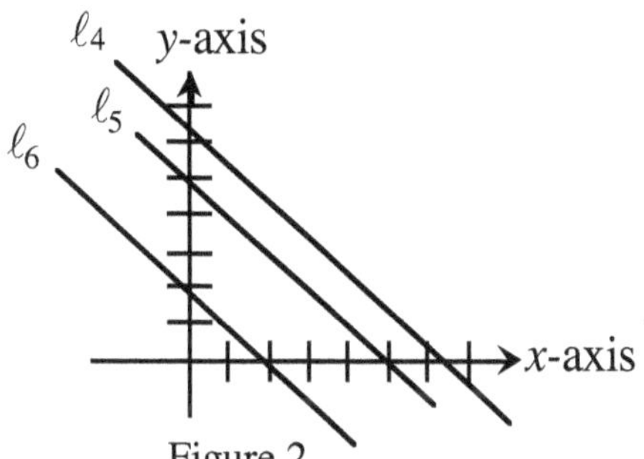

Figure 2

Lines ℓ_4, ℓ_5 and ℓ_6 have negative slopes. (Simply, these lines **lean** to the **left** in the page; or these lines fall as one moves ones head from the left to the right in the page.)

Note: The slope of a **horizontal line** is zero. The slope of a **vertical line** is undefined. (For the equations of horizontal and vertical lines, see p.68)

Example 1 Find an equation of the line whose graph is given below.

Method 1: Apply $\boldsymbol{y = -\frac{b}{a}x + b}$,

where a is the x-intercept and b is the y-intercept. (see also p.61)

Step 1: From the graph, we read the values of a and b: $a = 3$, $b = -4$

Step 2: Substitute $a = 3$, $b = -4$ in

$$y = -\frac{b}{a}x + b.$$

Then $y = -\frac{-4}{3}x + (-4)$

(Make sure you take into account the **minus sign** that comes with the formula.)

$y = +\frac{4}{3}x - 4$ (Two minus signs make the x-term positive)

An equation of the line is $y = \frac{4}{3}x - 4$.

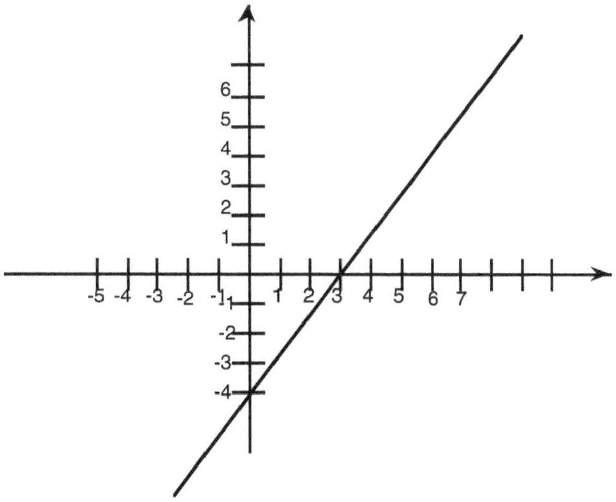

Method 2

Step 1: By inspection, the line has a positive slope. (See figure 1 above.)

Step 2: Slope, $m = \dfrac{\text{vertical change}}{\text{horizontal change}}$

$= \dfrac{\text{change in } y}{\text{change in } x}$

Pick any two convenient points on the line, say, $P_1(0,-4)$ and $P_2(3, 0)$ (see figure 2, below).

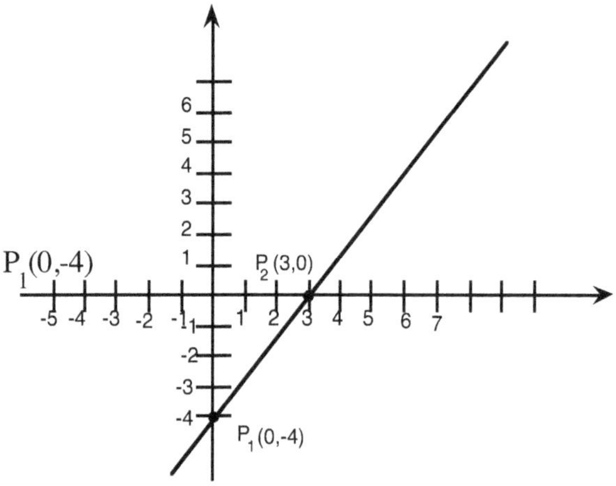

Figure 2

Finding the vertical and horizontal changes by counting.

We will assume that we want to go from the point P_1 to the point P_2, and count the number of equal intervals (units) we will travel vertically, and the number of equal intervals (units) we will travel horizontally. Then, we will go up vertically 4 equal intervals and horizontally to the right 3 equal intervals. Divide the vertical change, 4, by the horizontal change, 3 , to obtain the slope $\frac{4}{3}$

Note that the slope is positive. (see p.65, Figure 1).

Step 3: From the graph , the y-intercept, $b = -4$

Step 4: With $m = \frac{4}{3}$, $b = -4$, apply $y = mx + b$

then, $y = \frac{4}{3}x - 4$

Therefore, an equation of the line is

$y = \frac{4}{3}x - 4$

In fact, once we have been able to specify the points $P_1(0,-4)$, $P_2(3,0)$, we

can apply $m = \dfrac{y_2 - y_1}{x_2 - x_1} = \dfrac{0 - (-4)}{3 - 0} = \dfrac{+4}{3}$

to find the slope $m = \frac{4}{3}$,

and then apply $y - y_1 = m(x - x_1)$ to obtain

$y + 4 = \frac{4}{3}(x - 0)$ $\quad (x_1 = 0, y_1 = -4)$

$y + 4 = \frac{4}{3}x - 0$

$y = \frac{4}{3}x - 4$ < -------slope-intercept form of the equation of a straight line.

Example 2 Find an equation of the line whose graph is given below.

Method 1 Apply $y = -\frac{b}{a}x + b$ where a is the x-intercept and b is the y-intercept.

Step 1: From the graph, $a = 3, b = 6$

Step 2: Substitute $a = 3, b = 6$ in $y = -\frac{b}{a}x + b$

Then $y = -\frac{6}{3}x + 6$

$y = -2x + 6$ < ---slope-intercept form of the equation of a straight line.

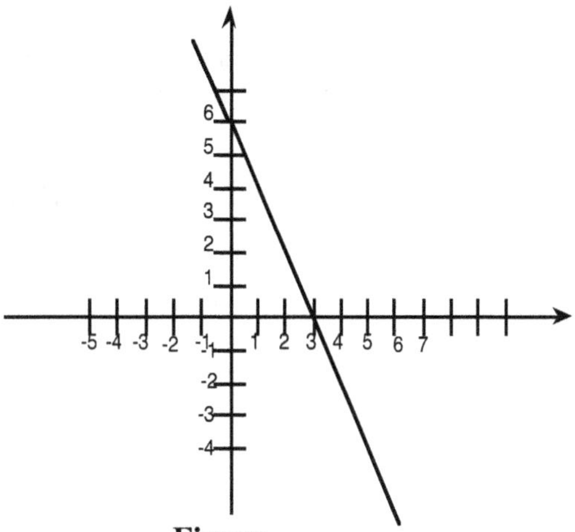

Figure

Method 2

Step 1: Note that the slope is negative. (See p.67, Figure 2.)

Step 2: Pick any two convenient points on the line, say P_1 and P_2

Step 3: Assume that we want to travel (vertically and horizontally only) from P_1 to P_2 .

Then, we will go down 6 units and then to the right 3 units.

The slope, $m = -\frac{6}{3} = -2$

(The slope is negative since the line leans to the left. See page 65, Fig..2)

Step 4: Read the y-intercept, $b = 6$

Step 5: Apply $y = mx + b$, with $m = -2, b = 6$

Then, an equation of the given graph is $y = -2x + 6$

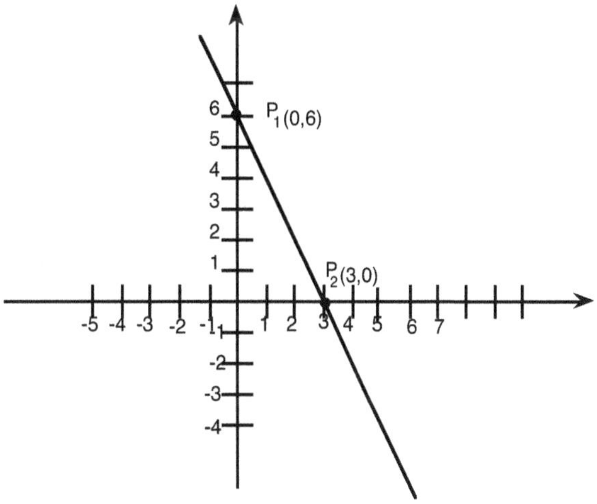

Figure

Method 3 From $P_1(0,6)$ and $P_2(3,0)$

Step 1: $m = \frac{y_2 - y_1}{x_2 - x_1}$ $(x_1 = 0, y_1 = 6)$

$= \frac{0 - 6}{3 - 0} = \frac{-6}{3} = -2$

$m = -2$

Step 2: Apply $y - y_1 = m(x - x_1)$

$y - 6 = -2(x - 0)$

$y - 6 = -2x + 0$

$y = -2x + 6$

An equation of the line is $y = -2x + 6$

Note that, in Method 3, the negative sign results solely from the calculation and we do not have to know in advance the sign of the slope)

In **Case 4**, the author recommends **Method 1.**

Special Cases of the Equations of Straight Lines and their Graphs

The equation $y = mx + b$, (with m defined and, $m \neq 0$) geometrically represents oblique lines (i.e., lines which are neither vertical nor horizontal). The special cases of the equation of a straight line are for horizontal and vertical lines. In these cases, either the x- or the y-term is missing.

The Equation and Graph of a Horizontal Line

The equation of a horizontal line is of the form $y = b$, where b is the y-intercept.

The slope, $m = 0$, since the vertical change is zero.

Substituting $m = 0$, in $y = mx + b$,

$$y = 0x + b$$

$$y = b \qquad (1)$$

Equation (1) means that as x varies, y remains unchanged.

Examples: Sketch the graphs of the following lines:

1. $y = 3$, **2.** $y = -4$. **3.** $y = 0$.

Solution: The line $y = 3$ is the horizontal line passing through (0,3). See Figure 1, below.

The line $y = -4$ is the horizontal line passing through (0,-4).

The line $y = 0$ is the horizontal line along the x-axis (i.e., the line $y = 0$ is the x-axis)

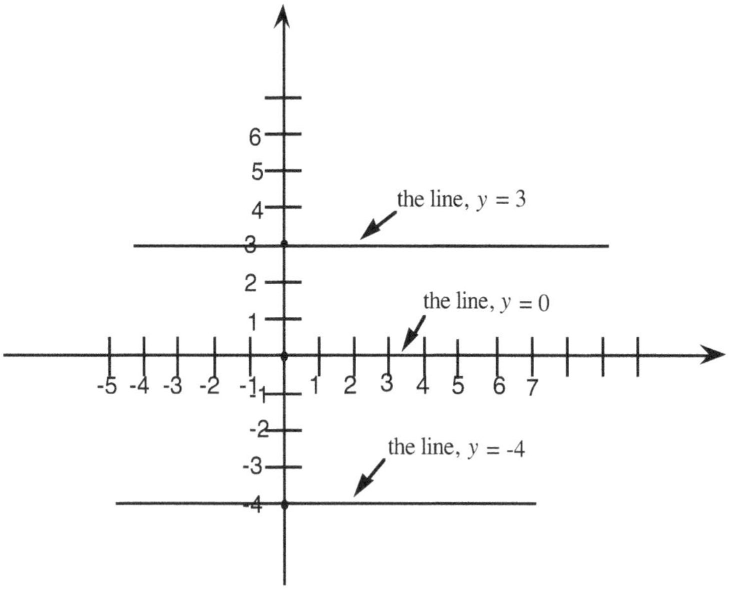

Figure 1

The Equation and Graph of a Vertical Line

The slope of a vertical line is undefined since the horizontal change is zero.
The equation of a vertical line is of the form $x = a$, where a is the x-intercept. This form of the equation means that as y varies, x remains unchanged.

Examples Sketch the lines with the following equations:

1. $x = 2$, **2.** $x = 0$, **3.** $x = -4$

Solution: See Figure 2, below.

1. The line $x = 2$ is the vertical line passing the point (2,0).
2. The line $x = 0$ is the vertical line along the y-axis (i.e., the line $x = 0$ is the y-axis).
3. The line $x = -4$ is the vertical line passing through the point (-4,0).
4. The line $x = -2$ is the vertical line passing through the point (-2,0).

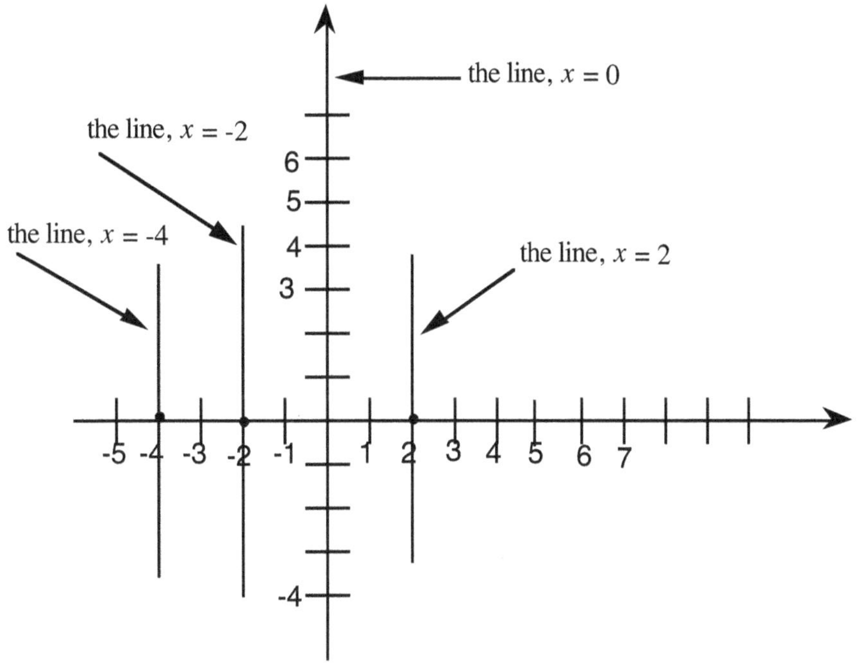

Figure 2

Lesson 15 Exercises

A 1. Find an equation of the line with slope 7 and y-intercept -2.

2. Find an equation of the line with slope $\frac{2}{3}$ and y-intercept 5.

3. A line has a y-intercept -3 and a slope of 6. Find an equation for this line.

Answers: 1. $y = 7x - 2$; **2.** $y = \frac{2}{3}x + 5$; **3.** $y = 6x - 3$

B 1. Find an equation of the line with slope -4 and passing through the point (3,-2).

2. A line passes through the point (-1,-7) and has a slope of 5. Find an equation for this line.

Answers: **1**. $y = -4x + 10$; **2**. $y = 5x - 2$

C 1. Find an equation of the line passing through the points (2,2) and (-5,6)

2. If the points (2,-5) and (-3,1) are on a certain line, find an equation for this line.

3. Find an equation of the line with slope -3 and y-intercept 8.

4. Find an equation of the line with slope 2 and passing through the point (1,6)

Answers: **1.** $y = -\frac{4}{7}x + \frac{22}{7}$; **2.** $y = -\frac{6}{5}x - \frac{13}{5}$; **3.** $y = -3x + 8$; **4.** $y = 2x + 4$.

D Find the equations (slope-intercept forms) of the lines ℓ_1 , ℓ_2, ℓ_3, ℓ_4

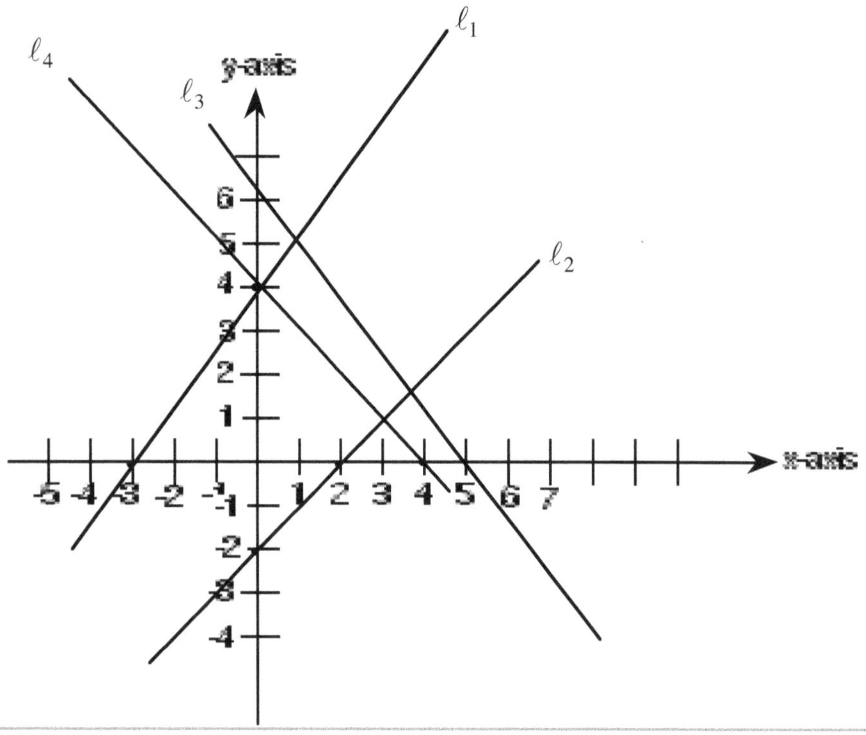

Answers: For ℓ_1: $y = \frac{4}{3}x + 4$. For ℓ_2: $y = x - 2$. For ℓ_3: $y = -\frac{5}{4}x + \frac{25}{4}$ For ℓ_4: $y = -x + 4$

E **1.** Find an equation of the horizontal line passing through the point (3,-4).

2. Sketch the graph the line in Problem 1.

3. Does the line in Problem 1 have a y-intercept ? If yes, find it.

4. Sketch the graph of the line $y = -5$.

5. Sketch the graph of the line $y = 2$.

Answers: **1.** $y = -4$; **2.** See Fig 1, above and imitate. ; **3.** Yes. y-intercept = - 4 ; **4.** & **5.** Imitate Fig. 1 above

F **1.** Find an equation of the vertical line passing through the point (2,-3).

2. Sketch the graph the line in Problem **1**.

3. Does the line in Problem **1** have a y-intercept ? If yes, find it.

Answers: **1.** $x = 2$; **3.** No.

G **1.** Draw the graph of the line $x = 4$. **2.** Draw the graph of the line $x = -1$.

H **1.** Find an equation of the line with slope -5 and y-intercept 2.

2. Find an equation of the line whose y-intercept is -3 and whose slope is $\frac{1}{2}$.

3. Find an equation of the line whose slope is -2 and which passes through the point (4, -5).

4. A line passes through the point (-3, 2) and has slope 4. Find an equation for this line.

5. Find an equation of the line passing through the points (5, 4) and (8, 6).

Answers: **1.** $y = -5x + 2$; **2.** $y = \frac{1}{2}x - 3$; **3.** $y = -2x + 3$; **4.** $y = 4x + 14$; **5.** $y = \frac{2}{3}x + \frac{2}{3}$

CHAPTER 6

Applications of Ratios and Proportion in Physics and Chemistry

Lesson 16

Boyle's Law

(An inverse proportion relationship)

Preliminaries:
Pressure = Force per unit area
Postulate: The pressure of a gas is the result of collisions of the molecules of the gas with the walls of the container.
Boyle's Law deals with the effects of pressure changes on the volume of a gas at constant temperature.

Boyle's Law:
Boyle's Law states that at constant temperature, the pressure exerted on a fixed quantity of a gas is inversely proportional to the volume of the gas. (i.e., pressure and volume vary indirectly).

The pressure being inversely proportional to the volume implies that when the volume increases, the pressure decreases, and when the volume decreases, the pressure increases.
Note: The product of pressure, P, and volume, V, is a constant. If the pressure decreases, the volume must increase the keep the product PV constant.

Boyle's Law in equation form

Let the initial pressure on a gas = P_1; and let the initial volume of the gas = V_1
Let the final pressure of the gas = P_2; and let the final volume of the gas = V_2

Then Boyle's Law states that $P_1V_1 = P_2V_2$ (Note: **product = product**, inverse proportion)

Example When the pressure on a given mass of a gas is 740 mm Hg, the volume is 4.7 L. Find the volume of the gas when the pressure is 720 mm Hg, assuming constant temperature.

Solution $P_1 = 740$ mm Hg, $V_1 = 4.7$ L
$P_2 = 720$ mm Hg, $V_2 = ?$

We are to find V_2. Substitute the known values in $P_1V_1 = P_2V_2$ (Boyle's Law)

$$740 \text{ mm Hg} \times 4.7 \text{ L} = 720 \text{ mm Hg} \times V_2$$

$$\frac{740 \text{ mm Hg} \times 4.7 \text{ L}}{720 \text{ mm Hg}} = V_2$$

$$4.83 \text{ L} = V_2$$

$$4.8 \text{ L} = V_2$$

$$V_2 = 4.8 \text{ L}.$$

When the pressure is 720 mm Hg, the volume is 4.8 L

Note above that we can also use any of the other methods (already covered) for solving inverse proportion problems. See Lesson 6.

Lesson 17
Charles' Law
(A direct proportion relationship)

Charles' Law deals with the effects of temperature changes on the volume of a gas. keeping the pressure constant.

Charles' Law: Charles' Law states that at constant pressure, the volume of a gas is directly proportional the temperature of the gas.
(i.e., volume and temperature vary directly)

The volume being directly proportional to the temperature implies that when the temperature i ncreases, the volume increases, and when the temperature decreases, the volume also decreases.

Charles' Law in equation form

Let the initial volume of a gas = V_1; and let the initial temperature of the gas = T_1
Let the final volume of the gas = V_2; and let the final temperature of the gas = T_2
(Note: T_1 and T_2 are in degrees Kelvin)

Then Charles' Law states that $\frac{V_1}{T_1} = \frac{V_2}{T_2}$.(Note: **quotient = quotient**, a direct proportion)

Example At constant pressure, when the volume of a given mass of a gas is 4.2 L, the temperature is 36°. Find the volume when the temperature is 24° C.

Solution

$$V_1 = 4.2L$$
$$T_1 = 36^\circ$$
$$= (36 + 273)^\circ K$$
$$T_1 = 309^\circ K$$
$$T_2 = 24^\circ$$
$$= (24 + 273)^\circ K$$
$$T_2 = 297^\circ K$$
$$V_2 = ?$$

We are to find V_2.

Substituting the above values in $\frac{V_1}{T_1} = \frac{V_2}{T_2}$

$$\frac{4.2 \text{ L}}{309^\circ} = \frac{V_2}{297^\circ K}$$
$$309^\circ \times V_2 = 4.2 \text{ L} \times 297^\circ\text{K}$$
$$V_2 = \frac{4.2 \text{ L} \times 297^\circ K}{309^\circ\text{K}}$$
$$= 4.036 \text{ L}$$
$$V_2 = 4.0 \text{ L}$$

When the temperature is 24° C, the volume is 4.0 L.

Similarly, as it is in the case of Boyle's Law, note that we can also use any of the other methods (already covered) for solving direct proportion problems. See Lesson 5.

..

Lesson 18
Gay-Lussac's Law
(A direct proportion relationship)

Gay-Lussac's Law deals with the effects of temperature changes on the pressure on a gas.

Gay-Lussac's Law:
Gay-Lussac's Law states that at constant volume, the pressure on a gas is directly proportional the temperature of the gas. (That is, the pressure and temperature vary directly)

The pressure being directly proportional to the temperature implies that when the temperature increases, the pressure also increases, and when the temperature decreases, the pressure also decreases.

Gay-Lussac's Law in equation form
Let the initial pressure on a gas = P_1; and let the initial temperature of the gas = T_1
Let the final pressure on the gas = P_2; and let the final temperature of the gas = T_2

Then Gay-Lussac's Law states that $\frac{P_1}{T_1} = \frac{P_2}{T_2}$ (Note: **quotient = quotient,** a direct proportion)

Example At constant volume, when the pressure on a given mass of a gas is 720 mm Hg, the temperature is 30°C. What is the pressure when the temperature is 36°C?

Solution

$$P_1 = 720 \text{ mm Hg};$$
$$T_1 = 30°$$
$$= (30 + 273)°K;$$
$$T_1 = 303°K$$
$$P_2 = ?$$
$$T_2 = 36°C$$
$$= (36 + 273)°K$$
$$T_2 = 309°K$$

Substituting the above values in $\frac{P_1}{T_1} = \frac{P_2}{T_2}$

$$\frac{720 \text{ mm Hg}}{303°\text{K}} = \frac{P_2}{309°K}$$
$$303°K \times P_2 = 720 \text{ mm Hg} \times 309°\text{K}$$
$$\text{P}_2 = \frac{720 \text{ mm Hg} \times 309°\text{K}}{303°\text{K}}$$
$$P_2 = 734.257 \text{ mm Hg}$$
$$= 734 \text{ mm Hg}$$

When the temperature is 36°C, the pressure is 734 mm Hg.

Note above that we can also use any of the other methods (already covered) for solving direct proportion problems. See Lesson 5.

Lesson 19
Combined Gas Law
(A combined proportion relationship)

The combined gas law is obtained from Boyle's Law, Charles' Law, and Gay-Lussac's Law.

In the combined gas law, pressure, volume, and temperature, all vary at the same time and none is constant.

The combined gas law states that the pressure on a given mass of a gas is inversely proportional to the volume and directly proportional to the temperature.

Combined Gas Law in equation form

Let the initial pressure = P_1, initial volume = V_1, and initial temperature = T_1

Let the final pressure = P_2, final volume = V_2, and final temperature = T_2

Then the combined gas Law states that $\frac{P_1V_1}{T_1} = \frac{P_2V_2}{T_2}$

Example When the pressure on a given mass of a gas is 720 mm Hg, and the volume is 6.2 L, the temperature is 28 °C. Find the temperature when the pressure is 740 mm Hg and the volume is 8.5 L

Solution Method 1

$P_1 = 720 \text{ mm Hg},\ V_1 = 6.2 \text{ L}$

$T_1 = 28°C$

$= (28 + 273)°K$

$T_1 = 301°K$

$P_2 = 740 \text{ mm Hg}.\ V_2 = 8.5 \text{ L}$

$T_2 = ?$

We are to find T_2.

Substituting in $\frac{P_1V_1}{T_1} = \frac{P_2V_2}{T_2}$

$$\frac{720 \text{ mm Hg} \times 6.2 \text{ L}}{301°\text{K}} = \frac{740 \text{ mm Hg} \times 8.5 \text{ L}}{T_2}$$

$$720 \text{ mm Hg} \times 6.2 \text{ L} \times T_2 = 740 \text{ mm Hg} \times 8.5 \text{ L} \times 301°\text{K}$$

$$T_2 = \frac{740 \text{ mm Hg} \times 8.5 \text{ L} \times 301°\text{K}}{720 \text{ mm Hg} \times 6.2 \text{ L}}$$

$$= 424.124°K$$

$$= (424.124 - 273)°C$$

$$= 151.124$$

$$= 151°C$$

When the pressure is 740 mm Hg, and the volume is 8.5 L, the temperature is 151°C.

Note above that we can also use any of the other methods (already covered) for solving combined proportion problems. See Lesson 7.

Lesson 20

Application of Ratios in Physics-B

Velocity

If a body travels a distance, *s*, in a straight line, in time *t*, the velocity, *v* ,is given by the **ratio,**

$$v = \frac{s}{t}$$

Acceleration

If a body starts with a velocity v_0 and changes to velocity v_f after a time interval *t*, the body's acceleration, *a*, is given by the **ratio,** $\frac{\text{change in velocity}}{\text{time interval}}$

$$\text{Thus } a = \frac{v_f - v_0}{t}$$

Newton's second law

From Newton's second law, if a net force *F* acts on a body of mass, *m* , the body's acceleration, *a* is given by the **ratio,** $\frac{\text{net force}}{\text{mass}}$

$$\text{Thus } a = \frac{F}{m}.$$

Friction

If the normal force of reaction holding two surfaces together is N, and the friction force is F, the coefficient of friction, μ , for the two surfaces is given by the **ratio** $\frac{F}{N}$.

$$\text{Thus } \mu = \frac{F}{N}.$$

Circular motion

The centripetal acceleration, a_c, of a body in uniform circular motion is given by the **ratio,** $\frac{v^2}{r}$, where *r* is the radius of its path, and. *v* is the tangential velocity..

$$\text{Thus } a_c = \frac{v^2}{r}$$

Power

Power is the rate of doing work. If the work performed is *W*, and the time interval is *t*, then power, *P*, is given the **ratio** $\frac{W}{t}$. Thus $P = \frac{W}{t}$.

Elasticity

Modulus of elasticity is defined as the **ratio,** $\frac{\text{stress}}{\text{strain}}$

Pressure:

If a force. *F*, acts perpendicular to a surface of area, *A*, the pressure, *P*, exerted on the surface is the **ratio** $\frac{\text{force}}{\text{area}}$.. Thus $P = \frac{F}{A}$

Mechanical Advantage

The mechanical advantage of a hydraulic press is given by the **ratio** $\frac{A_{out}}{A_{in}}$**,** where A_{out} is the area of the output piston, and A_{in} is the area of the input piston.

Kinematic Viscosity

kinematic coefficient of viscosity, $\nu = \frac{\mu}{\rho}$, where μ = absolute viscosity and ρ = mass density

Electric field

If a charge q at a given point is acted upon by a given force F, then the electric field, E, at the given point is defined as the **ratio** $\frac{\text{force}}{\text{charge}}$.

Thus $E = \frac{F}{q}$

Coulomb's Law

The force, F one charge exerts on another charge is given by

$F = k\frac{q_1 q_2}{r^2}$, where q_1, q_2 are the charges, and r is the distance between the charges.

Ohm's law

If the potential difference between the ends of a conductor is V, and the resistance of the conductor is R, then the current, I, in the conductor is given by the **ratio,** $\frac{V}{R}$.

Thus $I = \frac{V}{R}$.

Resistance

The resistance, R, of a conductor is the **ratio** $\frac{V}{I}$, where V is the potential difference across the conductor, and I is the current in the conductor.

Electric Current

If a quantity of charge, q, passes a given point in a conductor in the time interval t, the electric current, I, in the conductor is given by the **ratio,** $\frac{\text{charge}}{\text{time interval}}$

Thus $I = \frac{q}{t}$.

Capacitance

If the potential difference between the plates of a capacitor V, and the charge on either plates Q, then the capacitance, C, of the capacitor is the **ratio** $\frac{\text{charge}}{\text{potential difference}}$. Thus $C = \frac{Q}{V}$.

Magnetic Intensity

Magnetic intensity **H**, is defined as the **ratio** $\frac{\mathbf{B}}{\mu}$

where **B** is the magnetic field and μ is the permeability of medium

Thus $\mathbf{H} = \frac{\mathbf{B}}{\mu}$

Index of refraction The index of refraction, n, in a transparent medium is the ratio $\frac{c}{v}$, where c is the velocity of light in free space, and v is the velocity of light in the medium. Thus $n = \frac{c}{v}$

Index of refraction For a ray of light passing from medium A into medium B at an oblique angle, the index of refraction, n, is defined by the ratio $\frac{\text{velocity of light in A}}{\text{velocity of light in B}}$

Thus $n = \frac{\text{velocity of light in A}}{\text{velocity of light in B}}$

Illumination

The illumination E, of a surface of area A, is the ratio $\frac{F}{A}$, where F is the luminous flux.

Linear Magnification

The linear magnification of an optical system is the ratio, $\frac{\text{image height}}{\text{object height}}$

Snell's law

$\mu = \frac{\sin i}{\sin r}$, where μ = refractive index, i = angle of incidence, r = angle of refraction

Telescope

$M = \frac{f_o}{f_e}$, where M = magnifying power of objective at infinity, f_o = focal length of objective, f_e = focal length of eyepiece.

Lesson 21
Application of Ratios in Chemistry

Law of definite composition; Law of multiple proportions; Valence; Writing formulas; Balancing equations, Vapor density; Specific gravity, Molecular weight of gases, Hydrogen equivalent; Problems based on equations

Law of Definite composition

The law of definite composition states that when elements combine to form compounds, the masses of the elements involved in the combination are in a definite **ratio.**

or

The law of definite composition states that in a pure compound, the masses of the elements are always in definite **proportions.**

Example In water (H_2O) the **ratio** of the mass of hydrogen to the mass of oxygen is 1 to 8, irrespective of the source of the water.

Law of Multiple Proportions

When two elements A and B combine to form more than one compound, the masses of B which combine with a fixed mass of A are in the **ratio** of small whole numbers.

Illustration of the law of multiple proportions

Nitrogen and oxygen can combine to form two different substances, nitric oxide, and nitrogen pentoxide:

Composition ⇓	Fixed Mass of Nitrogen (from experiment) ⇓	**Mass of Oxygen** (from experiment) ⇓	**Ratio** of oxygen in different compounds ⇓
Nitric Oxide	28 g of nitrogen	32 g of oxygen	32 to 64 or 1 to 2
Nitrogen pentoxide	28 g of nitrogen	64 g of oxygen	

From the above table, the masses of oxygen, 32 g and 64 g, which combine with a fixed mass of nitrogen, namely 28 g are in the **ratio** 32 to 64 or 1 to 2 (small whole numbers).

Density: The density of a substance is the **ratio** $\frac{\text{mass of substance}}{\text{volume of substance}}$

Specific Gravity: For **solids** and **liquids**, specific gravity is the **ratio** $\frac{\text{density of substance}}{\text{density of water}}$

For **gases,** specific gravity is the **ratio** $\frac{\text{density of gas}}{\text{density of air (or other gases)}}$

Vapor density: The vapor density of a gas is defined as the **ratio** $\frac{\text{weight of any volume of gas}}{\text{weight of an equal volume of air}}$

Molecule: A molecule is a group of atoms bound together such that they behave as a single particle.

Molecular weight is the weight of one molecule of a substance.
Gram-molecular weight (GMW) = Sum of the atomic weights in a molecule of the substance.

Mole
1 gram mole contains 6.02×10^{23} molecules.

1 g mole = $\frac{\text{mass in g}}{\text{molecular weight}}$

Question How many grams of $C_6H_{12}O_6$ are in one molecule of $C_6H_{12}O_6$?

Solution We use the "Units Label Method" of solving proportion problems.
Note 1: 1 mole of $C_6H_{12}O_6 \sim$ 180 g of $C_6H_{12}O_6$
Note 2: 1 mole of $C_6H_{12}O_6$ contains 6.02×10^{23} molecules of $C_6H_{12}O_6$

$$1 \text{ molecule } C_6H_{12}O_6 \times \frac{1\text{mole of } C_6H_{12}O_6}{6.02 \times 10^{23} \text{molecules of } C_6H_{12}O_6} \times \frac{180 \text{ g } C_6H_{12}O_6}{1 \text{ mole } C_6H_{12}O_6}$$

$$= \cancel{1 \text{ molecule } C_6H_{12}O_6} \times \frac{\cancel{1\text{mole of } C_6H_{12}O_6}}{6.02\times 10^{23} \cancel{\text{molecules of } C_6H_{12}O_6}} \times \frac{180 \text{ g } C_6H_{12}O_6}{\cancel{1 \text{ mole } C_6H_{12}O_6}}$$

$= 2.99 \times 10^{-22}$ **g**
$= 3.0 \times 10^{-22}$.

Mole fraction: Mole fraction is the ratio $\frac{\text{moles of substance}}{\text{total number of moles}}$

Calculations based on Chemical reactions
Chemical Equations
The coefficients of a chemical equation indicate the **ratio** in which the moles of one substance reacts with the moles of another.
For example, in the equation, $2H_2 + O_2 \rightarrow 2H_2O$,
2 moles of hydrogen react with 1 mole of oxygen to form 2 moles of water
Also, 2 molecules of hydrogen react with 1 molecule of oxygen to form 2 molecules of water

Proportion Problem
Given the equation $2H_2 + O_2 \rightarrow 2H_2O$
(a) what is the ratio of moles of hydrogen to moles of oxygen?
(b) what is the ratio of moles of oxygen to moles of water?
(c) How many moles of water are produced from 8 moles of oxygen?

Valence (or Valency)

Memorize the entries in the following table:

Valence: Valence of an element or a radical is its combining capacity.

Some Common Cations and Anions with their Charges (Valences or valencies)

Monovalent**+1**	**Divalent: +2**	**Trivalent +3**
Li^{+1} (lithium)	Mg^{2+} (magnesium)	Al^{3+} (alumnum)
Na^{+1} (sodium)	Ca^{2+} (calcium)	Cr^{3+} (chromic; chromium III)
K^{+1} (potassium)	Ba^{2+} (Barium)	Mn^{3+}(Manganic; manganese III)
NH_4^{+1} (ammonium)	Zn^{2+} (zinc)	Fe^{3+} (ferric:Iron III)
HC_3^{+1}	Hg^{2+} (mercuric: mercury II	Co^{3+}(cobaltic: cobalt III)
Ag^{+1} (silver)	Fe^{2+} (Ferrous: Iron II)	Ni^{3+}(Nickel III or nickelic)
Au^{+1} (gold)	Ni^{2+} (nickel II: nickelous))	
Hg^{+1}(mercury I; mercurous)	Mn^{2+}(manganous:manganeseII)	
Cu^{+1}(cuprous: copper I)	Cu^{2+} (cupric: copper II)	
	Co^{2+} (cobaltous: cobalt I)	
	Sn^{2+} (:tin II: stannous)	
	Pb^{2+} (lead II; plumbous)	
	Cr^{2+} (chromous; chromium II)	
-1	**-2**	**-3**
F^- (flouride)	S^{2-} (sulfide)	PO_4^{3-}(phosphate)
Cl^- (chloride)	O^{2-}(oxide)	P^{3-} (phosphide)
Br^- (bromide)	O_2^{2-}(peroxide)	BO_3^{3-} (borate)
I^- (iodide)	SO_4^{2-}(sulfate)	
CN^- (cyanide)	SO_3^{2-}(sulfite)	
HCO_3^-(bicarbonate)	CO_3^{2-}(carbonate)	
HSO_4^-(bisulfate)	$Cr_2O_7^{2-}$ (dichromate)	
NO_3^- (nitrate)	CrO_4^{2-}(chromate)	
NO_2^- (nitrite)	$S_2O_3^{2-}$ (thiosulfate)	
OH^- (hydroxide)	$C_2O_4^{2-}$ (oxalate)	
ClO_4^- (perchlorate)		
ClO_3^- (chlorate)		
ClO_2^- (chlorite)		
ClO^- (hypochlorite)		
$H_2PO_4^-$(dihyrogen phosphate)		
MnO_4^-(permanganate)		
$C_2H_3O_2^-$(acetate)		

Extra: Valence , +4 Sn^{4+}(tin IV: stannic); Pb^{4+} (lead IV)

How to write chemical formulas using valences

Rule 1: We use the valences (disregarding the signs) as subscripts

Rule 2: The more electropositive element or radical (metal, metallic radical) is written first , followed by the non metallic (or the less electropositive element.

Rule 3: The valences of the electropositive and the electronegative elements or radicals are interchanged, but if the subscripts are the same, they are omitted.

Rule 4 : A radical should be enclosed in parenthesis if followed by a subscript other than 1.

Example 1 Write a formula for potassium chloride
Using K **(+1)** and Cl (-1), we obtain KCl

Example 2: Write a formula for Barium sulfate
Using Ba (+2) and SO_4(-2), we obtain $BaSO_4$

Example 3: Write a formula for Ferrous phosphate
Using Fe **(+2)** and PO_4(-3), we obtain $Fe_3(PO_4)_2$

Example 4: Write a formula for Ferric phosphate
Using Fe **(+3)** and PO_4(-3), we obtain $FePO_4$

CHAPTER 7

Lesson 22: **Nursing Math: Dosage Calculations**
Lesson 23: **Food Preparation, Recipes**

Lesson 22
Nursing Math: Dosage Calculations

The calculations covered in this lesson are those of direct proportion, and therefore, the methods of Lesson 5 are all applicable. However, we will cover only Method 2, namely, the "Units Label method" or the "Dimensions method"

Recall that in direct proportion, as one quantity increases, the other quantity also increases; or as one quantity decreases, the other quantity decreases. Do not use the approaches here to solve indirect or inverse problems. At the end of Chapter 2, it was commented that the Units Label or Dimensions method is a very efficient method for solving direct proportion involving three or more variables.

Units Label or Dimensions Method

In the units label method, we use the units to guide us to obtain the expression which on simplifying produces the correct result.

We begin with an everyday example

Example 1: If 2 dollars can buy 8 apples, how many dollars are needed to buy 24 apples?

Solution: We can consider this problem as converting 24 apples to dollars. The justification of this method is in parentheses. Note that 2 dollars and 8 apples are paired together, and we want to find the member to be paired with the 24 apples.

Step 1: $\frac{24 \text{ apples}}{1} \times \frac{?}{?}$

Step 2: Multiply $\frac{24 \text{ apples}}{1} \times \frac{?}{?}$ by a fraction formed by using the quantities 2 dollars and 8 apples as the terms of the fraction such that the denominator has the same units as the 24 apples (the quantity in the numerator), and under such conditions, we can divide out (cancel) the common units in the numerator and the denominator, leaving us the units, dollars.

Then, $\frac{24 \text{ apples}}{1} \times \frac{2 \text{ dollars}}{8 \text{ apples}}$

Step 3: $\frac{24 \cancel{\text{apples}}}{1} \times \frac{2 \text{ dollars}}{8 \cancel{\text{apples}}}$ (= Number of apples $\times$ cost per each apple)

= 6 dollars

Conclusion: 6 dollars are needed to buy 24 apples.

Tasks in Dosage Calculations

The two main tasks in dosage calculations are
1. Reading Labels correctly; and
2. Applying ratios and proportions to determine correct dosages.

Case 1: Simple proportion involving two ordered pairs (four quantities). Note that each ordered pair contains two quantities. A paired quantity may be in the form of a ratio (such as 2 mL to 6 gm, or in the form of a fraction such as $\frac{2 \text{ mL}}{6 \text{ gm}}$.

Why does a nurse or a pharmacist need to do basic dosage calculations?
Answer: What is available at the pharmacy or drug store from the manufacturer are of various strengths and quantities.
Sometimes, the tablet prescribed by a doctor one buys at the drug store is exactly of the same strength and quantity as was delivered by the manufacturer; and sometimes, what is prescribed by the doctor is different from what is on the market. A nurse or a pharmacist may therefore have to calculate what the patient needs using the information in the doctor's prescription and the drug manufacturer's labels.

Case 1: A one step calculation

Example 1 A drug label reads 100 mg per 5 ml. A doctor orders 140 mg for a patient. How many ml should be prepared?

Solution: We can consider this problem as converting 140 mg to ml.

Step 1: Begin with $\frac{140 \; mg}{1} \times \frac{?}{?}$

Step 2: Multiply $\frac{140 \; mg}{1} \times \frac{?}{?}$ by a fraction formed by using the paired quantities 100 mg and 5 ml as the terms of a fraction such that the denominator has the same units as the 140 mg". (the quantity in the numerator), and under such conditions, we can divide out (cancel) the common units in the numerator and the denominator, leaving us the units,

Tnen $\frac{140 \; mg}{1} \times \frac{5 \text{ ml}}{100 \text{ mg}}$ $\left(\frac{140 \; \cancel{mg}}{1} \times \frac{5 \text{ ml}}{100 \; \cancel{mg}}\right)$

$= 7 \text{ ml}$

7 ml should be prepared.

Case 2: A two step calculation, but can be done in a single line.

Example 2: **Intravenous flow rate calculation**

A doctor orders 5 units of insulin per hour.

Available: **1**. 25 units of insulin in 120 ml of saline solution, and

2. Administration set delivers 10 *gtt/ml* (tubing drop factor)

(*gtt* - from the Latin word, gutta, which means drop)

Find the flow rate in gtt/min. (drop per minute)

Solution: We can consider this problem as converting 5 units of insulin per hour to gtt/min

Step 1: Begin with $\frac{5\text{ units}}{60\text{ min}} \times \frac{?}{?}$ (1 hr = 60 min)

Step 2: Multiply $\frac{5\text{ units}}{60\text{ min}}$ by a fraction formed by using the quantities 25 units and 120 ml as the terms of a fraction such that the denominator has the same units as the 5 units". (the quantity in the numerator), and under such conditions, we can divide out (cancel) the common units in the numerator and the denominator, leaving us the units, gtt/min.

Then, $\frac{5\text{ units}}{60\text{ min}} \times \frac{120\text{ ml}}{25\text{ units}}$

Step 3: $\frac{5\,\cancel{\text{units}}}{60\text{ min}} \times \frac{120\text{ ml}}{25\,\cancel{\text{units}}} = \boxed{\frac{5}{60\text{ min}} \times \frac{120\text{ ml}}{25}}$

Step 4: Multiply the result in step 3 by a fraction using the 10 gtt and ml (from the given 10 *gtt/ml*) as a fraction, such that the denominator has the same units as the 200 ml in the numerator.

(That is multiply by $\frac{10\text{ gtt}}{\text{ml}}$)

$$\frac{5}{60\text{ min}} \times \frac{120\,\cancel{\text{ml}}}{25} \times \frac{10\text{ gtt}}{\cancel{\text{ml}}} = \boxed{\frac{5}{60\text{ min}} \times \frac{120}{25} \times \frac{10\text{ gtt}}{1}}$$

(Note: the only units left now $= \frac{\text{gtt}}{\text{min}}$, which is what we are looking for.)

$$\frac{5}{60\text{ min}} \times \frac{120}{25} \times \frac{10\text{ gtt}}{1} = \frac{4\text{ gtt}}{\text{min}}$$

The flow rate is $\frac{4\text{ gtt}}{\text{min}}$

(**Note** that we can in practice, write only the expression in Step 4. Steps 1-3 are for illustrative purposes, and after mastering the steps, one can show only a single expression in a line as in Step 4)

Lesson 23

Food Preparation, Nutrition & Recipes

A. How to Prepare an Ashanti Light Soup

Ingredients (for three people):

1. Three pounds of chicken (or lamb) 2. One bulb of fresh onion

3. Two medium "bulbs" of fresh tomatoes

4. Salt (sprinkle conservatively on the chicken. You can add more salt after cooking)

5. One small can of sardines in tomato sauce **6. Fresh carrots (about 10 1-inch pieces)**

Method

Step 1: At time **t = 0:** Cut pieces of chicken, onion, carrots, and some salt are placed in a cooking utensil. You may pour a thin film of vegetable oil on the chicken **Do not add any water at this time** The contents are boiled (for between 6 and 9 minutes using very low flame, stirring the contents every two minutes. The contents will turn gray. The first 9 minutes are critical; so watch the contents very carefully, otherwise, they would burn.

Step 2: At **time, t =10** minutes, Freshly cut tomato are added to the contents, and boiled for about 3 more minutes. Using a spoon, press the tomatoes against the walls of the cooking utensil to flatten them.

Step 3: At **time t =14 minutes**, ground fresh tomato or tomato paste, sardines in tomato sauce, are added; followed by with 3 to 4 cups of water, "Turn up the flame" now. The contents are boiled for another 35 minutes.

Step 4: At time, t = 55 minutes, turn off the fire, and an appetizing soup would be ready. You may also add 35 extra minutes in Step 4 if you want to thicken the soup

Extra: (In Step 1, you may include fresh crabs (regular crabs).

The soup may be served with boiled rice, yam, green plantains, mashed potato, boiled potatoes; and fufu (mashed potato plus potato starch) etc

Question: If the above recipe is for 3 people, then using proportion , determine the quantities of the ingredients required to prepare the soup for 9 people.

B. Breakfast Ingredients: 2 eggs, crackers, oats, sugar, milk, broccoli, salt, vegetable oil.

The author has seen seniors walking with canes. Why? Answer: Nutritional deficiency.

Solution: 1. 2 eggs, fresh tomato, fresh onion, salt, sardines in oil

2. Crich crackers (unsalted on surface, a product of Italy) instead of bread.

3. One minute preparation oats (such as Quaker Oats).

Step 1: At time t = 0, Some cooking vegetable oil (e.g. Mazola corn oil) is poured into a frying pan on a stove with very low flame. Cut pieces of onion, and tomato are placed in the frying pan containing the oil and some salt (very small amount) is sprinkled over the contents of the frying pan. Cover the contents and let them fry for about 5 minutes. Beat-up two fresh eggs in a saucepan to mix the yellow and white part of the eggs. Remove the cover and pour the egg mixture into to the contents and stir thoroughly, Cover the contents and fry for another five minutes.. Remove the cover and stir the contents to break-up the contents, Cover them and let contents fry for another 3 minutes. and turn off the fire. .Remove and place contents on a plate and drain off the oil in the contents by tilting the plate.

Step 2: Prepare the oats porridge. Add some sugar and milk.

Step 3: Boil about a cup of broccoli. Eat a cup of broccoli with breakfast and another cup for dinner.

Step 4: For breakfast, eat about $\frac{2}{3}$ of the contents from Step 1; and save the remaining $\frac{1}{3}$ for dinner with boiled rice. Do not eat butter or margarine of any brand. Your cholesterol level will be safe because of the tomato-onion combination with the eggs. Repeat the above for about a week, and note the changes and learn from them. Also note the **ratio** of the ingredients,

CHAPTER 8

Applications of Ratios in Engineering

Lesson 24: **Machine Design**
Lesson 25: **Model-Prototype Design**
Lesson 26 **More Science and Engineering Ratios**

Lesson 24

Machine Design

Gear Design
Gears are used to transmit power from one (axis) shaft to another and provide an exact **velocity ratio** .A velocity ratio of 2 to 1 is obtained by mating two gears (A and B, where A is the driven gear, and B is the driver gear) such that the number of teeth on A equals twice the number of teeth on B,

Gear trains:
If n is the number of revolutions per minute (rpm), and t is the number of teeth of the gear, then for a pair of gears, $n_1 t_1 = n_2 t_2$ **(inverse proportion)**

Factor of Safety:
If the limit-stress value is twice as large as the chosen design-stress, we say that the factor of safety, F, in the design is 2.. That is the **ratio**. $F = \frac{s_L}{s_D}$, where s_L is the limit stress, and s_D is the design stress.

Elasticity of a shaft:

The modulus of elasticity , G, is given by the **ratio,** $G = \frac{\text{stress}}{\text{strain}}$

Hollow Shaft design:
The **ratio** of the inside diameter to the outside diameter is usually 1:2 or 1:3

Mechanical advantage of a machine is defined as the **ratio,** $\frac{\text{output force}}{\text{input force}}$

Pulley Design:
In designing elliptical arms of a pulley, a recommended **ratio** of the major axis to the minor axis is 2 to 1.

Torque ratio: $\frac{\text{output torque}}{\text{input torque}}$
Used for designing mechanisms

Efficiency of a machine is defined as the **ratio** $\frac{\text{output work}}{\text{input work}}$

Mechanical Advantage of Lever = ratio $\frac{L_{in}}{L_{out}}$

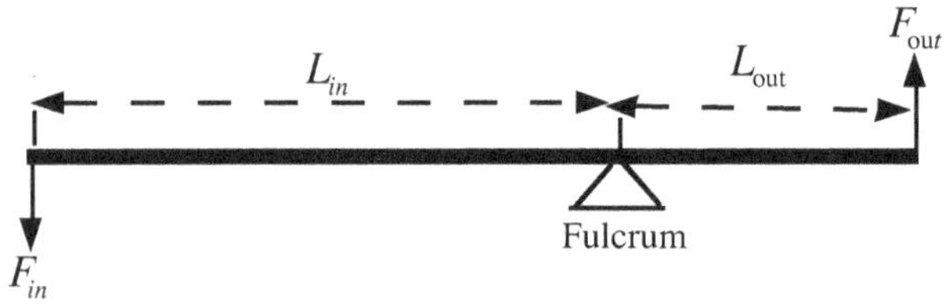

Mechanical Advantage of a hydraulic press $\frac{A_{out}}{A_{in}}$

(where *A* is the cross-sectional area of piston)

Lesson 25
Applications of Proportion in Engineering
Model-Prototype Design

A **model** is a small scale copy of something, and a **prototype** is generally the first actual full scale physical system from which others may be developed or copied. Models are therefore physically similar to the prototype, but differ only in scale (size).

If an engineer wants to build a new and untested power plant costing $180,000, what the engineer will do is first to design, build and test a similar but much smaller power plant, the **model** power plant, costing say, $5,000. After testing this model, if the engineer is satisfied with the performance of this model, the next step is to design and build the actual full scale $180,000 power plant, the **prototype p**ower plant. In this approach, the engineer saves a lot of money. In going from the model to the prototype, the principle of **similarity** between the model and prototype must exist..

The main types of similarity that must exist between model and prototype include the following:

1. Geometric similarity (proportionality of dimensions)
2. Mechanical similarity which includes
  (a) Static similarity (proportionality of time);
  (b) Dynamic similarity (proportionality of forces)
  (c) Kinematic similarity (proportionality of time); (Geometric similarity must also exist)
3. Thermal similarity (proportionality of temperature)
4. Chemical similarity (proportionality of concentration) in chemical engineering

Example
Geometric similarity will exist between model and prototype if the **ratio** of any pair of corresponding dimensions of the model and the prototype is equal to the **ratio** of any other pair of corresponding dimensions.

Note Models such as mathematical models, mechanical and electrical analogs are physically dissimilar from the prototype,

Other ratios include:
velocity **ratio,** acceleration **ratio**, discharge **ratio**, force **ratio,** pressure **ratio**,:
inertia force **ratio**, time **ratio,** surface tension **ratio,** gravity force **ratio**, elasticity force **ratio,** viscous force **ratio,** and power **ratio.**

Lesson 26

More Science and Engineering Ratios

Activation ratios	Displacement ratios	Point-hour ratio
Air-fuel ratio	Drive ratio	Poisson ratio
Air-liquid ratios	Effector-target ratios	Popped ratios
Air mixing ratios	Efficacy ratio	Pressure ratios
Amplitude ratios	Efficiency ratios	Pulley ratio
Area ratios	Electrooptical ratios	Radius ratio
Bedrock waves ratio	Engineering ratio	Ratio analysis
Bentonite-sand ratios	Extension/compression ratio	Reinforcement ratios
Brewing ratios	Extraction ratios	Resistance ratios
C/N ratios	Gear ratios	Resolution ratio
Coefficient ratios	Great ratios	Risk ratios
Consolidation ratios	H/V Spectratios	Roof Drifts ratios
Chemical ratios	Isotope ratios	Reactivity ratios
Coal-to- air ratios	Likelihood ratio	Reduction ratio
Concentration ratios	Linear ratios	Reactivity ratios
Compression ratio	Male-Female ratios	Reflux ratio
Critical Mix ratios	Mass to charge ratio	Reduction ratio
Damping ratios	Mixing ratios	Reflux ratio
Deformation ratios	Mean ratio	Settlement ratio
Delivery ratios	Mole ratios	Signal output ratio
Density ratio	Odds ratios	Signal-to-noise ratio
Depth-area ratios	Oxygen ratio	Solubility ratios
Dilution ratio	Power-weight ratios	Snow-air ratio
Dimensionless ratios	Phase Distribution ratio	Step-down ratios
Direction ratios	Producing gas-oil ratio	Stiffness-weight ratios

Science and Engineering Ratios (continued)	**Biology Ratios**	
Stoichiometric ratios	Birth Sex ratios	Mortality ratios
Surface area/volume ratio	Cell substrate/ adhesion ratio	Nurse-Patient ratios
Taguchi S/N ratios	Extinction ratio	Nutrient ratios
Tensile Strength ratios	Gender ratios	Offspring sex ratios
Torque-weight ratio	Gene expression ratio	Sex-ratio
Trigonometric ratios	Genotype/phenotype ratios	Sons/daughters ratio
Variable aspect ratio	Growth rates	Student/ faculty ratios
Water- cement ratios	Memory ratio	

CHAPTER 9
Lesson 27
Applications of Ratios in Business

Compound Interest:
At first glance, one might think that compound interest calculation has nothing to do with ratios, but observing the formula for compound interest, reveals that ratios are involved.

The **ratio** $\frac{\text{amount}}{\text{principal}} = \frac{A}{P} = (1+i)^n$ (from $A = P(1+i)^n$)

Compound interest depends on upon four factors, namely (a) the principal (b) the interest rate per conversion period (c) the duration of the money invested.
An interest rate of 7% percent means 7% per year. If the compounding period is not one year, then the period is specified. For example, an interest rate of 7 % compounded quarterly (4 times a year) means the rate per each quarter is $\frac{7\%}{4} = 1\frac{3}{4}\%$ per quarter. An interest rate of 8% compounded quarterly (that is four times a year) means the interest rate per each quarter is $\frac{8\%}{4} = 2\%$ per quarter.
If A = amount at the end of n compounding periods.
P = the principal invested
i = rate compounded per specified period, then amount A is given by

$$A = P(1+i)^n$$

If the interest is compounded k times a year for n years the amount A is given by

$$A = P(1+\tfrac{i}{k})^n$$

If $P = 1$ then $A = (1+i)^n$ becomes
$A = (1+i)^n$ and then we can use the binomial theorem or tables in computing the interest.

Simple Interest:

The simple interest on P dollars at a rate of $R\%$ for T years is given by $\frac{PRT}{100}$.

Price-Earnings (P.E.) Ratio is the ratio $\frac{\text{Current market price of company's stock}}{\text{Company's per share earnings for a 12–month period}}$

Earning's yield is the inverse of the price-earnings ratio. Example: If P.E = 4, then

earnings yield $= \frac{1}{4}$ or 25%

Rate: Many terms in economics are rates, each of which is a **ratio** of two quantities

Interest rate An interest rate of 5% means the **ratio** of interest to principal is 5:100.

Parity Ratio: $\frac{\text{prices received}}{\text{prices paid}}$ (in agriculture)

Cross Rates: Paired exchange **rates** among three or more bases.

Ratio scale: In using a **ratio scale** in graphing, one uses a logarithmic scale for one axis or both axes of the graph.

Income and Expenses

The ratio, $\frac{\text{Income}}{\text{Expenses}} > 1$: It means your income is greater than your expenses, and which means you can save some of your income.

The ratio, $\frac{\text{Income}}{\text{Expenses}} = 1$: It means your income equals your expenses. This means you cannot save any of your income.

The ratio, $\frac{\text{Income}}{\text{Expenses}} < 1$: It means your income is less than your expenses, and which means you are spending more than you earn; and therefore you are living beyond your means.

The ratio $\frac{\text{Debt}}{\text{Credit}} > 1$: It means your debt is greater than your credit, and which means you have exceeded your credit limit and are in financial trouble.

The ratio $\frac{\text{Debt}}{\text{Credit}} = 1$: It means your debt equals your credit and have used all your available credit.

The ratio $\frac{\text{Debt}}{\text{Credit}} < 1$: It means your debt is less than your credit and can use more credit. The smaller this ratio, the more the available credit.

More Business Ratio Terms (About 60)

Business Ratios

Accounting ratios	Debt Service ratio	Management ratios
Actual-Target Cost ratio	Debt/Equity ratio	Operating ratios
Bank Capital ratios	Digit ratios	Optimal ratios
Benefit/cost ratios	Effectiveness ratios	Parity ratio
Bidding ratios	Employment ratios	Payout ratio
Business ratio	Equilibrium ratios	PC ratio
Capital Market ratios	Equity ratios	Performance ratio
Cash Flow ratio	Equivalency ratios	Personnel ratios
Claims ratios	Export ratios	Price index
Company's P/E ratio	Financial ratios	Price ratios
competitor ratios	Growt rates	Price-sales ratio
Consumption ratios	Hedging ratios	Price/Earnings ratio
Contribution ratios	Host–guest ratios	Productivity ratios
Cost ratios	Income ratios	Profile ratios
Credit ratios	Inflation ratio	Profitability ratios
Cross ratio	Interim ratios	Turnover ratio
Current ratios	Investment ratios	
Damage ratios	L/B population ratios	
Debt/Credit ratio	Management efficiency ratios	

CHAPTER 10

Lesson 28

Miscellaneous Applications

Ratio test in calculus:

Let v_n and v_{n+1} be two successive terms of the series, $\sum_{n=1}^{+\infty} V_n$. If $\lim_{n\to\infty}\left|\frac{V_{n+1}}{V_n}\right| = L$, and $L < 1$, then the series is absolutely convergent, but if $L > 1$, or $L = +\infty$, the series diverges. However if $L = 1$, no conclusion can be drawn about convergence.

Combustion
Ratio of oxygen in air to the volume of air is 20:80 or 1:4

Statistics: Estimation of **proportions** of data. Here, proportion implies **ratio** (a fraction)

Fluid Mechanics
Reynolds number, Re,

is defined by the ratio $\frac{\rho DV}{\mu}$, where ρ= density of fluid, V,= average forward velocity, D = diameter of tube, μ = viscosity of fluid.

Streamlines of Fluid Flow (for relating fluid velocity components to the flow field)

$$\frac{dx}{u_x} = \frac{dy}{u_y} = \frac{dz}{u_z}$$ (streamline is tangent to the velocity vector component)

Partial Differential Equations (First order linear)
If $P(x,y,z)z_x + Q(x,y,z)z_y = R(x,y,z)$

$$\frac{dx}{P(x,y,z)} = \frac{dy}{Q(x,y,z)} = \frac{dz}{R(x,y,z)}$$ or simply, $\frac{dx}{P} = \frac{dy}{Q} = \frac{dz}{R}$

Mach number = the ratio, $\frac{\text{speed of object}}{\text{speed of sound}}$

Soil Mechanics:

Void Ratio $= \frac{\text{Volume of voids (air or liquids)}}{\text{Volume of soil particles}}$

Golden Ratio (Geometry) which is associated with beauty .

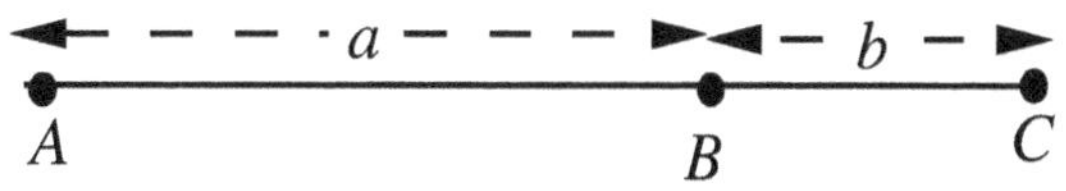

Two numbers a and b $(a > b)$ are in the golden ratio if $\frac{a}{b} = \frac{a+b}{a}$

$$\frac{a}{b} = \frac{a+b}{a}$$

Golden ratio $= \frac{1+\sqrt{5}}{2} \approx 1.618$

The golden ratio is also known as the golden number or the golden section

For a **golden rectangle,** the ratio, $\frac{\text{length}}{\text{width}} = 1.618$

Control Systems:
Closed-loop transfer function = control ratio (use a diagram)

Shear modulus of a solid = the ratio $\frac{\text{shear stress}}{\text{shear strain}}$

Viscosity = the ratio $\frac{\text{shear stress}}{\text{rate of shear strain}}$

Nusselt number, Nu = the ratio $\frac{hL}{k}$, where L = characteristic length,
h = convective heat transfer coefficient, k = thermal conductivity of fluid

Note the following:
For two positive numbers A and B,
if the ratio, $\frac{A}{B} > 1$, then $A > B$; if the ratio, $\frac{A}{B} = 1$, then $A = B$, but if the ratio, $\frac{A}{B} < 1$, then $A < B$.
Application: For an equation, to prove that $A = B$, prove that $\frac{A}{B} = 1$.

Appendix 1
Fractions and Mixed Numbers

Lesson 30
Definitions; Writing Fractions and Mixed Numbers in Words; Reducing Fractions to lowest terms; Equivalent Fractions

Fraction: A fraction is a number representing one or more of the equal parts into which a whole (thing) has been divided.

Unit fraction One part of a whole which has been divided (or broken) into equal parts.
The numerator of a unit fraction is 1. Examples of unit fractions are $\frac{1}{2}$, $\frac{1}{2}$ and $\frac{1}{100}$.

Terms and Meaning of a Fraction

Example In the fraction, $\frac{2}{3}$, the 2 and 3 are called the terms of the fraction; the 2 is called the numerator and the 3 is called the denominator.

In a fraction, the numerator indicates how many of the unit fraction the fraction represents; the denominator indicates into how many equal parts a whole has been divided; the denominator also indicates the size of the unit fraction: the smaller the denominator, the larger the size, and the larger the denominator, the smaller the size of the unit fraction.

Note that in addition to representing parts of a whole, a **fraction** is also used to express **division**, and also to **compare** numbers.

Types of Fractions

Proper Fraction: A fraction in which the numerator is smaller than the denominator.

A proper fraction is always less than 1. Examples are $\frac{1}{2}$, $\frac{2}{7}$ and $\frac{34}{45}$

Improper Fraction: A fraction in which the numerator is greater than or equal to the denominator.

An improper fraction is greater than or equal to 1. Examples are $\frac{8}{5}$, $\frac{9}{3}$ and $\frac{7}{7}$.

Mixed Number: A mixed number consists of a whole number part and a proper fraction part. We obtain a mixed number from an improper fraction when the denominator does not divide the numerator exactly (i.e. without a remainder).

Converting an Improper Fraction to a Mixed Number

Example Convert $\frac{91}{17}$ to a mixed number.

Solution $\frac{91}{17}$ = 5 remainder 6 (using long division) ←

$$\begin{array}{r} 5 \\ 17\overline{)91} \\ \underline{-85} \\ 6 \end{array}$$

Therefore, $\frac{91}{17} = 5\frac{6}{17}$

Converting a Mixed Number to an Improper fraction

Example Convert $6\frac{3}{4}$ to an improper fraction.

Step 1: $6 \times 4 = 24$
Step 2: $24 + 3 = 27$
Step 3: $6\frac{3}{4} = \frac{27}{4}$

(Normally, we do the conversion mentally and show only Step 3 (skipping Steps 1 & 2)).

Writing Fractions and Mixed Numbers in Words

Example 1 Write in words: $\frac{4}{5}$
Solution: Four fifths.
Example 2 Write in words: $7\frac{3}{10}$
Solution : Seven **and** three tenths.

Example 3 Write in words: $568\frac{9}{100}$
Solution: Five hundred sixty-eight **and** nine hundredths.

Example 4 Write in words: $43\frac{2}{7}$
Solution Forty-three **and** two sevenths.

Example 5: Write in words: $64\frac{102}{200}$
Solution: Sixty-four **and** one hundred two two-hundredths

A note about hyphenation

The author suggests that one should hyphenate only for clarity and to avoid confusion.. Unnecessary hyphenation sometimes leads to confusion in communication. The objective must be to communicate unambiguously.

Reduction of Fractions to Lowest terms

A fraction is in its lowest terms if the numerator and the denominator do not have any common factors (divisors) other than 1.

Examples of fractions in lowest terms: $\frac{2}{3}$, $\frac{5}{4}$ and $\frac{1}{6}$

Example Reduce $\frac{14}{21}$ to lowest terms.

Solution $\frac{14}{21}$

$= \frac{2 \times 7}{3 \times 7}$ <-------You may skip this step.

$= \frac{2}{3}$ (Dividing out or canceling the common factor, 7.)

Equivalent Fractions

Equivalent fractions are fractions that have the same value but differ in the numerators and the denominators.

Examples: $\frac{1}{2}$, $\frac{2}{4}$, and $\frac{5}{10}$ are equivalent fractions. On reducing equivalent fractions to their lowest terms, they become identical. In the above examples, $\frac{1}{2}$ is in the lowest terms; $\frac{2}{4}$ and $\frac{5}{10}$ are higher terms of $\frac{1}{2}$

Forming Equivalent Fractions

Given a fraction **in** its lowest terms, to form an equivalent fraction, multiply both the numerator and the denominator by the same nonzero number.

Example 1 $\mathbf{\frac{1}{2}} = \frac{1 \times 3}{2 \times 3} = \mathbf{\frac{3}{6}}$

Given a fraction **not in** its lowest terms, to form an equivalent fraction, multiply (to higher terms) or divide (to lower terms) both the numerator and the denominator by the same nonzero number.

Example 2 $\mathbf{\frac{8}{12}} = \frac{8 \times 2}{12 \times 2} = \mathbf{\frac{16}{24}}$ <-----------Going to higher terms.

Example 3 $\mathbf{\frac{8}{12}} = \frac{8 \div 2}{12 \div 2} = \mathbf{\frac{4}{6}} = \mathbf{\frac{2}{3}}$ **<-------**-Going to lower terms.

Example 4 Find the missing number: $\frac{8}{12} = \frac{?}{9}$

Solution

Step 1: Reduce to lowest terms.

$$\frac{8}{12} = \frac{2}{3} = \frac{?}{9}$$

Step 2: Form the required equivalent fraction.

$$\frac{2}{3} = \frac{2 \times 3}{3 \times 3} = \frac{6}{9}$$

The missing number is 6.

Note above (as in Example 4) that in some cases, if the given fraction is not in its lowest terms, first reduce to lowest terms before forming the equivalent fraction.

Example 5 Find x: $\frac{8}{12} = \frac{x}{9}$

Use exactly the same procedure as in Example 4. In the future, we will learn how to find x using algebraic methods.

Answer: $x = 6$.

Some applications of equivalent fraction formation

Equivalent fraction formation has a number of applications:

1. In adding and subtracting unlike fractions. (see p.145)

2. In forming proportions and solving proportion problems.(see p.12)

3. In rationalizing denominators of fractions involving radicals and complex numbers.

Lesson 30 Exercises

A. (a) Convert the following to mixed numbers: **1.** $\frac{22}{7}$; **2.** $\frac{17}{5}$; **3.** $\frac{79}{18}$.

(b) Verify the answers in **(a)** by converting your answers back to improper fractions.

(c) Change the following to improper fractions: .**1.** $8\frac{5}{6}$ **2.** $9\frac{3}{8}$

Answers: (a). **1.** $3\frac{1}{7}$; **2.** $3\frac{2}{5}$; **3.** $4\frac{7}{18}$; (c): **1.** $\frac{53}{6}$; **2.** $\frac{75}{8}$

B Write in words: **1.**. $\frac{5}{8}$; **2.** $63\frac{4}{100}$; **3.** $2\frac{5}{17}$ **4.** $\frac{24}{25}$; **5.** $9\frac{3}{5}$; **6.** 63,400.

Answers: 1. Five eighths; **2.** Sixty-three and four hundredths; **3.** Two and five seventeenths; **4.** Twenty-four twenty-fifths; **5.** Nine and three fifths; **6.** Sixty-three thousand, four hundred.

C. Write using numerals:: **1.** Eight and nine hundredths. **2.** Eight hundred nine. **3.** Eight hundred nine and nine tenths.

Answers: : **1** . $8\frac{9}{100}$; **2.** 809; **3.** $809\frac{9}{10}$

D. Reduce the following to lowest terms:

1. $\frac{21}{28}$ **2.** $\frac{24}{30}$ **3.** $\frac{20}{16}$ **4.** $\frac{512}{1024}$ **5.** $\frac{210}{1050}$

Answers: 1 . $\frac{3}{4}$; **2.** $\frac{4}{5}$; **3.** $\frac{5}{4}$); **4.** $\frac{1}{2}$; **5.** $\frac{1}{5}$

E 1. Convert $\frac{3}{4}$ to 28ths (That is, change $\frac{3}{4}$) to an equivalent fraction whose denominator is 28.)

2. Convert $\frac{3}{2}$ to 16ths.

3. Change $\frac{5}{6}$ to 36ths

4. Find the missing number: $\frac{15}{16} = \frac{?}{32}$

5. Find x: $\frac{15}{16} = \frac{x}{32}$

6. Convert $\frac{11}{12}$ to 24ths.

7. Change $\frac{6}{12}$ to 20ths

8. Find the missing number: $\frac{?}{56} = \frac{5}{8}$

9. Find x: $\frac{x}{56} = \frac{5}{8}$

Answers: Answers: **1.** $\frac{21}{28}$; **2.** $\frac{24}{16}$; **3.** $\frac{30}{36}$; **4.** 30 ; **5.** 30; **6.** $\frac{22}{24}$) ; **7.** $\frac{10}{20}$; **8.** 35; **9.** 35.

Lesson 31

Addition and Subtraction of Fractions; Addition of Mixed Numbers

Addition of Fractions

Like Fractions (Fractions with the same denominator).

Example: Add $\frac{5}{8} + \frac{4}{8}$

Solution 1. Add the numerators and keep the common denominator.
2. Reduce the resulting fraction to lowest terms (if reducible) and change any improper fractions to mixed numbers.

$$\frac{5}{8} + \frac{4}{8} = \frac{9}{8}$$
$$= 1\tfrac{1}{8}$$

Unlike fractions (Fractions with different denominators).

Example 1 Add: $\frac{2}{7} + \frac{5}{9}$

In this case, we change the unlike fractions to like fractions, and then add the numerators

Format 1 (Sideways Format)

Step 1: Find the LCD of the fractions.
(The LCD of the fractions is the LCM of the denominators.
The LCD of the fractions is 63.

Step 2: Change the fractions to equivalent fractions with LCD of 63.

$$\frac{2}{7} = \frac{2 \times 9}{7 \times 9} = \frac{18}{63}$$

$$\frac{5}{9} = \frac{5 \times 7}{9 \times 7} = \frac{35}{63}$$

Step 3: Add the numerators and keep the LCD, 63.

$$\frac{18}{63} + \frac{35}{63} = \frac{53}{63}$$

Therefore, $\frac{2}{7} + \frac{5}{9} = \frac{53}{63}$.

Format 2 (Vertical Format)

$$\frac{2}{7} + \frac{5}{9}$$

Step 1: Find the LCD of the fractions
The LCD = 63

Step 2: Form equivalent fractions using the same LCD of 63 to obtain

$$\frac{2 \times 9}{7 \times 9} + \frac{5 \times 7}{9 \times 7}$$
$$= \frac{18}{63} + \frac{35}{63}$$

Step 3: Add the numerators and keep the LCD, 63. to obtain

$$\frac{53}{63}$$

Therefore, $\frac{2}{7} + \frac{5}{9} = \frac{53}{63}$.

Example 2 Add: $\frac{5}{6}+\frac{1}{2}+\frac{3}{4}$

Solution

Method 1

Example 2 Add: $\frac{5}{6}+\frac{1}{2}+\frac{3}{4}$

Format 1 (Sideways Format)

Step 1: Find the LCD of the fractions
The LCD = 12

Step 2: Form equivalent fractions using the same LCD of 12.

Then $\frac{5\times 2}{6\times 2}=\frac{10}{12}$

$\frac{1\times 6}{2\times 6}=\frac{6}{12}$

$\frac{3\times 3}{4\times 3}=\frac{9}{12}$

Step 3: Add the numerators and keep the common denominator to obtain

$$\frac{10+6+9}{12}=\frac{25}{12}=2\frac{1}{12}$$

Format 2 (Vertical Format)

$\frac{5}{6}+\frac{1}{2}+\frac{3}{4}$

Step 1: Find the LCD of the fractions
The LCD = 12

Step 2: Form equivalent fractions using the same LCD of 12 to obtain

$=\frac{5\times 2}{6\times 2}+\frac{1\times 6}{2\times 6}+\frac{3\times 3}{4\times 3}$

$=\frac{10}{12}+\frac{6}{12}+\frac{9}{12}$

$=\frac{25}{12}=2\frac{1}{12}$

Therefore $\frac{5}{6}+\frac{1}{2}+\frac{3}{4}=2\frac{1}{12}$

Method 2 Using prime factorization of the denominators followed by making the denominators of the fractions the same

Add: $\frac{5}{6}+\frac{1}{2}+\frac{3}{4}$

Step 1: Factor the denominators into primes.

$$\frac{5}{2\times 3}+\frac{1}{2}+\frac{3}{2\times 2}$$

Step 2: Make the denominators have the same factors by looking for missing factors and multiplying accordingly; but note that whenever, you multiply the denominator by any factor, you must also multiply the numerator by the same factor (in order to keep the value of the original fraction unchanged).

$\frac{5\times 2}{2\times 3\times 2}+\frac{1\times 3\times 2}{2\times 3\times 2}+\frac{3\times 3}{2\times 2\times 3}$ <---- The denominator of the first fraction needs a "2." The denominator of the second fraction needs a "2" and a "3". The denominator of the third fraction needs a "3".

$=\frac{10}{12}+\frac{6}{12}+\frac{9}{12}$

$=\frac{10+6+9}{12}$

$=\frac{25}{12}$

$=2\frac{1}{12}$

Subtraction of Fractions

The process in the subtraction of fractions is exactly like that of the addition of fractions, except that, in this case, we **subtract** the numerators (instead of adding) of the like fractions.

Addition of Mixed Numbers

Procedure: 1. Add the whole number parts.
2. Add the fractional parts and simplify.

Example Add: $2\frac{3}{4}+3\frac{7}{8}$

Format 1 (Sideways Format)	Format 2 (Vertical Format)
$2\frac{3}{4}+3\frac{7}{8}$	$2\frac{3}{4}+3\frac{7}{8}$
$2\frac{3}{4}=2\frac{6}{8}$	$=2\frac{3\times2}{4\times2}+3\frac{7}{8}$
$3\frac{7}{8}=3\frac{7}{8}$	$=2\frac{6}{8}+3\frac{7}{8}$
Scrapwork	$=5\frac{6+7}{8}$
$5\frac{13}{8}=5+1\frac{5}{8}$ 6 + 7 =13	$=5\frac{13}{8}$
$=5+1\frac{5}{8}$	$=5+1\frac{5}{8}$
$=6\frac{5}{8}$	$=6\frac{5}{8}$

Lesson 31 Exercises

A Add the following: **1.** $\frac{5}{7}+\frac{4}{7}$; **2.** $\frac{5}{12}+\frac{7}{9}$ **3.** $\frac{1}{6}+\frac{3}{5}$; **4..** $\frac{3}{10}+\frac{4}{15}$
5. $\frac{3}{16}+\frac{5}{14}+\frac{7}{8}$; **6.** $\frac{8}{21}+\frac{1}{3}+\frac{9}{14}$; **7.** $\frac{2}{45}+\frac{5}{18}+\frac{7}{30}$

Answers 1. $1\frac{2}{7}$; **2.** $1\frac{7}{36}$; **3.** $\frac{23}{30}$; **4.** $\frac{17}{30}$; **5.**$1\frac{47}{112}$) ; **6.** $1\frac{5}{14}$; **7.** $\frac{5}{9}$

B Subtract: **1..** $\frac{9}{11}-\frac{5}{33}$; **2.** $\frac{3}{7}-\frac{2}{7}$; **3.** Subtract $\frac{1}{16}$ from $\frac{11}{48}$;. **4.** $\frac{7}{8}-\frac{5}{12}$;
5. $\frac{13}{18}-\frac{5}{12}$; **6.** $\frac{5}{18}-\frac{2}{9}$; **7.** $\frac{17}{48}-\frac{5}{16}$

Answers **1.** $\frac{2}{3}$; **2.** $\frac{1}{7}$; **3.** $\frac{1}{6}$; **4.** $\frac{11}{24}$; **5.** $\frac{11}{36}$; **6.** $\frac{1}{18}$) ; **7.** $\frac{1}{24}$

C Add : **1.** $5\frac{2}{3}+7\frac{3}{5}$; **3.** $6\frac{4}{5}+7\frac{1}{5}$ **5.** $4\frac{5}{8}+7\frac{3}{4}$; **7.** $4\frac{8}{11}+2\frac{1}{5}$
2. $4\frac{5}{12}+9\frac{5}{8}$; **4.** $3\frac{2}{5}+6\frac{7}{15}$; **6.** $16\frac{3}{4}+5\frac{1}{6}$; **8.** $2\frac{3}{4}+3\frac{7}{8}$

Answers 1. $13\frac{4}{15}$; **2.** $14\frac{1}{24}$; **3.** 14; **4.** $9\frac{13}{15}$ **5.** $12\frac{3}{8}$; **6.** $21\frac{11}{12}$; **7.** $6\frac{51}{55}$; **8.** $6\frac{5}{8}$.

Lesson 32

Comparison of Fractions; Subtraction of Mixed Numbers

Comparison of Fractions

Example 1 Which fraction is the larger:

$$\frac{2}{5} \text{ or } \frac{4}{7} ?$$

There are a number of methods for comparing fractions. We will use a method which we can call "cross-multiplication of the numerators and denominators". We will multiply the numerator of one fraction by the denominator of the other fraction. The fraction whose numerator (in the multiplication) results in the larger product is the larger of the two fractions; and the fraction whose numerator results in the smaller product is the smaller of the two fractions.

$7 \times 2 = 14$ $\quad$ $5 \times 4 = 20$

$$\frac{2}{5} \times \frac{4}{7}$$

(**Note** that the **directions** of the **arrows** are from the denominators to the numerators.)

Since 20 is larger than 14, $\frac{4}{7}$ is the larger of the two fractions (but of course, $\frac{2}{5}$ is the smaller of the two fractions). **Note** that $\frac{2}{5} = \frac{14}{35}$

and $\frac{4}{7} = \frac{20}{35}$

Example 2 Which is the larger:

$$\frac{7}{16} \text{ or } \frac{3}{5} ?$$

$5 \times 7 = 35$ $\quad$ $16 \times 3 = 48$

$$\frac{7}{16} \times \frac{3}{5}$$

Since 48 is larger than 35, $\frac{3}{5}$ is the larger fraction (It is the fraction whose **numerator** when multiplied by the denominator of the other fraction resulted in the larger 48).

By using the above method , we do not have to find the least common denominator. We however, produce equivalent fractions, in which the common denominator may or may not be the least common denominator.

The objective of the above method was to find a fast method of comparing fractions.

Note: In some cases, if the LCD of the fractions can easily be found, we can form equivalent fractions using the same LCD, and then compare the numerators.

Determining the Smallest or Largest Fraction

Example 3 Determine the smallest fraction.

$$\frac{2}{3}, \frac{3}{5}, \frac{5}{7}, \frac{10}{13}$$

We will compare two fractions at a time, and after any comparison, we will keep the smaller fraction (since we want to determine the smallest fraction) and compare it with the other fractions.

Step 1:

$5 \times 2 = 10 \qquad 3 \times 3 = 9$

$$\frac{2}{3} \times \boxed{\frac{3}{5}}$$

(We "box "the fraction to keep for the next step)

$\frac{3}{5}$ is the smaller (since 9 is smaller than 10) and we keep $\frac{3}{5}$ for the next step.

Step 2:

$7 \times 3 = 21 \qquad 5 \times 5 = 25$

$$\boxed{\frac{3}{5}} \times \frac{5}{7}$$

(bringing down the next fraction, $\frac{5}{7}$, for comparison)

$\frac{3}{5}$ is the smaller (since 21 is smaller than 25)

Step 3:

$13 \times 3 = 39 \qquad 5 \times 10 = 50$

$$\boxed{\frac{3}{5}} \times \frac{10}{13}$$

$\frac{3}{5}$ is the smaller (since 39 is smaller than 50)

The smallest fraction is $\frac{3}{5}$.

Note, above, that after every comparison, we 'box" the fraction to keep for the next comparison.

Example 4 For the last example, we will determine the largest fraction

$$\frac{2}{3}, \frac{3}{5}, \frac{5}{7}, \frac{10}{13}.$$

Bear in mind that in this example, we will always keep the larger fraction after every comparis on (and compare it with the other fractions) since we want to determine the largest fraction.

Step 1: $5 \times 2 = 10 \qquad 3 \times 3 = 9$

$$\boxed{\frac{2}{3}} \quad \frac{3}{5}$$

(We "box" the fraction to keep for the next step)

$\frac{2}{3}$ is the larger

Step 2: $7 \times 2 = 14 \qquad 3 \times 5 = 15$

$$\frac{2}{3} \quad \boxed{\frac{5}{7}}$$

$\frac{5}{7}$ is the larger

Step 3: $13 \times 5 = 65 \qquad 7 \times 10 = 70$

$$\frac{5}{7} \quad \boxed{\frac{10}{13}}$$

$\frac{10}{13}$ is the larger

The largest fraction is $\frac{10}{13}$

Arranging Fractions in Descending (decreasing) or Ascending (increasing) Order

To arrange in descending order: Determine the largest fraction first, remove it from the list and write it down; then determine the largest of the remaining fractions, and similarly remove it from the list and write it after the largest fraction. Repeat the process until only one fraction remains, and write this fraction last. (It is the smallest.)

To arrange in ascending order: The process here is similar to the descending order case, but in this case, we determine the smallest fraction first, followed by the next smallest and so on. The last remaining fraction becomes the largest, and it is written last.

Subtraction of Mixed Numbers

Case 1: We do **not** have to borrow

Example 1 Subtract $9\frac{5}{8} - 2\frac{1}{4}$

In this problem, because the fractional part being subtracted is smaller than the fractional part from which we are subtracting, we do **not** have to borrow from the whole number part; (Of course, if you borrow, you can still arrive at the correct answer, at the expense of extra simplification work).

$$\begin{array}{r} 9\frac{5}{8} = 9\frac{5}{8} \\ -2\frac{1}{4} = -2\frac{2}{8} \\ \hline 7\frac{3}{8} \end{array}$$

Case 2: We have to borrow

Example 2 **Subtract:**

$14\frac{3}{7} - 4\frac{5}{8}$

A quick check (see p.148) indicates that $\frac{5}{8}$ is larger than $\frac{3}{7}$ $(8 \times 3 = 24;\ 7 \times 5 = 35)$

Therefore we borrow 1 from the 14. (**The fractional part of the subtracted mixed number is larger than the fractional part of the mixed number from which we are subtracting.**)

Then, $14\frac{3}{7} = 13\frac{7}{7} + \frac{3}{7} = 13\frac{10}{7}$

Now, we subtract: $13\frac{10}{7} - 4\frac{5}{8}$

LCD = 56

$$\begin{array}{r} 13\frac{10}{7} = 13\frac{80}{56} \\ -4\frac{5}{8} = -4\frac{35}{56} \\ \hline 9\frac{80-35}{56} = 9\frac{45}{56} \end{array}$$

Case 3: We have to borrow

Example 3 $12 - 4\frac{2}{5}$

↓ ↓

$11\frac{5}{5} - 4\frac{2}{5}$ (we borrow 1 from the 12 and change it to $\frac{5}{5}$)

$$\begin{array}{r} 11\frac{5}{5} \\ -4\frac{2}{5} \\ \hline 7\frac{3}{5} \end{array}$$

(To check, add $4\frac{2}{5}$ and $7\frac{3}{5}$ to obtain 12)

Case 4: We do **not** have to borrow. **Example 4** Subtract $25\frac{4}{7} - 10$

$$\begin{array}{r} 25\frac{4}{7} \\ -10 \\ \hline 15\frac{4}{7} \end{array}$$

In this case we do not have to borrow, since the number to be subtracted is a whole number (i.e., does not have a fractional part)

Lesson 32 Exercises

A. Determine the larger fraction:

1. $\frac{2}{7}$ or $\frac{4}{15}$; **2.** $\frac{5}{8}$ or $\frac{7}{11}$; **3.** $\frac{7}{8}$ or $\frac{8}{11}$; **4.** $\frac{3}{7}$ or $\frac{4}{9}$; **5.** $\frac{5}{6}$ or $\frac{9}{11}$

Answers 1. $\frac{2}{7}$; **2.** $\frac{7}{11}$; **3.** $\frac{7}{8}$; **4.** $\frac{4}{9}$; **5.** $\frac{5}{6}$

B Determine the smallest fraction: **1.** $\frac{3}{5}, \frac{4}{7}, \frac{9}{13}, \frac{8}{11}.$; **2.** $\frac{3}{4}, \frac{2}{3}, \frac{11}{15}, \frac{19}{30}.$

Answers 1. $\frac{4}{7}$; **2.** $\frac{19}{30}$

C Determine the largest fraction:

1. $\frac{3}{5}, \frac{4}{7}, \frac{9}{13}, \frac{8}{11}$. ;**2.** $\frac{3}{4}, \frac{2}{3}, \frac{11}{15}, \frac{19}{30}$; **3.** $\frac{3}{8}, \frac{5}{16}, \frac{1}{4}$

Answers **1.** $\frac{8}{11}$; **2.** $\frac{3}{4}$; **3.** $\frac{3}{8}$

D Subtract the following:

1. $6\frac{5}{12} - 2\frac{1}{4}$;
2. $15\frac{2}{9} - 6\frac{5}{18}$;
3. $48 - 7\frac{3}{8}$
4. $76\frac{3}{4} - 23$
5. $3\frac{1}{4} - 1\frac{5}{8}$
6. $51\frac{4}{9} - 7\frac{1}{6}$
7. $89 - 5\frac{3}{7}$
8. $43\frac{2}{5} - 15$

Answers **1.** $4\frac{1}{6}$; **2.** $8\frac{17}{18}$; **3.** $40\frac{5}{8}$; **4.** $53\frac{3}{4}$; **5.** $1\frac{5}{8}$; **6.** $44\frac{5}{18}$; **7.** $83\frac{4}{7}$; **8.** $28\frac{2}{5}$

E

Arrange the fractions in the above, B, and C in descending order; and in ascending order.

Word Problems

F

1. What must be added to 2 to make 5.?

2. What must be added to $2\frac{1}{4}$ to make $6\frac{5}{12}$.?

3. Maria needs $6\frac{5}{12}$ pounds of sugar to make some cakes, At present, she has only $2\frac{1}{4}$ pounds of sugar. How many more pounds of sugar does she need to have in order to make the cakes?

4. Mary had $6\frac{5}{12}$ pints of milk in her refrigerator this morning. In the afternoon she used $2\frac{1}{4}$ pints of this milk. How much milk is left in the refrigerator?

Answers: **1.** 3; **2.** $4\frac{1}{6}$ **3.** $4\frac{1}{6}$ pounds **4.** $4\frac{1}{6}$ pints

Lesson 33

Multiplication and Division: of Fractions and Mixed Numbers

Multiplication of Fractions

Procedure:
Step 1: Divide out (cancel) **any** common factors in the numerators and the denominators.

Step 2: Multiply the remaining numerators, and multiply the remaining denominators.

Example 1 Multiply $\frac{3}{5} \times \frac{5}{16} \times \frac{2}{9}$

Solution:
Step 1: Canceling the common factors, we obtain

$$\frac{\cancel{3}^{\,1}}{\cancel{5}_{\,1}} \times \frac{\cancel{5}^{\,1}}{\cancel{16}_{\,8}} \times \frac{\cancel{2}^{\,1}}{\cancel{9}_{\,3}}$$

Step 2: Multiply the remaining factors.

$$\frac{1 \times 1 \times 1}{1 \times 8 \times 3}$$
$$= \frac{1}{24}$$

Example 2 Multiply: $\frac{2}{3} \times \frac{2}{5}$

Solution: There are no common factors to cancel.
Therefore, we multiply the numerators, and then multiply the denominators.

$$\frac{2}{3} \times \frac{2}{5}$$
$$= \frac{2 \times 2}{3 \times 5}$$
$$= \frac{4}{15}$$

Division of Fractions

Reciprocals: The reciprocal of a real number A is $\frac{1}{A}$

Examples: The reciprocal of $\frac{2}{3}$ is $\frac{3}{2}$. The reciprocal of 4 is $\frac{1}{4}$. The reciprocal of $\frac{1}{4}$ is 4.
Thus, to find the reciprocal of a number, invert the number (or interchange the numerator and the denominator)
The reciprocal of number is also known as the multiplicative inverse of that number.
The product of a number and its reciprocal is 1. Example $\frac{1}{4} \times \frac{4}{1} = 1$

Example 1 $\frac{4}{9} \div \frac{2}{15}$ ($\frac{4}{9}$ is the dividend; $\frac{2}{15}$ is the divisor)

Procedure: Invert the divisor and multiply. (or multiply the dividend by the reciprocal of the divisor)

Then $\frac{4}{9} \div \frac{2}{15}$
$$= \frac{4}{9} \times \frac{15}{2}$$

$$= \frac{\overset{2}{\cancel{4}}}{\underset{3}{\cancel{9}}} \times \frac{\overset{5}{\cancel{15}}}{\underset{1}{\cancel{2}}}$$

$$= \frac{2 \times 5}{3 \times 1}$$

$$= \frac{10}{3}$$

$$= 3\tfrac{1}{3}$$

(Always, cancel any common factors in the numerators and the denominators before multiplying)

Division of Complex Fractions

Example Simplify: $\dfrac{\frac{3}{4}}{\frac{16}{27}}$

Solution: $\dfrac{\frac{3}{4}}{\frac{16}{27}} = \dfrac{3}{4} \div \dfrac{16}{27}$

$= \dfrac{3}{4} \times \dfrac{27}{16}$ (inverting the divisor and multiplying)

$= \dfrac{81}{64}$

$= 1\tfrac{17}{64}$

Multiplication of Mixed Numbers

Procedure: Step 1: Change each mixed number to an improper fraction.
Step 2: Multiply the resulting fractions.

Example $2\tfrac{4}{5} \times 2\tfrac{1}{7}$

Step 1: $= \dfrac{14}{5} \times \dfrac{15}{7}$

Step 2: $= \dfrac{\overset{2}{\cancel{14}}}{\underset{1}{\cancel{5}}} \times \dfrac{\overset{3}{\cancel{15}}}{\underset{1}{\cancel{7}}}$

$= 6$

Division of Mixed Numbers

Example $5\tfrac{2}{3} \div 4\tfrac{1}{2}$ ($4\tfrac{1}{2}$ is the divisor)

Procedure: Change each mixed number to an improper fraction and divide.

$5\tfrac{2}{3} \div 4\tfrac{1}{2}$

Step 1: $= \dfrac{17}{3} \div \dfrac{9}{2}$

Step 2: $= \dfrac{17}{3} \times \dfrac{2}{9}$ (inverting the divisor and multiplying)

$= \dfrac{34}{27} = 1\tfrac{7}{27}$

Lesson 33 Exercises

A. Multiply:: **1.** $\frac{21}{32}\times\frac{8}{15}$; **2.** $\frac{3}{7}\times\frac{35}{36}$; **3.** $\frac{4}{5}\times\frac{6}{7}$;; **4.** $12\times\frac{2}{3}$; **5.** $\frac{9}{8}\times\frac{4}{45}$
6. $\frac{3}{5}\times\frac{5}{16}\times\frac{2}{9}$

Answers **1**. $\frac{7}{20}$; **2.** $\frac{5}{12}$; **3.** $\frac{24}{35}$; **4**. 8 ; **5.** $\frac{1}{10}$; **6.** $\frac{1}{24}$

B. Divide: **1.** $\frac{4}{5}\div\frac{3}{7}$; **2.** $\frac{3}{8}\div\frac{5}{9}$; **3.** $\frac{5}{12}\div\frac{5}{4}$; **4.** $\frac{3}{4}\div\frac{4}{3}$; **5.** $\frac{7}{15}\div\frac{21}{20}$; **6.** $\frac{7}{8}\div\frac{3}{16}$

Answers: **1.** $1\frac{13}{15}$; **2.** $\frac{27}{40}$; **3.** $\frac{1}{3}$; **4.** $\frac{9}{16}$; **5.** $\frac{4}{9}$; **6.** $4\frac{2}{3}$

C. Multiply: (a) $3\frac{3}{4}\times6\frac{2}{3}$; (b) $1\frac{2}{5}\times1\frac{1}{2}$; (c) $1\frac{7}{9}\times\frac{3}{4}$
(d) $2\frac{1}{3}\times5\frac{1}{4}$; (e) $3\times4\frac{1}{2}$; (f) $\frac{4}{5}\times2\frac{1}{7}$; (g) $3\frac{3}{7}\times\frac{3}{4}\times14$

Answers: (*a*) 25 ; (*b*) $2\frac{1}{10}$; (*c*) $1\frac{1}{3}$; (*d*) $12\frac{1}{4}$; (*e*) $13\frac{1}{2}$; (*f*) $1\frac{5}{7}$; (*g*) 36

D. Divide: (a) $5\frac{1}{2}\div3\frac{1}{7}$; (d) $2\frac{2}{3}\div5\frac{1}{3}$; (g) $7\div4\frac{2}{3}$
(b) $12\frac{3}{4}\div2\frac{1}{8}$; (e) $1\frac{7}{8}\div2\frac{3}{16}$; (h) $3\frac{1}{2}\div1\frac{4}{5}$
(c) $14\frac{1}{4}\div3$; (f) $2\frac{1}{8}\div12\frac{3}{4}$

Answers (*a*) $1\frac{3}{4}$; (*b*) 6 ; (*c*) $4\frac{3}{4}$; (*d*) $\frac{1}{2}$; (*e*) $\frac{6}{7}$; (*f*) $\frac{1}{6}$; (*g*) $1\frac{1}{2}$; (*h*) $1\frac{17}{18}$

Appendix 2
DECIMALS

Lesson 34: **Definitions, Addition and Subtraction; Multiplication and Division by Powers of 10, Multiplication of Decimals; Comparison of Decimals ; converting decimals to fractions**

Lesson 35: **Complex Decimals; Dividing Decimals; Converting Fractions to Decimals**

Lesson 34

Definitions, Addition and Subtraction of Decimals; Multiplication and Division by Powers of 10; Multiplication of Decimals; Comparison of Decimals ; Converting Decimals to Fractions

Definitions

Powers of 10

Examples: $10^1 = 10$

$10^2 = 100$

$10^3 = 1{,}000$

$10^4 = 10{,}000$

A **decimal fraction** is a fraction whose denominator is a power of 10.

Examples are $\frac{3}{10}$, $\frac{5}{100}$, and $\frac{23}{1000}$. .

A **decimal** is the symbol for an equivalent decimal fraction.

Examples: $\frac{3}{10} = .3$ ($\frac{3}{10}$ is the decimal fraction; .3 is the decimal)

$\frac{5}{100} = .05$

$\frac{23}{10000} = .0023$

Some Common Fractions, Decimal Fractions, Decimals and Names

Common Fraction	Decimal Fraction	Decimal	Name
$\frac{1}{2}$	$\frac{5}{10}$	.5	Five tenths
$\frac{1}{4}$	$\frac{25}{100}$	.25	Twenty-five hundredths
$\frac{61}{100}$	$\frac{61}{100}$	.61	Sixty-one hundredths
$\frac{490}{1000}$	$\frac{490}{1000}$	.490	Four hundred ninety thousandths

Number of decimal places or decimal digits: The number of decimal places is the number of digits to the right of the decimal point.

Examples: (a) 0.62 has two decimal places; (b) 46.3 has one decimal place;
(c) 53.005 has three decimal places.

Also note for example that 0.62 = .62 (The zero preceding the decimal point on the left-hand side is used to remind the reader that a decimal point follows.)

Equivalent decimals

Equivalent decimals are decimals that have the same value but differ in the number of decimal place
(see also p.143 for a similar definition of equivalent fractions)

Example: .5, .50, and .500 are equivalent decimals, but note that .5 is the simplest form
(in much the same way as $\frac{1}{2}$ is the lowest terms of the equivalent fractions $\frac{2}{4}$ and $\frac{5}{10}$).

Similarly, 4.2 and 4.20 and 4.2000 are equivalent decimals..

Addition of Decimals

Example Add: 13.23 + 7.1 + 9

Procedure: Indicate the decimal point explicitly in all the whole numbers.
Arrange the numbers vertically, while lining up the decimal points.
Note: 9 = 9.

Then we obtain: ↓ ——— decimal points in line

```
 13.23    (Add as is done in whole
  7.10      number addition)
  9.00
 -----
 29.33
```

Subtraction of Decimals

Example: Subtract: 15 - 2.43

Indicating the decimal point in the 15 and arranging the numbers vertically, while lining up the decimal points, we obtain:

```
 15.00<-------------------------(attach zeros)
 -2.43
 -----
 12.57
```

Multiplying decimals or whole numbers by a power of 10

Procedure: Move the decimal point to the right as many places as there are zeros in the power of 10. Examples of powers of 10: $10^1 = 10$, $10^2 = 100$, $10^3 = 1000$

Example 1 $3.45 \times 10 = 34.5$

Example 2 $3.45 \times 10^2 = 345.$

Example 3 $3.45 \times 10^3 = 3450.$

Example 4 $23 \times 10000 = 230000.$ or 230000
(In Examples 3 and 4, we wrote zeros to hold places)

Dividing decimals or whole numbers by a power of 10

Procedure: Move the decimal to the left as many places as there are zeros in the power of 10.

Example 1 $53.4 \div 10 = 5.34$

Example 2 $53.4 \div 100 = .534$

Example 3 $53.4 \div 1000 = .0534$

Example 4 $534 \div 10 = 53.4$

Example 5 $\frac{672.89}{100} = 6.7289$

Example 6 $\frac{89675}{100} = 896.75$

Multiplication of a decimal by a decimal or by a whole number

Step 1: Ignore the decimal points and multiply the whole numbers.

Step 2: Count the total number of decimal places (decimal digits) in the decimals being multiplied.

Step 3: Insert the decimal point in the product from Step 1, so that the number of decimal places equals the total number of decimal places counted in Step 2. (We may have to write zeros to hold places in some cases.)

Example Multiply 4.21 by 3.5 **Note**: $4.21 \times 3.5 = \frac{421}{100} \times \frac{35}{10} = \frac{14735}{1000} = 14.735$

```
  4.2 1
x  3.5      (The total number of decimal places is 3)
  2105
 1263
-------
 14.735
```

Comparison of Decimals

Example Determine which decimal is the smallest: .234, .098, .0725

There are a number of methods for comparing decimals.
First, we will use a method which we can call " comparison of place-value digits", and then we will discuss another method.

Method 1

Step 1: Arrange the numbers vertically while lining up the decimal points as if one were to add the numbers.

.2340 (we can attach zeros)
..0980
.0725

Step 2: We will compare the digits from left to right (tenths, hundredths, and thousandths places etc).
In the tenths' column, 2 is the largest, and therefore .2340 cannot be the smallest.
We cross out .2340 or ignore it.

Step 3: Compare .0980
.0725

Since the tenths' places are the same, we go to the next column (the hundredths' place) and there, 7 is smaller than 9 and therefore .
.0725 is smaller than .098. Hence .0725 is the smallest decimal.
(of course, .2340 is the largest decimal.)

Method 2

Multiply each decimal by a power of 10, so that all the decimals become whole numbers. In multiplying by a power of 10, move the decimal point to the right as many places as there are zeros in the power of 10. The power of 10 to use is based on the decimal with the most number of decimal places.
In the above example, .0725 has the most number of decimal places, four decimal places.

Proceeding,

$$.2340 \times 10000 = 2340$$
$$..0980 \times 10000 = 980$$
$$.0725 \times 10000 = 725$$

Since .0725 gave the smallest whole number 725, .0725 is the smallest decimal.

Arranging Decimals in Descending (decreasing) or Ascending (increasing) Order

To arrange in descending order: Determine the largest decimal first, remove it from the list and write it down; then determine the largest of the remaining decimals, and similarly remove it from the list and write it after the largest decimal. Repeat the process until only one decimal remains, and write this decimal last. (It is the smallest decimal.)

To arrange in ascending order: The process here is similar to the descending order case, above, but in this case, we determine the smallest decimal first, write it down, followed by the next smallest, and so on. The last remaining decimal becomes the largest, and it is written last.

Converting a Decimal to a Fraction

Example Change .96 to a fraction in its lowest terms.

Step 1: Write .96 as a decimal fraction .

$$.96 = \frac{96}{100}$$

(The number of zeros in the denominator equals the number of decimal places in the decimal)

Step 2: Reduce the fraction to its lowest terms.

Then, $\frac{\cancel{96}^{\cancel{48}\ 24}}{\cancel{100}_{\cancel{50}\ 25}} = \frac{24}{25}$

$$\therefore\ .96 = \frac{24}{25}$$

Lesson 34 Exercises

A. Write as decimals: (*a*) $\frac{36}{1000}$; (*b*) $\frac{40}{100}$; (*c*) $\frac{9}{100}$

Answers (*a*) .036 ; (*b*).40 ; (*c*) .0 9

B. Add the following:: (*a*) 13.456 + 4.03 + 9 + .463; (d) 19 + .018 + 298
(*b*) 43.050 + 73.4 + 7 + .0304 + 29 ; (e) .5 + .54 + .6407
(*c*) 7.8 + 47 + 5.605 + .006; (*f*) What must be added to 3.48 to make 9?.

Answers (*a*). 26.949; (*b*) 152.4804 ; (*c*) 60.411 ; (*d*) 317.018 ; (*e*) 1.6807 ; (*f*) 5.52

C. Subtract: **(*a*)** 60.54 - 3.679**;** **(*b*)** 28 - 16.603**;** **(*c*)** 46.07 - 9; **(*d*)** 8.23 - 5.025
(*e*) Subtract 37.5 from 76.4; **(*f*)** From 8, subtract 5.96; **(*g*)** Subtract .65 from 1.1.

Answers **(*a*)** 56.861; **(*b*)** 11.397 ; **(*c*)** 37.07; (*d*) 3.205; **(*e*)** 38.9 ; **(*f*)** 2.04 ; **(*g*)** .45

D. Multiply: (*a*) 65.8×100; (*b*) 4.306×10^2; (*c*) $.006 \times 10000$; (*d*) 26×1000

Answers (*a*) 6580. or 6580; (*b*) 430.6 ; (*c*) 60. or 60; (*d*) 26,000.

E. Divide:(*a*) $74.68 \div 100$; (*b*) $34.07 \div 10^4$; (*c*) $68953 \div 100$; (*d*) $\frac{245.6}{10000}$

Answers (*a*) .7468 ; (*b*) .003407 ; (*c*) 689.53 (*d*) .02456

F. Which decimal is the larger? (*a*) .0764 or .04685 (*b*) 3.245 or .986753
(*c*) 4 or .8954

Answers (*a*) .0764 ; (*b*) 3.245 ; (*c*) 4

G. Determine the smallest decimal: (*a*) .067, .0096, .12402; (*b*) 3.008, .094, .4509
(*c*) 1.007, .753, .087

Answers (*a*) .0096; (*b*) .094; (*c*) .087

H. Use the inequality symbol " > " to compare the following:
(*a*) .0764 or .04685; (*b*) 3.245 or .986753; (*c*) 4 or .8954

Answers: *a*) .0764 > .04685; (*b*) 3.245 > .986753; (*c*) 4 > .8954

I. Order the following using the symbol " < ": (*a*) .067, .0096, .12402;
b) 3.008, .094, .4509; (*c*) 1.007, .753, .087

Answers: (*a*) .0096 < .067 < .12402; (*b*) .094 < .4509 < 3.008 (c) .087 < .753 <1.007,

J. 1. Arrange the decimals in **G** and **H** (above) in descending order..
2. Arrange the decimals in **G** and **H** (above) in ascending order..

K. Change each of the following to a fraction in its lowest terms:
(*a*) .64 ; (*b*) 6.4 ; (*c*) 0.036 ; (*d*) 1.36 ; (*e*) .63; (*f*) 0.0825

Answers (*a*) $\frac{16}{25}$) ; (*b*) $\frac{32}{5}$; (*c*) $\frac{9}{250}$; (*d*) $\frac{34}{25}$; (e) $\frac{63}{100}$; (*f*) $\frac{33}{400}$

Lesson 35
Complex Decimals; Dividing Decimals; Converting Fractions to Decimals

Complex Decimals

A complex decimal consists of a decimal part and a common fraction part, Examples are $.11\frac{5}{7}$ and $.117\frac{1}{7}$.

Using complex decimals allows one to write the exact quotient of a division problem in which no round-off place for the quotient is specified. Example: .82 divided by 7 = $.11\frac{5}{7}$ or $.117\frac{1}{7}$

In a complex decimal, the **number of decimal places** equals the number of digits to the right of the decimal point, **ignoring** the fractional part.

Examples **1.** $.8\frac{2}{3}$ has **one** decimal place. **2.** $.14\frac{1}{3}$ has **two** decimal places.

3. $.3\frac{1}{3}$ has **one** decimal place. **4.** $.33\frac{1}{3}$ has **two** decimal places.

5. $.166\frac{2}{3}$ has **three** decimal places.

Converting a Complex Decimal to a Fraction

Example 1 Convert $.8\frac{2}{3}$ to a fraction in its lowest terms.

$$.8\frac{2}{3} = \frac{8\frac{2}{3}}{10}$$ ($.8\frac{2}{3}$ has one decimal place; and note, for example, that $.8 = \frac{8}{10}$)

$$= \frac{26}{30}$$ $\left(\frac{26}{3} \div \frac{10}{1} = \frac{26}{3} \times \frac{1}{10} = \frac{26}{30}\right)$

$$= \frac{13}{15}$$

Example 2 Convert $.14\frac{1}{3}$ to a fraction in its lowest terms.

$$.14\frac{1}{3} = \frac{14\frac{1}{3}}{100}$$ ($.14\frac{1}{3}$ has two decimal places)

$$= \frac{43}{300}$$

How to divide a Decimal by a Whole number

Procedure Align the decimal point vertically in the quotient and divide as in whole number division. You may attach zeros (one at a time) if needed to continue the division process.

Case 1: Round-off place for the quotient is specified

Example 1 Divide .82 by 7 and round-off quotient to the nearest hundredth.

$$\begin{array}{r} .117 \approx \mathbf{.12} \\ 7\overline{)\,.82} \\ \underline{-7} \\ 12 \\ \underline{-7} \\ 50 \\ \underline{-49} \\ 1 \end{array}$$

(We divided to the thousandths place before rounding-off)

Conclusion: To the nearest hundredth, $\frac{.82}{7} \approx \mathbf{.12}$

Case 2: No Round-off place for the quotient is specified

In this case, we will leave the answer as a complex decimal, since this will be equivalent to the exact quotient. Note that when we round-off, the quotient obtained is an approximation.

Example 2 Divide .82 by 7. (Round-off-place is **not** specified)

$$\begin{array}{r} .11\frac{5}{7} \\ 7\overline{)\,.82} \\ \underline{-7} \\ 12 \\ \underline{-7} \\ 5 \end{array}$$

(dividing to **two** decimal places)

or

$$\begin{array}{r} .117\frac{1}{7} \\ 7\overline{)\,.820} \\ \underline{-7} \\ 12 \\ \underline{-7} \\ 50 \\ \underline{-49} \\ 1 \end{array}$$

((dividing to **three** decimal places)

To check: Let us convert $.11\frac{5}{7}$ and $.117\frac{1}{7}$ back to the original division problem.

$$.11\tfrac{5}{7} = \frac{82}{700} = \frac{.82}{7} \qquad .117\tfrac{1}{7} = \frac{820}{7000} = \frac{82}{700} = \frac{.82}{7}$$

Therefore, $.11\frac{5}{7}$ and $.117\frac{1}{7}$ are equivalent.

We will make the following agreement: In computations, unless otherwise instructed, we will leave the decimal as a complex decimal of two decimal places. This will produce sufficient decimal places for most applications and yet provide additional information for either obtaining more decimal places or for rounding-off.

To **round-off** a complex decimal by **dropping only** the **fractional part,** add 1 to the digit immediately to the left of the fractional part if the numerator of the fractional part is greater than or equal to half of the denominator; but if the numerator is less than half of the denominator, the digit to the left of the fractional part remains unchanged.

Examples (a): $.11\frac{5}{7}$ rounded-off to the nearest hundredth becomes .12 $\left(5 > \frac{7}{2}\right)$

(b) $.117\frac{1}{7}$ rounded-off to the nearest thousandth becomes .117 $\left(1 < \frac{7}{2}\right)$

Note however that $.117\frac{1}{7}$ rounded-off to the nearest hundredth becomes .12

Dividing a Number by a Decimal

Procedure

Step 1: Make the divisor a whole number by moving the decimal point to the right as many places as the number of decimal places in the divisor, and at the same time, move the decimal point in the dividend to the right as many places as the decimal point was moved in the divisor. Note however, that for a divisor such as .20 or .200, we need to move the point only one place to the right to make it a whole number, since 2 is equivalent to 2.0 or 2.00.

Step 2: Rewrite the dividend and the divisor and divide as by a whole number.

Example 1 Divide 1.23 by .8 and round-off quotient to the nearest hundredth.

Step 1: $.8\overline{)1.23} = .8\overline{)1.23}$ (moving the decimal point 1 place to the right in the divisor and dividend)

Step 2: $8\overline{)12.3}$

Step 3:

$$\begin{array}{r} 1.537 \approx \mathbf{1.54} \\ 8\overline{)12.300} \\ \underline{-8} \\ 43 \\ \underline{-40} \\ 30 \\ \underline{-24} \\ 60 \\ \underline{-56} \\ 4 \end{array}$$

Conclusion: To the nearest hundredth, $\frac{1.23}{.8} \approx 1.54$

Example 2 Divide .86 by .20

Step 1: $.20\overline{).86} = .20\overline{).86}$ (We need to move the decimal point only one place to make it a whole number)

Step 2: $\begin{array}{r} 4.3 \\ 2\overline{)8.6} \end{array}$

Writing Fractions as Decimals

When a fraction (a rational number) is converted to a decimal, the decimal is either terminating or repeating

Examples of terminating decimals: **1.** $\frac{1}{4}=.25$; **2.** $\frac{1}{8}=.125$

Examples of repeating decimals: **1.** $\frac{1}{3}=.\overline{3}$ or .33...; **2.** $\frac{3}{11}=.\overline{27}$ or .2727...

Converting a Fraction to a Decimal

Case 1: Round-off place of the decimal is specified

Example Change $\frac{3}{7}$ to a decimal, rounding off the quotient (answer) to the nearest hundredth (two decimal places).

Procedure: Using long division, divide the numerator by the denominator, attaching zeros to the dividend during the division process. We will carry the division to three decimal places (that is, one extra place) before rounding off the quotient. Do **not** just stop at the hundredth place.

$$\begin{array}{r} 0.428 \\ 7\overline{)3.000} \\ -2\,8 \\ \hline 20 \\ -14 \\ \hline 60 \\ -56 \\ \hline 4 \end{array}$$

Then, 0.428 to the nearest hundredth becomes 0.43.

Note: To round off quotient to the nearest tenth, we carry out the division to two decimal places before rounding off; to the nearest thousandth, the division is to four decimal places.

Case 2: Round-off place of the decimal is **not** specified

In this case, we will leave the decimal as a complex decimal of two decimal places, since this will produce sufficient decimal places for most applications and yet provide additional information for either obtaining more decimal places or for rounding-off.

Example 2 Change $\frac{3}{7}$ to a decimal. (Round-off place is **not** specified)

Procedure: Using long division, divide the numerator by the denominator, attaching zeros to the dividend during the division process. No round-off place is specified and therefore we will carry the division to two decimal places, and write the remainder divided by the divisor as a fraction.

$$\begin{array}{r} 0.42\tfrac{6}{7} \\ 7\overline{)3.000} \\ -2\,8 \\ \hline 20 \\ -14 \\ \hline 6 \end{array}$$ ← (complex decimal with two decimal places)

Thus $\frac{3}{7}=0.42\frac{6}{7}$ (a complex decimal with two decimal places)

Example 3: Change $\frac{2}{3}$ to a decimal rounded-off to two decimal places.

$$\begin{array}{r} 0.666 \approx .67 \\ 3\overline{)\,2.000} \\ -\underline{1\,8} \\ 20 \\ -\underline{18} \\ 20 \\ -\underline{18} \\ 2 \end{array}$$

Example 4: Change $\frac{2}{3}$ to a decimal.

$$\begin{array}{r} 0.66\tfrac{2}{3} \\ 3\overline{)\,2.00} \\ -\underline{1\,8} \\ 20 \\ -\underline{18} \\ 2 \end{array}$$

← (complex decimal with two decimal places)

Note: In Example 4, we could write the quotient as $0.\bar{6}$ or 0.6... instead of $0.66\frac{2}{3}$. However, in computations, it is much easier to convert $0.66\frac{2}{3}$ to a fraction than to convert $0.\bar{6}$ or 0.6...to a fraction. The general method for converting $0.\bar{6}$ or 0.6...to a fraction is algebraic.

Lesson 35 Exercises

A. Determine the number of decimal places

1. $.2\frac{2}{3}$; 2. $.02\frac{2}{3}$; 3. $1.2\frac{1}{3}$; 4. $.4\frac{3}{11}$; 5. $43.4\frac{2}{7}$

Answers: 1. 1 ; 2. 2; 3. 1 ; 4. 1 ; 5. 1 .

B. Convert to a fraction in its lowest terms.

1. $.2\frac{2}{3}$; 2. $.02\frac{2}{3}$; 3. $1.2\frac{1}{3}$; 4. $.4\frac{3}{11}$; 5. $43.4\frac{2}{7}$

Answers: 1. $\frac{4}{15}$; 2. $\frac{2}{75}$; 3. $\frac{37}{30}$; 4. $\frac{47}{110}$; 5. $\frac{304}{7}$

C. Divide: **1.** $0.72 \div 5$; **2.** $4.32 \div 12$; **3.** $72.1 \div 0.20$; **4.** $0.738 \div 0.6$; **5.** $30.03 \div 2.1$

Divide and round-off quotient to the nearest hundredth.

6. $0.463 \div 0.8$; **7.** $62.82 \div 14$; 8. $56.6536 \div 23$; 9. $46.95 \div 1.1$; **10.** $38.93 \div 0.08$

Answers: **1.** 0.144; **2.** 0.36; **3.** 360.5; 4, 1.23; **5.** 14.3; **6.** 0.58; **7.** 4.49; **8.** 2.46; **9.** 42.68; **10.** 486.63

D. **1.** Change $\frac{6}{7}$ to a decimal, rounding off the quotient to the nearest hundredth.

2. Change $\frac{5}{9}$ to a decimal, rounding off quotient to the nearest tenth.

Answers 1. $0.857 \approx 0.86$ **2.** $0.55 \approx 0.6$

Appendix 3

Lesson 36: **Percent (%) and Inter-conversions**
Lesson 37: **Calculations Involving Percent**
Lesson 38: **More Applications Involving Percent**

Lesson 36
Percent (%) and Inter-conversions

Some interpretations of the percent symbol "%":

1. **Over hundred**: For example, 20% means $\frac{20}{100}$ (Twenty over hundred or 20 divided by 100)
2. **For each hundred**: For example, a savings account with a 5% interest rate pays the depositor \$5 for each \$100 in the account. (Five for each hundred)
3. **Hundredths:** For example, 20% means $\frac{20}{100}$ or .20 (Twenty hundredths)
4. **As a number out of 100**: For example, a grade of 80% on a test means a student got 80 points out of 100 points.

Changing a decimal (or any fraction) to percent

Procedure: Multiply by 100%. (i.e., multiply by 100 and attach the percent symbol "%".

To change a decimal to percent, move the decimal point two places to the right and attach the percent symbol. (Note that moving the decimal point two places to the right is equivalent to multiplying by 100)

Example 1 Convert .74 to percent.
Solution .74 = 74% .

Example 2 Convert .008 to percent.
Solution .008 = .8%

Example 3 Convert 1 to percent.
Solution 1 = 100%

Example 4 Convert 12 to percent.
Solution 12. = 1200%

Example 5 Change $\frac{1}{4}$ to percent

Method 1 $\frac{1}{4} \times \frac{100\%}{1} = 25\%$

Method 2 $\frac{1}{4} = .25 = .25 \times 100\% = 25\%$ (Changing to decimal first, and then changing to percent)

Changing a percent to a decimal

Procedure: Divide by 100%. (i.e., drop the percent symbol "%" and divide by 100) or apply the meaning of the percent symbol to change the percent to a decimal fraction and then easily to a decimal.

To change a percent to a decimal, drop the percent symbol and move the decimal point two places to the left . (**Note** that moving the decimal point two places to the left is equivalent to dividing by 100)

Case 1: Example 1 74% = .74 $\left(74\% = \frac{74}{100} = .74\right)$

Example 2 .8% = .008 $\left(\text{or } .8\% = \frac{.8}{100} = .008\right)$

Example 3 145% = 1.45

Example 4 14.5% = .145

Example 5 .145% = .00145

Case 2: Example 6 Convert $84\frac{2}{3}\%$ to a decimal

$$84\tfrac{2}{3}\% = \frac{84\frac{2}{3}}{100}$$

$= .84\frac{2}{3}$ (moving the decimal point two places to the left)

Example 7 Convert $84\frac{2}{3}\%$ to a decimal rounded-off to the nearest **hundredth.**

$84\frac{2}{3}\% = 0.846... \approx 0.85$ $\left(\frac{254}{3}\% = 84.6...\% = .846... \approx .85\right)$

Example 8 Convert $8\frac{1}{4}\%$ to a decimal.

$8\frac{1}{4}\% = 0.08\frac{1}{4}$ <-------- Complex decimal.

However, since $\frac{1}{4} = 0.25$, a terminating decimal,

$8\frac{1}{4}\% = 0.08\frac{1}{4} = 0.0825$ (Also, $8\frac{1}{4}\% = \frac{33}{4}\% = 8.25\% = \frac{8.25}{100} = .0825$.)

Example 9 Convert $4\frac{2}{3}\%$ to a decimal

$$4\tfrac{2}{3}\% = \frac{4\frac{2}{3}}{100}$$

$= .04\frac{2}{3}$ (moving the decimal point two places to the left and writing a zero to hold place)

To check: Let us convert $.04\frac{2}{3}$ to a percent.

$.04\frac{2}{3} = .04\frac{2}{3} \times 100\%$

$= 4\frac{2}{3}\%$ (moving the decimal point two places to the right and attaching the percent symbol)

Changing a Percent to a Fraction (in its lowest terms)

Example 1. Convert 25% to a fraction in its lowest terms.

Solution: $25\% = \frac{25}{100} = \frac{1}{4}$.

Example 2. Convert 23% to a fraction in its lowest terms.

Solution: $23\% = \frac{23}{100}$.

Example 3. Convert 74% to a fraction in its lowest terms

Solution: $74\% = \frac{74}{100} = \frac{37}{50}$.

Example 4 Convert $4\frac{2}{3}\%$ to a fraction

$$4\tfrac{2}{3}\% = \frac{4\frac{2}{3}}{100} = \frac{14}{300} = \frac{7}{150}$$

Example 5 Convert $.4\frac{2}{3}\%$ to a fraction

$$.4\tfrac{2}{3}\% = \frac{.4\frac{2}{3}}{100} = .4\tfrac{2}{3} \div 100 = \frac{14}{30} \div 100 = \frac{14}{3000} = \frac{7}{1500}$$

Note: Attaching the % symbol is equivalent to dividing by 100; and dropping the % symbol is equivalent to multiplying by 100.

Lesson 36 Exercises

A. Convert to a decimal**: 1.** 23%**; 2.** 8.25%; **3.** 0.8%; **4.** 10%**; 5.** 10.5%**; 6.** 8%

Answers: **1.** 0.23 ; **2.** 0.0825 **3.** 0.008; **4.** 0.10 or 0.1; **5.** 0.105; **6.** 0.08

B. Convert to a decimal**: 1.** $2\frac{2}{3}\%$**; 2.** $25\frac{2}{7}\%$; **3.** $4.3\frac{5}{11}\%$; **4.** $0.16\frac{2}{3}\%$**; 5.** 8.2%**; 6.** $64\frac{3}{11}\%$

Answers: 1. $0.02\frac{2}{3}$; **2.** $0.25\frac{2}{7}$; **3.** $0.043\frac{5}{11}$; **4.** $0.0016\frac{2}{3}$. **5.** 0.082; **6.** $0.64\frac{3}{11}$

C. Convert to a fraction in its lowest terms

1. 24% **; 2.** 63% ; **3.** 96%; **4.** 8.2%

Answers: 1. $\frac{6}{25}$; **2.** $\frac{63}{100}$; **3.** $\frac{24}{25}$; **4.** $\frac{41}{500}$.

D. Convert to a fraction in its lowest terms

1. $2\frac{2}{3}\%$; **2.** $25\frac{2}{7}\%$; **3.** $4.3\frac{5}{11}\%$; **4.** $0.16\frac{2}{3}\%$.

Answers: 1. $\frac{2}{75}$; **2.** $\frac{177}{700}$; **3.** $\frac{239}{5500}$; **4.** $\frac{1}{600}$.

Lesson 37
Calculations Involving Percent (%)

In calculations involving percent, three main quantities are involved, namely the percentage, the base, and the rate percent. Some authors call the percentage the amount.
In these problems, you are usually given two of these quantities and you are asked to find the third quantity.

***Percentage:** This is what is obtained when a percent is taken of a number.

Base: This is the number **of** which a percent is taken.

Rate: This the **percent** that is taken of a number.

There are formulas relating these three quantities:

1. **percentage = base × rate%**
2. **base = percentage ÷ rate %**
3. **rate% =** $\frac{\textbf{percentage}}{\textbf{base}}$ **× 100%** (i.e. rate% = (the ratio of percentage to base) × 100%)

You do not need to memorize the first formula, provided you note that "**of**" implies multiply. **Memorize** the second and the third formulas (even though some of the methods discussed below do not need the recall of these formulas). *Some authors call the percentage the **amount** and suggest the proportion

$$\frac{r}{100} = \frac{\text{Percentage}}{\text{base}} = \frac{r}{100} = \frac{A}{B}$$, where r = rate, A = Amount and B = Base.

Finding the Percentage

Example 1 Find 20% of 72 (i.e. **Finding the Percentage**)

(Note: "%" means over 100. Example: $20\% = \frac{20}{100}$

20% of 72 ("of" means multiply)

Step 1: $= \frac{20}{100} \times \frac{72}{1}$ <---------Simplify this by any of the methods discussed below.

Step 2:

Method 1: $\frac{20}{100} \times \frac{72}{1} = .20 \times 72 = 14.40 = 14.4$ (Using decimals)

Method 2: $\frac{20}{100} \times \frac{72}{1} = \frac{20 \times 72}{100} = \frac{1440}{100} = 14.40$ or 14.4 (Multiplying numerators and dividing by 100)

Method 3: $\frac{\cancel{20}^{\,1}}{{}_5\cancel{100}} \times \frac{72}{1} = \frac{72}{5} = 14\frac{2}{5}$ or 14.4 (using cancellation)

The most convenient method will depend on the type of numbers involved, and whether we want the answer as a decimal, as a fraction, or as a mixed number.
For example, if there are common factors in the numerator and the denominator, cancellation, (Method 3) may be more convenient; but if there are no common factors , use Method 1 or Method 2.
In any case, it is a good practice to set up the problem as in Step 1. The next example will show the usefulness of setting up the problem before proceeding to simplify.

Example 2 Find $4\frac{2}{7}\%$ of 28000.

Step 1 : Translating: $4\frac{2}{7}\%$ of 28000

$$= \frac{4\frac{2}{7}}{100} \times \frac{28000}{1}$$

Step 2: $= \frac{30}{700} \times \frac{28000}{1}$ <-----------Simplify this by any method.

The easiest method is by cancellation, since there are common factors in both the numerator and the denominator.

$$\frac{30}{\cancel{700}_{1}} \times \frac{\cancel{28000}^{40}}{1} = 30 \times 40 = 1200$$

Note that steps 1 and 2 are important. Do not round off $4\frac{2}{7}\%$ as .04% because you will not get the exact answer. If the rate % were say, 70%, then you could immediately write 70% = .70 and then similarly write $70\frac{1}{4}\%$ = .7025 ; but better, $70\frac{1}{4}\% = \frac{281}{400}$.

Finding the Base (Finding the Original Number)

Note that the next problem is different from the last two examples both in the wording of the problem and how we solve it. In the last two problems, we multiplied. In the next problem; we will divide, but, we must note which number is the divisor.

Example If 20% of a number is 64, what is the number?

There are a number of methods for solving this problem. We will discuss five methods, which include algebraic methods. You may skip the algebraic methods if you do not have the algebraic background to allow you to follow the procedure. Later, in Chapter 7 (percent problems) we will repeat the algebraic method.

Method 1: Using Formula

Base = Percentage ÷ Rate %

In the above problem, 64 is the percentage. (The percentage is sometimes called the " is number". It is the number that (usually) immediately follows or precedes the word "is " in the word problem.

20% is the rate%

base = percentage ÷ rate %

$= 64 \div 20\%$

$= \frac{64}{1} \div \frac{20}{100}$ <-----you can simplify this by approach 1 or 2 below.

Approach 1. $\frac{64}{1} \times \frac{\overset{5}{\cancel{100}}}{\cancel{20}} = 64 \times 5 = 320$

Approach 2. $64 \div .20$ (by long division) $20\overline{)6400}$ = 320

The number is 320.

Method 2 Using Algebra

Let the number be x.

Then, "20% of the number is 64" translates to $\frac{20x}{100} = 64$

i.e. $\frac{20x}{100} = 64$

Solve for x: $\frac{\overset{1}{\cancel{(100)}}}{\underset{1}{\cancel{(20)}}}\ \frac{\overset{1}{\cancel{20}}x}{\cancel{100}} = 64\frac{\overset{5}{\cancel{(100)}}}{\underset{1}{\cancel{(20)}}}$ (or $20x = 100 \times 64$

$x = 320$ (or $x = \frac{100 \times 64}{20} = 320$)

Again, the number is 320 .

Note also that since 20% of the number is 64, the number must be greater than 64.

Method 3 If 20% of the number = 64

then 1% of the number $= \frac{64}{20}$

and 100% of the number $= \frac{64}{20} \times \frac{100}{1}$

$= 320.$

Method 4 "**Ratio Method**" **(Arithmetic)**

(This method follows from method 3.)

If 20% of a number = 64

then, 100% of the number $= \frac{64}{1} \times \frac{\cancel{100\%}^{5}}{\cancel{20\%}}$

$= 320$

In method 4, we used a very useful principle which states that " If more, the smaller divides, and if less, the larger divides". That is, since we expect 100% of the number to be greater than 64, in forming the fraction involving 20% and 100%, the smaller of 20% and 100% is the divisor (smaller divides), and hence we used the fraction,

$\frac{100\%}{20\%}$ <----- " smaller divides" (We multiplied by an improper fraction)

However, if we had expected the number (answer) to be less than 64, we would have used the fraction $\frac{20\%}{100\%}$ <----- "larger divides" (i.e., we would have multiplied by a proper fraction)

Method 5 **Using Proportion**

20% is to 64 as 100% is to x, where 100% of the original number (the base) is x.

Translating the proportion,

$\frac{20\%}{64} = \frac{100\%}{x}$ (or $\frac{.20}{64} = \frac{1}{x}$)

Solve for x: $20\%\, x = 100\%\,(64)$ (or $.20x = 64$)

$x = \frac{100\%}{20\%}(64)$ (or $x = \frac{64}{.20}$

$x = 320.$

Note: Methods 3, 4 and 5 are basically the same.

Note that 20% as a fraction is $\frac{20}{100} = \frac{1}{5}$.

Therefore, in the above problem, instead of asking "if 20% of a number is 64, what is the number?", we could have asked "if $\frac{1}{5}$ of a number is 64, what is the number? "

Example If $\frac{1}{5}$ of a number is 64, what is the number? We can solve this by any of the methods discussed above. Let us use the algebraic method.

Solving Algebraically:

Let the number be x.

Then $\frac{x}{5} = \frac{64}{1}$ (Note that $\frac{1}{5}x = \frac{x}{5}$)

$(5)\frac{x}{5} = \frac{64}{1}(5)$ or $x \times 1 = 5 \times 64$ (by cross-multiplication)

$x = 320$ or $x = 320$.

The number is 320.

(See also Method 1 of the preceding example)

Go over the last two problems, and note the differences between how the questions are worded and how they are solved. You **must remember** how each problem is worded and how to proceed to solve it.

Note also that the previous example was a direct proportion problem. Direct proportion involves a relationship between two quantities whereby as one quantity **increases** the other quantity also **increases.** See p.7 for more examples on direct proportion and also examples on inverse (indirect) proportion whereby as one quantity **increases** the other quantity **decreases** and vice versa.

Example Mary spends 20% of her weekly income on food. If she spends \$64 on food every week, what is her weekly income?

Solution : We could reword this problem in the familiar form as " If 20% of a number is 64, what is the number?"; and then use exactly the same method as in the example, above.

Answer: \$320 (numerically, the same answer as for the last example).

Finding the Rate %

Example 1 What rate % of 24 is 15?

Method 1

$$\text{By formula: Rate\%} = \frac{\text{Percentage}}{\text{Base}} \times 100\%$$

In this problem , the percentage is 15. The percentage is sometimes referred to as the " is number". It is the number that follows or precedes the word "is" if the problem is worded in the above form. The base is 24. The base is sometimes referred to as the "of number". It is the number that (usually) follows the word "of" if the problem is worded in the above form.

$$\text{Then, rate \%} = \frac{15}{24} \times \frac{100\%}{1}$$
$$= \frac{\overset{5}{\cancel{15}}}{\underset{8}{\cancel{24}}} \times \frac{100\%}{1}$$
$$= .625 \times 100\%$$
$$= 62.5\% \text{ or } 62\tfrac{1}{2}\%$$

Method 2

Step 1: Form a fraction using the mnemonic device " $\frac{\text{is number}}{\text{of number}}$ "

Then , we obtain $\frac{15}{24}$

Step 2: Change $\frac{15}{24}$ to percent by multiplying by 100 and attaching the '"%" symbol

$$\frac{\overset{5}{\cancel{15}}}{\underset{8}{\cancel{24}}} \times \frac{100\%}{1} = \frac{500\%}{8}$$
$$= 62\tfrac{1}{2}\% \text{ or } 62.5\%$$

Method 3 Using algebra <--You may skip this method if you do not have the algebraic background to allow you to follow the procedure. Later, in Chapter 7, we will repeat this method.

Let $x\%$ be the rate %. We will write an equation in terms of x, and solve for x.
From the wording, "$x\%$ of 24 is 15" translates to:

$$\frac{x}{100} \times \frac{24}{1} = 15$$

$$\frac{24x}{100} = \frac{15}{1}$$

$24x = 100 \times 15$ (by cross-multiplication or by multiplying both sides of the equation by 100)

$x = \frac{100 \times 15}{24}$ (by dividing both sides of the equation by 24)

$x = 62\frac{1}{2}$

$\therefore$ the required rate = $62\frac{1}{2}\%$.

Example 2 What rate % of 18 is 44?

Method 1 base = 18 (the "of" number)
percentage = 44 (the "is" number)

$$\text{rate } \% = \frac{\text{percentage}}{\text{base}} \times 100\%$$

$$= \frac{\overset{22}{\cancel{44}}}{\underset{9}{\cancel{18}}} \times \frac{100\ \%}{1}$$

$\therefore$ the required rate is $244\frac{4}{9}\%$

Note in the above problem that, the larger number does NOT have to be in the denominator.
The base (the "of" number) must always be in the denominator. Therefore, in forming the fraction, ignore the relative sizes of the numbers.

Method 2 (Using algebra)
Let the required rate be $x\%$.
Then, "$x\%$ of 18 is 44" translates to:

$$\frac{x}{100} \times \frac{18}{1} = 44$$

$$\frac{18x}{100} = \frac{44}{1}$$

$18x = 100 \times 44$ (by cross-multiplication or by multiplying both sides of the equation by 100)

$x = \frac{100 \times 44}{18}$ (by dividing both sides of the equation by 18)

$x = 244\frac{4}{9}$

$\therefore$ the required rate is $244\frac{4}{9}\%$

Example 3 A family's annual income last year was \$20,000. This year, the income is \$33 ,000. What is the percent increase in the annual income?

Solution

Step 1: The increase in income = \$33,000 - \$20,000
= \$13,000

Step 2: The percent increase in income $= \frac{13000}{20000} \times 100\%$ (Finding the rate percent)
$= 65\%$

Note: In Step 2 , the question could have been posed as: What percent of 20,000 is 13.000 ?

Example 4 On a class test , out of 20 questions, a student answered 16 questions correctly. What was the student's grade in percent?

Step 1: Fraction of questions answered correctly $= \frac{16}{20}$

Step 2: Change $\frac{16}{20}$ to percent by multiplying by 100 and attaching the % symbol.

$$\frac{16}{20} \times 100\% = 80\%$$

The student's grade was 80%.

Lesson 37 Exercises

A. (*a*) Find 80% of 25; (*b*) Find 23% of 60; (*c*) Find $5\frac{2}{9}\%$ of 1800

Answers (*a*) 20; (*b*) 13.8 or $13\frac{4}{5}$; (*c*) 94

B. 1. If 20% of a number is 72, what is the number?

2. 53 is 25% of what number?

3. 16% of what number is 82?

4. If 25% of a number is 140, what is the number?

5. If 120% of a number is 103.2, what is the number?

Answers **1.** 360 ; **2.** 212; **3.** 512.5 or $512\frac{1}{2}$; **4.** 560 ; **5.** 86.

C. 1. The monthly rent for Maria's apartment is $800. If Maria spends 25% of her monthly income on this rent, what is her monthly income?

2. A homeowner borrowed money from a bank at the interest rate of 12% per year. If the homeowner pays the bank an interest of $6, 000 per year, how much money did the homeowner borrow?

3. If $3\frac{2}{7}$ of a number is 46, What is the number?

Answer: **1.** $3,200; **2.** $50,000; **3.** 14.

D. 1. What rate percent of 32 is 12?

2. What rate% of 20 is 80?

3. On a math test, there were 40 questions. James answered 32 questions correctly. What was his grade in percent?

Answers **1.** 37.5% or $37\frac{1}{2}\%$; **2.** 400% ; **3.** 80%

Lesson 38

More Applications Involving Percent: Discount , Salary Change and Sales Tax Problems

Applications of Base Finding and Percentage Finding

In these applications, the questions are **not** worded in forms such as "find 20% of a number; "if 30% of a number is 45, what is the number?". A good approach is to reword the problem in any of these familiar forms and then proceed accordingly.

Note **1. The** fraction involved in the problem may very likely be the **rate** (but it may **not** be the rate).

Note for example that $\frac{3}{5} = 60\%$

2. The quantity following the word "of" may very likely be the **base**.
If you know any two of the three quantities, then third quantity is easily deduced.

Example 1 3 out of 5 students at a certain college study Biology.

(a) If 600 students at this college study Biology, how many students are there at this college?

(b) If one were to collect a sample of 200 students at this college, how many of these students would study Biology?

Solution:

There are a number of approaches for solving this problem.
We will cover two methods.

Note that 3 out of 5 means $\frac{3}{5}$

Part (a): We can reword this part of the question as: if $\frac{3}{5}$ of a number is 600, what is the number?

Method 1 Let the number be x

$$\text{Then } \frac{3x}{5} = 600$$
$$3x = 3000$$
$$x = 1000$$

There are 1000 students at this college.

Method 2 (see also p. 171) Divide 600 by $\frac{3}{5}$

$$600 \div \frac{3}{5} = \frac{600}{1} \times \frac{5}{3}$$
$$= 1000$$

Part (b) We can reword this part of the question as: Find $\frac{3}{5}$ of 200.

$$\frac{3}{5} \text{ of } 200 = \frac{3}{5} \times \frac{200}{1} = 120$$

That is, of a sample of 200 students, 120 students would study Biology.

Example 2

Note that since $\frac{3}{5} = 60\%$, the above problem could have been posed as 60% of students at a certain college study Biology.

(a) If 600 students study Biology, how many students are there?

(b) If one were to collect a sample of 200 students, how many would study Biology?
Solution: Proceed exactly as in Example 1 above.

Example 3 At a certain college, 53% of students registered for chemistry, and 24% registered for Biology. If there are 500 students at this college, how many students did not register for Chemistry or Biology?

Solution: We will use two methods to solve this problem.

Method 1
Step 1: Percent of students registering for Chemistry or Biology is
$(53\% + 24\%) = 77\%$

Step 2: Percent of students **not** registering for Chemistry or Biology is
$100\% - 77\% = 23\%$

Step 3: Number of students who did not register for Chemistry or Biology is

$$\begin{aligned} &23\% \text{ of } 500 \\ &= \frac{23}{100} \times 500 \\ &= 115. \end{aligned}$$

Method 2
Step 1: Percent of students registering for Chemistry or Biology is
$(53\% + 24\%) = 77\%$

Step 2 : Find 77% of 500 and subtract the result from 500.

i.e. $\frac{77}{100} \times 500 = 385$

Then, the number of students not registering for Chemistry or Biology is

Step 3: $500 - 385 = 115$

By either Method 1 or Method 2, the number of students not registering for Chemistry or Biology is 115.

Discount Problem

Example A bag originally sold for \$85.00

The selling price was reduced by 30%
What is the new selling price?

Method 1

Step 1: Find 30% of \$85.00

i.e. $\frac{30}{100} \times \frac{85}{1} = 25.50$ or $\begin{array}{r} 85 \\ \times.30 \\ \hline 25.50 \end{array}$

Step 2: New Price = \$85.00 - \$25.50
= \$59.50

Method 2 Assuming a 100% rate % for the original price,

Step 1: Subtract 30% from 100%
i.e. 100% - 30% = 70%

Step 2 : Find 70% of \$85.00

i.e. $\frac{70}{100} \times \frac{85}{1} = \$59.50$ or $\begin{array}{r} 85 \\ \times.70 \\ \hline 59.50 \end{array}$

Salary Change Problem

Example

A year ago, Mary's annual salary was \$45,000. This year she received a 15% raise. What is her new salary?

The approach in solving the problem is similar to that of the above discount problem, except that in this case, we add any change (in salary).

Method 1 Step 1: Find 15% of 45,000

$$\frac{15}{100} \times \frac{45000}{1} = 6{,}750$$

The increase (raise) in salary is \$6,750

Step 2: New Salary = Original Salary + Increase in salary
= \$45,000 + \$6,750
= \$51,750

Method 2

Step 1: Add 15% to 100%.
15% + 100 % = 115%

Step 2: Find 115% of 45,00

$$\frac{115}{100} \times \frac{45000}{1} = \$51{,}750$$

Sales Tax Problem

Example A book sells for $50.00
and the sales tax is 8%.
How much does the purchaser pay?

Method 1

Step 1: Find 8% of $50

$\frac{8}{100} \times \frac{50}{1} = 4$ or $\begin{array}{r} 50 \\ \times .08 \\ \hline 4.00 \end{array}$

the sales tax is $4.00

Step 2: Total cost = $50.00 + $4.00
=$54.00
∴ The purchaser pays $54.00

Method 2 Step 1: Add 8% to 100%
i.e. 8% + 100% = 108%

Step 2: Find 108% of $50.00

$\frac{108}{100} \times \frac{50}{1} = \$54.00$

Lesson 38 Exercises

1. A new math textbook sells for $32.00. However, a used edition of this book is sold at a 6% discount. What is the selling price of the used edition?

Answer $30.08

2, Last year , the president of a corporation was earning $160,000 per year. This year, because of financial problems, the annual salary is to be reduced to $140,000.
What is the percent decrease in salary?

Answer 12.5% or $12\frac{1}{2}\%$

3. A book sells for $65.00 and the sales tax is 8.25%.
What is the total price?

Answer $70.36

Appendix 4

First Degree Equations Containing One Variable (Simple Linear Equations)

Lesson 39: **Axioms for Solving Equations**
Lesson 40: **Solving First Degree Equations**

Lesson 39

Axioms for Solving Equations

Axioms are general mathematical statements that we accept as true, without any proof, in order to deduce other less obvious statements. The following are very useful in constructing proofs and solving equations.

1. A quantity is equal to itself. (reflexive property of equality, also identity principle)
$a = a$

2. An equality may be reversed. (symmetric property of equality)
If $a = b$, then
$b = a$

3. Quantities equal to the same quantity are equal to each other.(transitive property of equality)
If $a = b$, and $b = c$, then
$a = c$.

4. A quantity may be substituted for its equal in any expression or equation. (substitution axiom)

5. A whole equals the sum of all its parts. (partition axiom)

6. If equal quantities are added to equal quantities, the sums are equal (addition axiom)

If $a = b$, then $a + c = b + c$ **ALSO** (If $a = b$, and $c = d$, then $a + c = b + d$)

7. If equal quantities are subtracted from equal quantities, the differences are equal.(subtraction axiom)

If $a = b$, then $a - c = b - c$ **ALSO** (If $a = b$, and $c = d$, then $a - c = b - d$)

8. If equal quantities are multiplied by equal quantities, the products are equal. (multiplication axiom)

If $a = b$, then $ac = bc$ **ALSO** (If $a = b$, and $c = d$, then $ac = bd$.)

9. If equal quantities are divided by equal quantities (not zero), the quotients are equal. (division axiom)

If $a = b$, then $\frac{a}{c} = \frac{b}{c}$ **ALSO** (If $a = b$, and $c = d$, then $\frac{a}{c} = \frac{b}{d}$)

10. Like powers of equals are equal . (powers axiom)

Example: If $a = 3$, then

$a^2 = 3^2$ or $a^2 = 9$.

11. Like roots of equals are equal. (roots axiom)

Example: if $a^3 = 8$, then

$\sqrt[3]{a^3} = \sqrt[3]{8}$

$a = 2$.

Lesson 40
Solving First Degree Equations

An equation is a statement of the equality between two expressions.

Example: $2x + 3 = x - 10$

The equality symbol "=' breaks up an equation into two sides or members , namely the left-hand side and the right-hand side of the equation. To solve an equation involving a single variable, say x, means we are to find values of x which satisfy the equation. **A value of x is said to satisfy an equation if this value when substituted in the equation makes both the left-hand side and the right-hand side of the equation equal to each other.**
A value of x which satisfies a given equation is said to be a solution or a root of the given equation.
To obtain a value for the variable, x, we will get x by itself on one side of the equation .
We agree that a value of x has been obtained if we have x by itself on one side of the equation and all the other quantities on the other side of the equation do **not** involve x. To get x by itself, we will use inverse operations. Addition and subtraction are inverse operations. Multiplication and division are inverse operations. Inverse operations "undo" each other. For example, to undo multiplication by a number, we will use division by the same the number. We will keep the above discussion in mind when we solve linear equations.

Example 1 Solve for x:

1. $x - 5 = 11$

To get x by itself on the left-hand side of the equation, we will remove the "-5" by adding its opposite to both sides of the equation.(This is the same as adding + 5 to both sides of the equation)

$$\begin{array}{r} x - 5 = 11 \\ \underline{\quad +5 \quad +5} \\ x = 16 \end{array}$$

The solution is 16.

Example 2 Solve for x: $x + 8 = -11$

To get x by itself on the left-hand side of the equation, we remove the "+8" by adding -8 to (that is, subtracting 8 from) both sides of the equation.

$$\begin{array}{r} x + 8 = -11 \\ \underline{\quad -8 \quad -8 \quad .} \\ x = -19 \end{array}$$

(by stressing on **addition** of - 8, the right-hand side of the equation is done without confusion)

Note that to remove (undo) any number, use the inverse (opposite) operation.

Example 3 Solve for x: $5x = 20$

To obtain x by itself, we will undo (remove) the "5". We use division by 5 (since the "5" is multiplying the x)

$$\frac{5x}{5} = \frac{20}{5}$$

$$x = 4$$

The solution is 4.

Example 4 Solve for x:

$$-6x = 24$$

$$\frac{-6x}{-6} = \frac{24}{-6}$$

$$x = -4$$

The solution is -4.

Example 5 Solve for x: $\frac{x}{6} = 4$

To remove the "6" which is a divisor, we will multiply both sides of the equation by 6

$$(6)\frac{x}{6} = (6)4$$

$$\overset{1}{\cancel{(6)}}\frac{x}{\underset{1}{\cancel{6}}} = (6)4$$

$x = 24$. The solution is 24.

Example 6 Solve for x: $\frac{-x}{5} = 2$ (or $\frac{x}{5} = -2 \text{---}> x = 5(-2) = -10$)

$$\overset{1}{\cancel{(5)}}\frac{(-x)}{\underset{1}{\cancel{5}}} = (5)2$$

$-x = 10$ <- ------------------------

$(-1)(-1x) = (-1)10$ **Note:** $-x = -1x$

$x = -10$ OR from [this step], just change the sign of the left-hand side and change the sign of the right-hand side.

The solution is -10.

Example 7 Solve for x

$2x - 6 = 10$

We will remove the "- 6 " first, and then remove the "2" (i.e., we will always undo multiplication and division last)

$$\begin{array}{r} 2x - 6 = 10 \\ \underline{+6 \quad +6} \\ 2x = 16 \end{array}$$

$$\frac{2x}{2} = \frac{16}{2}$$

$x = 8$. The solution is 8.

Example 8 Solve for x: $5x - 6 = x + 2$

In this case, we will collect all the x's on one side by removing either the x-term on the right-hand side or the x-term on the left-hand side.
Let us remove the x-term on the right-hand side.

$$\begin{array}{r} 5x - 6 = x + 2 \\ \underline{-x \qquad -x \quad} \\ 4x - 6 = +2 \\ \underline{+6 \quad +6} \\ 4x = 8 \end{array}$$

$$\frac{4x}{4} = \frac{8}{4}$$

$x = 2$. The solution is 2.

Example 9 Solve for x

$7x + 2 - 3x = 2x + 18 + 3x$

In this case, first, simplify each side of the equation, and then proceed.

Step 1: $7x + 2 - 3x = 2x + 18 + 3x$

$4x + 2 = 5x + 18$

Step 2:

$$\begin{array}{r} \underline{-5x \qquad -5x \quad} \\ -x + 2 = 18 \\ \underline{-2 \quad -2} \\ -x = 16 \end{array}$$

$x = -16$. The solution is -16.

Example 10 Solve for x

$$3(x + 2) = 5(x - 6)$$

First, remove the parentheses on both sides of the equation

$$\begin{aligned}
3(x + 2) &= 5(x - 6) \\
3x + 6 &= 5x - 30 \\
-5x \quad &\quad -5x \\
\hline
-2x + 6 &= -30 \\
-6 \quad &\quad -6 \\
\hline
-2x &= -36 \\
\frac{-2x}{-2} &= \frac{-36}{-2} \\
x &= 18
\end{aligned}$$

The solution is 18.

Checking Solutions of Equations

Example (a) Determine if - 5 is a solution of the equation

$$4x + 18 - x = 8x + 32 + 2x.$$

- 5 will be a solution of the given equation if this value when substituted in the given equation makes the left-hand side of the equation equal to the right-hand of the equation.

Checking: Replace x by -5 in the equation.

$$\begin{aligned}
\text{Then, } 4(-5) + 18 - (-5) &\stackrel{?}{=} 8(-5) + 32 + 2(-5) \\
-20 + 18 + 5 &\stackrel{?}{=} -40 + 32 - 10 \\
-2 + 5 &\stackrel{?}{=} -8 - 10 \\
3 &= -18 \quad \text{False}
\end{aligned}$$

The left-hand side of the equation and the right-hand side of the equation are **not** equal. Therefore, -5 is **not** a solution of the given equation.

(b) Determine if -2 is a solution of $4x + 18 - x = 8x + 32 + 2x$.

Checking: Replace x by -2 in the equation.

$$\begin{aligned}
\text{Then, } 4(-2) + 18 - (-2) &\stackrel{?}{=} 8(-2) + 32 + 2(-2) \\
-8 + 18 + 2 &\stackrel{?}{=} -16 + 32 - 4 \\
10 + 2 &\stackrel{?}{=} 16 - 4 \\
12 &= 12 \quad \text{True}
\end{aligned}$$

Since the left-hand side of the equation equals the right-hand side of the equation, -2 is a solution of the given equation.

More Examples on Solutions of Linear Equations

In the following examples, attempt the problems before reading the solutions:

Example 1 Solve for x: $5x - 9 = 6$

Solution

$$\begin{aligned} 5x - 9 &= 6 \\ +9 \quad &\ \ +9 \\ \hline 5x + 0 &= 15 \\ \frac{5x}{5} &= \frac{15}{5} \\ x &= 3 \end{aligned}$$

Check: $5(3) - 9 \stackrel{?}{=} 6$

$15 - 9 \stackrel{?}{=} 6$

$6 = 6$ True

Example 2 Solve for x: $6x - 3 = 2x + 8$

Solution :

$$\begin{aligned} 6x - 3 &= 2x + 8 \\ -2x \quad &\ \ -2x \\ \hline 4x - 3 &= 0 + 8 \\ +3 \quad &\ \ +3 \\ \hline 4x &= 11 \\ \frac{4x}{4} &= \frac{11}{4} \\ x &= \frac{11}{4} \end{aligned}$$

Check: $6(\frac{11}{4}) - 3 \stackrel{?}{=} 2(\frac{11}{4}) + 8$

$33 - 6 \stackrel{?}{=} 11 + 16$

$27 = 27$ True

Example 3 Solve for x:

$$\frac{x}{6} = \frac{11}{12}$$

Method 1 To get x by itself on the left-hand side of the equation, we undo the "6" (which is a divisor) by multiplying each side of the equation by 6.

$$(6)\frac{x}{6} = (6)\frac{11}{12_2}$$

$$\cancel{(6)}^{1}\frac{x}{\cancel{6}_1} = \cancel{(6)}^{1}\frac{11}{\cancel{12}_2}$$

$$x = \frac{11}{2}$$

Method 2 Multiply each term of the equation by the LCD of the fractions. The LCD is 12.

$$(12)\frac{x}{6} = (12)\frac{11}{12}$$

$$2\,\cancel{(12)}\frac{x}{\cancel{6}_1} = \cancel{(12)}^{1}\frac{11}{\cancel{12}_1}$$

$$2x = 11$$

$$\frac{2x}{2} = \frac{11}{2}$$

$$x = \frac{11}{2}$$

Method 3 **Cross-multiplication**

$$\frac{x}{6} = \frac{11}{12}$$

$$12x = 6(11)$$

$$\frac{\cancel{12}^{1}x}{\cancel{12}_{1}} = \frac{\cancel{6}^{1}(11)}{\cancel{12}_{2}}$$

$$x = \frac{11}{2}$$

Note: We can cross-multiply if there is only a single fraction on the left-hand side of the equation, and only a single fraction on the right-hand side of the equation. For instance, in the following example, we cannot cross-multiply immediately, but rather, we will have to combine the right-hand side into a single fraction before cross-multiplying.

Example 4 Solve for x: $\frac{x}{4} = \frac{x}{20} + \frac{2}{5}$

Method 1 **The LCD method**. Multiply each term of the equation by the LCD of the fractions. The LCD is 20.

$$(20)\frac{x}{4} = (20)\frac{x}{20} + (20)\frac{2}{5}$$

$$\overset{5}{\cancel{(20)}}\frac{x}{\cancel{4}_{1}} = \overset{1}{\cancel{(20)}}\frac{x}{\cancel{20}_{1}} + \overset{4}{\cancel{(20)}}\frac{2}{\cancel{5}_{1}}$$

$$5x = x + 8$$
$$\underline{-x \quad -x \quad .}$$
$$4x = 8$$

$$\frac{4x}{4} = \frac{8}{4}$$

$$x = 2$$

The solution is 2.

Method 2 By cross-multiplication

To apply cross-multiplication to the above problem, first, combine the two terms on the right-hand side into a single fraction. (We have a single fraction if we have a single denominator.)

continuing, $\frac{x}{4} = \frac{x}{20} + \frac{2}{5}$

$$\frac{x}{4} = \frac{x}{20} + \frac{2(4)}{5(4)}$$

$$\frac{x}{4} = \frac{x}{20} + \frac{8}{20}$$

$$\frac{x}{4} = \frac{(x + 8)}{20}$$

Now, we cross-multiply:

$20x = 4(x + 8)$ <------ The parentheses are important

$$20x = 4x + 32$$
$$\underline{- 4x \quad - 4x \quad .}$$
$$16x = 0 + 32$$

$$\frac{16x}{16} = \frac{32}{16}$$

$$x = 2$$

Method 3 You may undo one denominator at a time.

$$\frac{x}{4} = \frac{x}{20} + \frac{2}{5}$$

Step 1: Undo the "4" :

$$\frac{{}^{1}\cancel{(4)}x}{\cancel{4}_{1}} = \frac{{}^{1}\cancel{(4)}x}{\cancel{20}_{5}} + \frac{(4)2}{5}$$

(Multiply each term by 4)

$$x = \frac{x}{5} + \frac{8}{5}$$

Step 2: To undo the "5" multiply by 5:

$$(5)x = \frac{{}^{1}\cancel{(5)}x}{\cancel{5}_{1}} + \frac{\cancel{(5)}8}{\cancel{5}_{1}}$$

$$5x = x + 8$$
$$4x = 8$$
$$x = 2$$

Note: In method 3, multiplying the equation first by 4 and then by 5 is equivalent to multiplying the equation by 20 (as was done in Method 1).

Lesson 40 Exercises

A

Solve for x: **1.** $x - 7 = 12$; **2.** $x + 6 = 9$; **3.** $4x = 20$; **4.** $\frac{x}{6} = 8$; **5.** $4x + 3 = 35$

6. $7x - 6 = 2x + 24$; **7.** $5x - 7 + 3x + 9 = 6x + 5 + 8x$; **8.** $\frac{x}{6} + 4 = 9$;

9. $3(2x - 5) + 4 = -4(x + 7)$

Answers: **1.** $x = 19$; **2.** $x = 3$; **3.** $x = 5$; **4.** $x = 48$; **5.** $x = 8$; **6.** $x = 6$; **7.** $x = -\frac{1}{2}$

8. $x = 30$; **9.** $x = -\frac{17}{10}$

B Solve and check:

1. $x + 4 = 6$;	**2.** $x + 5 = -7$	**3.** $x - 3 = 9$
4. $x - 6 = -11$	**5.** $x - \frac{1}{3} = \frac{1}{4}$	**6.** $\frac{x}{2} = 8$
7. $\frac{x}{-2} = 7$	**8.** $\frac{-x}{4} = 9$	**9.** $3x = 18$
10. $-4x = 20$	**11.** $6x = 22$	**12.** $7x = -28$
13. $\frac{x}{4} - 6 = 14$	**14.** $\frac{-x}{3} + 5 = 7$	**15.** $2x + 4 = 9$
16. $-3x - 5 = 10$	**17.** $5x - 4 + x = 4x + 6 - 3x$	
18. $2x - 5 + 3x + 2 = 7x - 5 - 3x$	**19.** $2(x - 4) + 3 = 15$;	**20.** $-3(x + 5) + 1 = 5(x + 4) - 6$

Answers: **1.** 2 ; **2.** -12 ; **3.** 12 ; **4.** -5 ; **5.** $\frac{7}{12}$; **6.** 16 ; **7.** -14 ; **8.** - 36 ; **9.** 6 ; **10.** - 5 ; **11.** $\frac{11}{3}$;

12. - 4 ; **13.** 80 ; **14.** - 6 ; **15.** $\frac{5}{2}$; **16.** -5 ; **17.** 2 ; **18.** -2 ; **19.** 10 ; **20.** $-\frac{7}{2}$

C

1. Is 2 a solution of $3x + 1 = 7$? **2**. Is -3 a solution of $4x - 1 = 15$?

3. Is 2 a solution of $5x + 2 = x - 4$?

Answers: **1**. Yes ; **2**. No ; **3**. No.

Solve for x: **1.** $\frac{x}{6} + \frac{x}{3} = \frac{3}{4}$ **2.** $\frac{x}{8} = \frac{x}{2} - 6$; **3.** $\frac{x}{2} + 1 = \frac{x}{3}$; **4.** $\frac{x-2}{4} + \frac{2}{3} = \frac{1}{2}$

Answers : **1.** $x = \frac{3}{2}$; **2.** $x = 16$; **3.** $x = -6$; **4.** $x = \frac{4}{3}$

E

Solve for x: **1.** $\frac{x+3}{4} + \frac{x-2}{8} = 2$ **2.** $\frac{x-2}{3} + 1 = \frac{x}{5}$ **3.** $\frac{x}{4} + \frac{1}{2} = \frac{x}{5} + 1$

4. $\frac{x}{3} = \frac{x}{6} + \frac{3}{5}$ **5.** $\frac{x}{2} + \frac{2}{3} = \frac{x}{4}$

Answers: **1**. 4 ; **2**. $-\frac{5}{2}$; **3**. 10 ; **4**. $\frac{18}{5}$; **5**. $-\frac{8}{3}$

To show that if $\frac{a}{b} = \frac{c}{d} = \frac{e}{f}$, then $\frac{a}{b} = \frac{c}{d} = \frac{e}{f} = \frac{a+c+e}{b+d+f}$

We proceed in two steps. In Step 1, we consider $\frac{a}{b} = \frac{c}{d}$. In Step 2, We consider the result from Step 1 and $\frac{e}{f}$.

Step 1: $\frac{a}{b} = \frac{c}{d}$ (Given)

$\frac{a}{b} + \boxed{\frac{c}{b}} = \frac{c}{d} + \boxed{\frac{c}{b}}$ (Adding $\frac{c}{b}$ to both sides of the equation)

$\frac{a+c}{b} = \frac{bc+cd}{bd}$ (Adding on the LHS and on the RHS of the equation)

$\frac{a+c}{b} = \frac{c(b+d)}{bd}$ (Factoring the RHS)

$\frac{a+c}{b+d} = \frac{bc}{bd}$ (Dividing both sides by $b+d$ and multiplying both sides by b)

$\frac{a+c}{b+d} = \frac{c}{d}$ (Dividing out the b on the RHS)

Step 2: $\frac{a+c}{b+d} = \frac{e}{f}$ (Since $\frac{a}{b} = \frac{c}{d} = \frac{e}{f}$, replace $\frac{c}{d}$ on the RHS by $\frac{e}{f}$)

$\frac{a+c}{b+d} + \boxed{\frac{e}{b+d}} = \frac{e}{f} + \boxed{\frac{e}{b+d}}$ (Adding $\frac{e}{b+d}$ to both sides of the equation)

$\frac{a+c+e}{b+d} = \frac{e(b+d)+ef}{f(b+d)}$ (Adding on the LHS and on the RHS of the equation)

$\frac{a+c+e}{b+d} = \frac{be+de+ef}{f(b+d)}$

$\frac{a+c+e}{b+d} = \frac{e(b+d+f)}{f(b+d)}$ (Factoring out the e on the RHS)

$\frac{a+c+e}{b+d+f} = \frac{e(b+d)}{f(b+d)}$ (Dividing both sides by $b+d+f$ and multiplying by $b+d$)

$\frac{a+c+e}{b+d+f} = \frac{e}{f}$ (Dividing out the $b+d$ on the RHS)

Since $\frac{a}{b} = \frac{c}{d} = \frac{e}{f}$, (Given)

$\frac{a}{b} = \frac{c}{d} = \frac{e}{f} = \frac{a+c+e}{b+d+f}$

QED

Appendix 5
About Measurements

Standard Unit, Error, Rounding-off Numbers, Significant Digits, Scientific Notation

To determine the size of a physical quantity, we compare its size with a standard quantity called a unit. Example: To determine the length of the cover of a book in inches or in meters, we can use a ruler with its scale in inches or in meters.
A measurement is the ratio of the magnitude of a physical quantity to that of a standard unit

Standard unit
A standard unit is a measure with which other quantities are compared. A standard unit of measure is defined by a legal authority (such as the US Bureau of Weights and Measures) or by a conference of scientists.

Some universally accepted standards:
1. For mass, the standard (primary standard) is the kilogram (kg).
2. For length, the standard is the meter (m).
3. For time, the standard is the second (s)

Some devices for taking measurements
Examples: Rulers (for length), chemical balances (for mass), stop watches (for time), ammeters, (for electric current) voltmeters (for electric voltage) , thermometers (for temperature) and barometers (for pressure).

Experimental Errors (or Uncertainties)
There are two main types of errors, namely, systematic errors, and random errors.

Systematic Errors (constant errors):
These errors are due to faulty measuring devices. Systematic errors make the measurements either too small or too large:
Examples of faulty devices:
1. Instruments with needles off the zero mark; 2. Faulty clocks (stop clock)
3, Corroded weights; 4. Faulty thermometers. Heat leaking equipment

Random errors (accidental errors or indeterminate errors)
Random errors may be due to chance variations of the physical quantity being measured, or chance variations in the measuring device. Random errors may also be due to failure to take into account variables such as temperature fluctuations; and environmental effects. We can reduce random errors by making a large number of measurements and taking the average of the measurements.

Error (or uncertainty)

Error = Experimental value - accepted (true value)
Example: In an experiment to determine the acceleration due to gravity, g, the experimental value of g, was 986 cm/s^2. The accepted value of g is 980 cm/s^2. Find the error.

Solution Experimental value = 986 cm/s^2

Accepted value = 980 cm/s^2
Error = experimental value - accepted value
= (986 - 980) cm/s^2
= 6 cm/s^2

The positive value indicates that the experimental value is greater than the accepted value.

Note: If the experimental value = 964 cm/s^2

The error = (964 - 980) cm/s^2

= -16 cm/s^2

The negative value means that the experimental value is less than the accepted value.

Relative error (or relative uncertainty)

Relative error = $\frac{\text{error}}{\text{accepted value}}$

Example 1: If the error = 6 cm/s^2, and

the accepted value = 980 cm/s^2,

Relative error = $\frac{\text{error}}{\text{accepted value}}$

Then relative error = $\frac{6\ cm/s^2}{980\ cm/s^2}$

= 0.00612

Example 2: If the error = -16 cm//s^2 and

the accepted value = 980 cm/s^2

Then relative error = $\frac{-16\ cm/s^2}{980\ cm/s^2}$

= - 0.00612

= - 0. 0163

Note: If an accepted value is not known, and we have two or more experimental values, then the average of the experimental values would be used as the "accepted value" in calculations.

Rounding-Off Numbers

The rules for rounding-off a number may differ slightly depending upon the field. For instance, in accounting the rule may be slightly different from the rule in chemistry.

1. If the digit or group of digits to be dropped is more than 500...,hen drop that digit or group and add 1 to the last digit retained.

2. If the digit or group of digits to be dropped is less than 500...,then drop that digit or group and leave the last digit retained unchanged.

3. If the digit or group of digits to be dropped is exactly 500..., then drop that portion and add 1 to the last digit retained if this digit is odd but if this digit is even, then this digit remains unchanged.

Rounding off Whole numbers

Procedure:

Step 1: Locate the digit in the round-off place.
(The round-off place is the place to which we want to round-off the number)

Step 2: Drop all digits to the right of the round-off place, and if the digit immediately to the right of the round-off place is more than 5 or is 5 followed by non-zero digits, , add 1 to the round-off place digit (i.e. we round-up); but if the digit immediately to the right of the round-off place is less than 5, the round-off place digit remains unchanged. However, if the digit immediately to the right of the round-off place digit is 5 or 5 followed by zeros, we add 1 to the round-off place digit if it is odd, but if it is even, it remains unchanged.(i.e. we round-down). Also, replace each digit dropped by a zero.

Rounding-off Decimals

The procedure is the same as that for rounding-off whole numbers, except that after the decimal point, we do not replace any digits dropped by zeros.

Procedure:

Step 1: Locate the digit in the round-off place.
(The round-off place is the place to which we want to round-off the number)

Step 2: Drop all digits to the right of the round-off place, and if the digit immediately to the right of the round-off place is more than 5 or is 5 followed by non-zero digits, add 1 to the round-off place digit (i.e., we round-up); but if the digit immediately to the right of the round-off place is less than 5, the round-off place digit remains unchanged. However, if the digit immediately to the right of the round-off place digit is 5 or 5 followed by zeros, we add 1 to the round-off place digit if it is odd, but if it is even, it remains unchanged.(i.e., we round-down).

Rounding-off (Alternatively)

When we round-off a number, we drop some of the digits explicitly or implicitly specified. We must distinguish between rounding-off to a specified number of decimal places (or significant digits) and the implicit rounding-off which we must determine from the numbers involved in the calculation.

The rules for rounding-off a number may differ slightly depending upon the field. For instance, in accounting the rule may be slightly different from the rule in chemistry.

1. If the digit or group of digits to be dropped is more than 500...then drop that digit or group and add 1 to the last digit retained.

2. If the digit or group of digits to be dropped is less than 500...then drop that digit or group and leave the last digit retained unchanged.

3. If the digit or group of digits to be dropped is exactly 500..., then drop that portion and add 1 to the last digit retained if this digit is odd but if this digit is even, then this digit remains unchanged.

Example: The following have been rounded-off to three decimal places.

(1) .4398. ≈ **.440**

(2) .43652 ≈ **.437**

(3) .43637 ≈ **.436**

(4) .43750 ≈ **.438**

(5) .43650 ≈ **.436**

(6) .43946 ≈ **.439**

(7) .43650001 ≈ **.437**

(8) .4365001 ≈ **.437**

Example We round off **85376.7463** to the following places, using the simple "5 or greater or less than 5 rule"

1. 85376.7463 to the nearest **thousandth** becomes **85376.746** (We do **not** replace the 3 dropped by a zero)
2. 85376.7463 to nearest **hundredth** becomes **85376.75** (We added 1 to the digit in the round-off place)
3. 85376.7463 to the nearest **tenth** becomes **85376.7** (The 7 is unchanged since the 4 dropped is less than 5)
4. 85376.7463 the nearest **unit** becomes **85377**. (Adding 1 to the 6)
5. 85376.7463 to the nearest **ten** becomes **85380**. (Replacing the 6 dropped by a zero)
6. 85376.7463 to the nearest **hundred** becomes **85400.** (Replacing the digits (6 and 7) dropped by zeros)
7. 85376.7463 to the nearest **whole number** becomes **85377**. (same as to the nearest unit)

Estimation

In estimation, we round-off the numbers before carrying out the operations of addition, subtraction, multiplication , division etc. For convenience, we will round-off each number to the first non-zero digit, unless specified.

Approximate Numbers, Significant Digits, Scientific Notation,

A measurement consists of a numerical value and a unit of that measurement. Example: 4 kilograms, where the 4 is the numerical value and the kilograms is the unit of the measurement.

Numbers obtained from a measurement are never exact (i.e., are approximate) due to the limitations of the measuring instrument as well as the skill of the person making the measurement. As such, when one records a measurement, one should indicate the reliability of the measurement. All measurements may be assumed to have an uncertainty in at least one unit in the last digit of the measurement, since in making a measurement, we usually estimate the last digit.

Results obtained from calculations using measurements are also as uncertain as the measurements themselves.
In summary, the numbers that we deal with in calculations are obtained from observations. Some of the numbers are exact and some are approximate, The approximate numbers are those numbers obtained from making measurements.

Exact Numbers: An exact number is a number that contains no uncertainties. It is assumed to be infinitely accurate. We can obtain exact numbers from definitions and from direct count. For example, the number of students in a math class by count is 25. In this case, there is no uncertainty, since we know that there are exactly 25 students. Similarly, when one counts 200 dollars, one knows that one has exactly 200 dollars, and there is therefore, no uncertainty. Also by definition, 60 minutes = 1 hour; 2.54 centimeters = 1 inch. Since these numbers are defined, this 60 and 2,54 are exact and contain no uncertainties. We can also add that this 60 has an infinite number of significant digits, (we can write 60 as 60.000...) and therefore the zero in the 60 is significant. However if you make your own measurement and by coincidence obtain 2.54 centimeters, then this 2.54 would not be exact. We can generalize that all the conversion factors (from tables) are exact.

Significant Digits (Significant Figures), Digits obtained in a measurement

A significant digit (or figure) is one which is known to be reasonably reliable (or correct).
When we make measurements, the digits we read and estimate on a scale are also called significant digits (or significant figures). These digits include digits that we are certain of, and one additional digit that we are uncertain of. This uncertain digit is obtained by the estimation of the fractional part of the smallest subdivision on the scale being used. As such, the rightmost digit is assumed to be uncertain.

Significant figure notation is an approximate method of indicating the uncertainty of a measurement. when recording a measurement.
We agree to the following:

1. The digits 1, 2, 3, 4, 5, 6, 7, 8, 9 are always significant.

2. The digit zero, 0. may or may not be significant according to its position in the number as follows:

(a) Zeros before the first non-zero digits are **not** significant.

For example: (i) .0450 has **three** significant digits: The first zero is not significant; but the last zero is significant since if it were not we would not write it.

(ii) .0012 has **two** significant digits. The first two zeros are not significant

. (b) Zeros between non-zero digits **are** significant.

Example: 3.045 has **four** significant digits. The zero between 3 and 4 is significant.
40.240 has **five** significant digits. The last zero is significant because if it were not we would not write it.

More examples: The numbers referred to below are assumed to have been obtained from measurements.
23.00 has four significant digits: the zeros in this case are significant since we do not have to write the zeros if they were not significant. If the zeros were not significant we would have written 23.
2300. has two significant digits, and 600 has one significant digit; however, in each of these two examples, the number of significant digits is sometimes ambiguous.
It is suggested that when the number of significant digits is in doubt , the maximum number of significant digits is to be assumed. Also, in recording data, if we know the number of significant digits, we will use the scientific notation.
Note above that if 600 and 2300 had been obtained by counting, or by definition, all the zeros would have been significant. As a reminder, significant notation generally pertains to numbers obtained from measurements.
Using **scientific notation** avoids all ambiguities with respect to the number of significant digits.. For instance, if we know that the above number, 600 were measured to two significant digits, then we would write 6.0×10^2. If the measurement were to three significant digits, we would write 6.00×10^2
When we deal with very large or very small numbers we prefer to write the numbers in scientific notation form. In this form, the significant digits (digits) including the zeros can be unambiguously indicated. In scientific notation:

(1) 125000 would be written 1.25×10^5

(2) 1467 would be written 1.467×10^3

(3) .032500 would be written 3.2500×10^{-2}

(4) .0325 would be written 3.25×10^{-2}

(5) If 125000 were known to four significant digits it would be written 1.250×10^5.

Note: Some authors indicate which zeros are significant by underlining the last significant zero. For example, in 230$\underline{0}$0 the first two zeros are significant but the last zero is not significant. Note also that in some books, a decimal point placed after the last zero makes all the zeros significant. For example, 23000. has five significant digits, but 23000 has two significant digits. However, but there may still be ambiguity if the number is at the end of a sentence.

Accuracy and Precision in Measurements

Two contributions to uncertainty in measurement are limitations of precision, and limitations of accuracy.
Accuracy indicates how close a measured value is to the true value but **precision** indicates how close two measurements of the same quantity are close to each other. Generally, more precision implies more accuracy. However, there are instances in which numbers may be more precise, but may not be more accurate.. For example, if a measuring device is incorrectly calibrated (having incorrect scale).

Accuracy and Precision in Calculations

With respect to significant digits, **accuracy** refers to the number of significant digits but **precision** refers to the number of decimal places. The larger the number of significant digits in a number, the more accurate the number. The larger the number of decimal places, the more precise the number.

Note: In calculations, the approximate numbers determine how the rounding-off is done

Rounding-off to Significant Digits or Figures in Arithmetic Operations

(Implicit Specification)

1. In **multiplication and division** involving significant digits, the product or quotient (answer) should be rounded-off so that the number of significant digits in the answer is equal to the number of significant digits of the number with the least number of significant digits.
(In other words, the answer should not be more accurate than the number with the least accuracy)

Example 1: Multiply 2.34 cm by 5.6 cm

Step 1: $2.34 \times 5.6 = 13.104$

Step 2: The number with the fewest number of significant digits is 5.6 and it has two significant digits. Therefore, the product (answer) should contain only two significant digits.

Thus, 13.104 becomes 13.

Answer: 13 cm^2

In the above case the "4" in 2.34 is considered to be reasonably reliable. The " 6" in 5.6 is considered to be reasonably reliable.

Example 2: Maria determined the length of a piece of wood to be 6.47 yards. What is the length of this wood in feet?

Solution

By definition, 3 feet = 1 yard

6.47 yards is used to determine the number of significant digits in the answer,

(The 3 feet is exact and has an infinite number of significant digits)

$$\frac{6.47 \text{ yards}}{1} \times \frac{3 \text{ feet}}{1 \text{ yard}}$$

= 19.41 feet

= 19.4 feet (6.47 has three significant figures)

2. In addition and subtraction, we shall round-off so that the number of decimal place in the answer equals the number of decimal places in the number with the least number of decimal places.(In other words, the answer should not be more precise than the number with the least precision)

Example 1: Add: 143.54, 172.3, and 64.62

Solution

143.54 < - - - - has two decimal places
172.3 < - - - - -has one decimal place (determines the number of decimal places in answer)
64.62 < - - - - - has two decimal places
380.46
Answer: 380.5 (has one decimal place)

Example 2 Subtract 12.4 from 143.63

Solution

143.63 < - - - - has two decimal places
12.4 < - - - - -has one decimal place (determines the number of decimal places in answer)
131.23
Answer: 131.2 (has one decimal place)

3. **In finding powers and roots,** the root or power should be rounded-off so that the number of significant digits in the answer is equal to the number of significant digits in the number.

Example 1: Find $\sqrt{26.9}$

Radicand has three significant digits , and therefore the root should have three significant digits.

From a calculator, $\sqrt{26.9} = 5.1865$

$\sqrt{26.9} = 5.19$ (has three significant digits. Same as in the radicand)

Note: When two or more different operations are involved, the final operation determines how the final result is rounded-off.

Example Simplify: $38.3 + 12.9(3.58)$

Solution According to order of operations, we multiply 12.9 by 3.58 first and then add 38.3

$38.3 + 12.9(3.58)$

$= 38.3 + 46.182$

$= 84.485$

$= 84.5$ (has one decimal place as in 38.3)

Since addition was the last step, we use precision (the number of decimal places) to round-off.

Addition and Subtraction Involving Scientific Notation

Before adding or subtracting the numbers must have the **same powers** of 10. We will rewrite the expression so that the power of 10 is that of the highest power in the expression

Example: **1.** $4 \times 10^2 + 3 \times 10^2 = (4+3) \times 10^2$
$= 7 \times 10^2$

Example: **2.** $4 \times 10^2 + 2 \times 10^3 = 0.4 \times 10^3 + 2 \times 10^3$
$= (0.4 + 2) \times 10^3$
$= 2.4 \times 10^3$

or

$4 \times 10^2 + 2 \times 10^3 = 4 \times 10^2 + 20 \times 10^2$
$= (4 + 20) \times 10^2$
$= 24 \times 10^2$
$= 2.4 \times 10^3$ (Again, we obtain the same result)

Order of Magnitude (for comparing relative sizes using powers of 10).

The order of magnitude is the power of 10 closest to the given number.
(It is an approximation to the number. Note the sequence, ..., $10^{-2}, 10^{-1}, 10^0, 10^1, 10^2$,...)

If a given quantity is 1000 times another quantity, the given quantity is larger by three orders of magnitude.

Examples:

1. The order of magnitude of 123 is 10^2, since 123 is closer to 100 than to 1000.

2. Find the order of magnitude of 0.00352.

Solution

Step 1: Write the number in scientific notation.
$0.00352 = 3.52 \times 10^{-3}$.

Step 2: Since the integer before the decimal point, 3, is less than 5, we replace 3.52 by 1 (since this is closer to 1 or 10^0, than it is to 10 or 10^1)

Step 3: 3.52×10^{-3}
$= 10^0 \times 10^{-3}$
$= 1 \times 10^{-3}$
$= 10^{-3}$

The order of magnitude of 0.00352 is 10^{-3}. (since by definition, the order of magnitude is the power of 10 closest to the given number.).
We can use the order of magnitude in estimation by rounding-off to the orders of magnitude.

More examples: Round-off to the nearest order of magnitude.

1. 1.32×10^2
2. 8.02×10^4
3. 0.0009
4. 0.0302

Solution:

1. Since 1.32 is closer to 1 than to 10,

1.32×10^2
$= 10^0 \times 10^2$
$= 1 \times 10^2$
$= 10^2$

The order of magnitude of 1.32×10^2 is 10^2

2. 8.02 is closer to 10 than to 1

8.02×10^4
$= 10^1 \times 10^4$
$= 10^5$

The order of magnitude is 10^5

3. $0.0009 = 9 \times 10^{-4}$
$= 10^1 \times 10^{-4}$
$= 10^{-3}$

The order of magnitude is 10^{-3}.

4. $0.0201 = 2.01 \times 10^{-2}$
$= 10^0 \times 10^{-2}$
$= 1 \times 10^{-2}$
$= 10^{-2}$

The order of magnitude is 10^{-2}

Summary for rounding-off a number to the order of magnitude.

Step !: Write the number in scientific notation

Step 2: Ignoring the power of 10, if the integer before the decimal point is 5 or greater, replace the non-power of 10 part by 10 ((i.e. 10^1); but if the integer is less than 5, replace the non-power of 10 part by 1 (10^0)

Step 3: Simplify

International System of Units

The International System of Units (SI) has adopted a set of seven base (or primary) units.

Quantity	Unit	Symbol
Length	meter	m
Mass	kilogram	kg
Time	seconds	s
Electric current	ampere	A
Temperature	Kelvin	K
Amount of substance	mole	mol
Luminous Intensity	candela	cd

Derived units

In addition to the seven base units, there are derived units which are combinations of the base units

Example:

From the SI base unit, m (meter). for length, the unit for area is $m \times m = m^2$,
Since area = length $\times$ width, and the unit of length is m and the unit of with is m.

For more practical or convenient units, we use prefixes and multiplication factors (in powers of 10) to express other units.

Example: 1 kilometer = $10^3 m$, where kilo = 10^3
1 km = 10^3m or I km = 1000m.

Appendix 6
P vs NP: Solutions of NP Problems
Abstract

The simplest solution is usually the best solution---Albert Einstein

Best news. After over 30 years of debating, the debate is over. Yes, P is equal to NP. For the first time, NP problems. including the classic traveling salesman problem have been solved in this paper. The general approach to solving the different types of NP problems are the same, except that sometimes, specific techniques may differ from each other according to the process involved in the problem. Another type of NP problems covered is the division of items of different sizes, masses, or values into equal parts. The techniques and formulas developed for dividing these items into equal parts are based on an extended Ashanti fairness wisdom as exemplified below. If two people A and B are to divide items of different sizes which are arranged from the largest size to the smallest size, the procedure would be as follows. In the first round, A chooses the largest size, followed by B choosing the next largest size. In the second round, B chooses first, followed by A. In the third round, A chooses first, followed by B and the process continues up to the last item. To abbreviate the sequence in the above choices, one obtains the sequence "AB, BA AB". Let A and B divide the sum of the whole numbers, 10, 9, 8, 7, 6, 5, 4, 3, 2, 1 as equally as possible, by merely always choosing the largest number. Then A chooses 10, B chooses 9 and 8, followed by A choosing 7 and 6; followed by B choosing 5 and 4; followed by A choosing 3 an 2; and finally, B chooses 1. The sum of A's choices is 10 +7+ 6 + 3 + 2 = 28; and the sum of B's choices is 9 + 8 + 5 + 4 + 1 = 27, with error, plus or minus 0.5 . Observe the sequence "AB, BA, AB, BA, AB". Observe also that the sequence is **not** "AB, AB, AB, AB, AB as one might think. The reason why the sequence is "AB, BA AB, BA, AB" is as follows. In the first round, when A chooses first, followed by B, A has the advantage of choosing the larger number and B has the disadvantage of choosing the smaller number. In the second round, if A were to choose first, A would have had two consecutive advantages, and therefore, in the second round, B will choose first to produce the sequence AB, BA. In the third round, A chooses first, because B chose first in the second round. After three rounds, the sequence would be AB, BA, AB. When his technique was applied to 100 items of different values or masses, by mere combinations, the total value or mass of A's items was equal to the total value or mass of B's items. Similar results were obtained for 1000 items. By hand, the techniques can be used to prepare final exam schedules for 100 or 1000 courses. A new approach to solving the traveling salesman problem was used to determine the shortest route to visit nine cities and return to the starting city. The technique covered eliminates a shortcoming of the nearest neighbor approach as well as that of the grouping of the cities. The distances involved were arranged in increasing order and by inspection, ten distances were selected from a set of the shortest 14 distances, instead of the overall set of 45 distances involved. The selected distances were used to construct the shortest route. Confirmed is the notion that an approach that solves one of these problems can also solve other NP problems. Since six problems from three different areas have been solved, all NP problems can be solved. If all NP problems can be solved, then all NP problems are P problems and therefore, P is equal to NP. The CMI Millennium Prize requirements have been satisfied.

Solutions of NP Problems

The following sample problems will be solved and analyzed. They are based on the suggested sample problems from the Wikipedia (Simple English) website. Many Thanks to Wikipedia.

Basis of the method used in solving the NP problems: Ratios

Method 2 below is the method used for the solutions of the **NP** problems.

Example 1: Divide \$12 between A and B in the ratio 1: 2

Method 1 (Usual arithmetic method)

Step 1: Fraction of the money A receives = $\frac{1}{1+2} = \frac{1}{3}$

Fraction of the money B receives = $\frac{2}{1+2} = \frac{2}{3}$

Step 2: Amount A receives $= \frac{1}{3} \times \frac{12}{1} = 4$

Amount B receives $= \frac{2}{3} \times \frac{12}{1} = 8$

Therefore, A receives \$4, and B receives \$8

(**Method 1** above is from the author's book entitled "Power of Ratios" by A. A. Frempong, and published by Yellowtextbooks.com.)

Method 2 (The process method)

The ratio 1:2 means whenever A receives \$1, B receives \$2.

Step 1: In the first round, A receives \$1, and B receives \$2.
After the first round, the amount of money remaining is \$12 - (\$1 + \$2) = \$9.

Step 2: In the second round, from this \$9, A receives \$1 and B receives \$2.
After the second round, the amount of money remaining = \$9 - (\$1+ \$2) = \$6

Step 3: In the third round, A receives \$1 and B receives \$2.
The amount remaining = \$6 - (\$1 + \$2) = \$3

Step 4: In the fourth and final round, A receives \$1 and B receives \$2.
The amount remaining = \$3 - (\$1 + \$2) = 0

Step 5: A's total = \$1 + \$1 + \$1 + \$1 = \$4
B's total = \$2 + \$2 + \$2 + \$2 = \$8

Example 2: Divide \$12 between A and B in the ratio 1: 1

Method 1 (Usual arithmetic method)

Step 1: Fraction of the money A receives = $\frac{1}{1+1} = \frac{1}{2}$

Fraction of the money B receives = $\frac{1}{1+1} = \frac{1}{2}$

Step 2: Amount A receives $= \frac{1}{2} \times \frac{12}{1} = 6$

Amount B receives $= \frac{1}{2} \times \frac{12}{1} = 6$

Therefore, A receives \$6, and B receives \$6.

Method 2 (The process method)

The ratio 1:1 means whenever A receives \$1, B receives \$1.

Step 1: In the first round, A receives \$1, and B receives \$1.
After the first round, the amount of money remaining is \$12 - (\$1 + \$1) = \$10

Step 2: In the second round, from this \$10, A receives \$1 and B receives \$1.
After the second round, the amount of money remaining = \$10 - (\$1+ \$1) = \$8

Step 3: In the third round, A receives \$1, and B receives \$1.
The amount remaining = \$8 - (\$1 + \$1) = \$6

Step 4: In the fourth round, A receives \$1 and B receives \$1
The amount remaining = \$6 - (\$1 + \$1) = 4

Step 5: In the fifth round, A receives \$1 and B receives \$1.
The amount remaining = \$4 - (\$1 + \$1) = 2

Step 6: In the sixth and final round, A receives \$1 and B receives \$1.
The amount remaining = \$2 - (\$1 + \$1) = 0

Step 7: A's total = \$1 + \$1 + \$1 + \$1 + \$1 + \$1 = \$6.
B's total = \$1 + \$1 + \$1 + \$1 + \$1 + \$1 = \$6.

Case 1: Only two devisors A and B

Example 1 (Preliminaries)
By always choosing the largest number, A and B will divide the following set of numbers equally or nearly equally.

14,13,12,11,10,9,8,7,6,5,4,3,2,1.

Solution
For communication purposes, one will call the numbers to be divided the "dividends"; and one will call A and B the "divisors". Let the sum of A's choices be Q_A, and let the sum o f B's choices be Q_B.

Step 1: Check to ensure that the numbers are arranged in decreasing order.
One will apply the wisdom method of the introduction.
That is, one applies "AB, BA, AB, BA, AB, BA, AB"

Method 1 Using braces
Step 2: A chooses the first element, 14
Step 3: B chooses the next two elements, 13 and 12.,
Step 4: A chooses the next two elements 11, and 10, and the alternating consecutive choices continue to the end.

$$\underbrace{14}_{A},\ \underbrace{13,\ 12}_{B},\ \underbrace{11,\ 10}_{A},\ \underbrace{9,\ 8}_{B},\ \underbrace{7,\ 6}_{A},\ \underbrace{5,\ 4}_{B},\ \underbrace{3,\ 2}_{A},\ \underbrace{1}_{B} \qquad (1)$$

Step 5: Add the choices for A and add the choices for B.

$$Q_A = 14+11+10+7+6+3+2$$
$$= 53$$
$$Q_B = 13+12+9+8+5+4+1$$
$$= 52$$

The sum for A = 53; and the sum for B = 52.

Method 2 (Tabular form)
Step 1: List the dividends as shown below

14	13	12	11	10	9	8	7	6	5	4	3	2	1

Step 2: Write the divisors A, BB, AA, BB, AA, etc, above the numbers,
This is the choosing step.

A	B	B	A	A	B	B	A	A	B	B	A	A	B
14	13	12	11	10	9	8	7	6	5	4	3	2	1

Note: A over [14] means A chooses 14. B over [13] means B chooses 13.

Step 3: Collect and add the corresponding (dividends) choices

Q_A	Q_B
14	13
11	12
10	9
7	8
6	5
3	4
2	1
Total: 53	52

Mathematical formulas for choosing the elements

Let $a_1 = 14,\ a_2 = 13,\ a_3 = 12,\ a_4 = 11,\ a_5 = 10,$
$a_6 = 9,\ a_7 = 8,\ a_8 = 7,\ a_9 = 6,\ a_{10} = 5,\ a_{11} = 4,\ a_{12} = 3,\ a_{13} = 2,\ a_{14} = 1$

By experimentation, one obtains the following formulas for A and B.

$$Q_A = a_1 + \sum_{n=2,4,6,}^{6} a_{2n} + a_{2n+1} \qquad (Q_A = a_1 + a_4 + a_5 + a_8 + a_9 + a_{12} + a_{13})$$

$$Q_B = \sum_{n=1,3,5.}^{5} a_{2n} + a_{2n+1} + a_{14} \qquad (Q_B = a_2 + a_3 + a_6 + a_7 + a_{10} + a_{11} + a_{14})$$

Apply the formulas to above (The above formulas are valid for only **two** divisors.)

$$\begin{aligned} Q_A &= a_1 + \sum_{n=2,4,6}^{6} a_{2n} + a_{2n+1} \\ &= 14 + 11 + 10 + 7 + 6 + 3 + 2 \\ &= 53 \end{aligned}$$

$$\begin{aligned} Q_B &= \sum_{n=1,3,5.}^{5} a_{2n} + a_{2n+1} + a_{14} \\ &= 13 + 12 + 9 + 8 + 5 + 4 + 1 \\ &= 52 \end{aligned}$$

Note that the above formulas using the sigma notation are valid for only two divisors, A and B. For three divisors A, B, and C, different formulas would have to be derived, based on the solutions of the problem.

Example 2a

Consider the existence of dollar bills with denominations \$100, \$99, \$98,...\$2, down to \$1. Suppose the bills are on a table with the \$100 bill at the top, followed by the \$99, \$98, \$97 bills, and so on with the \$1 bill at the bottom of the stack. Now, by mere grabbing in turns always from the top of the stack, the total value of these dollar bills is to be divided equally between A and B.

Method 2a: Using the numerical values and braces

Apply, "AB, BA, AB, BA, AB, BA,..." (as in Method 1 of Example 1)

Step 1: A chooses the first \$100 bill. (Only a single item is removed).
Step 2: B chooses the next two bills, the \$99 and \$98 bills, (two items removed consecutively)
Step 3: A chooses the next two bills, the \$97 and \$96 bills, and the alternating removal continues to the end.

100, 99 98, 97, 96, 95, 94, 93, 92 91, 90, 89, 88, 87 86, 85, 84, 83, 82, 81, 80 79, 78,
A B A B A B A B A B A B

77, 76,, 75 74, 73, 72, 71, 70, 69, 68 67, 66, 65, 64, 63 62, 61, 60, 59, 58, 57, 56 55, 54,
A B A B A B A B A B A B

53, 52, ,51, 50, 49 48, 47, 46, 45, 44, 43, 42 41, 40, 39, 38, 37 36, 35, 34, 33, 32, 31, 30
A B A B A B A B A B A B

29, 28,, 27, 26, 25 24, 23, 22, 21, 20, 19, 18,17, 16, 15, 14,13, 12, 11, 10, 9, 8, 7, 6, 5, 4
A B A B A B A B A B A B A

3, 2, 1,
B A

Step 4: Collect and add the choices (dividends) for A and B

$$Q_A = 100+97+96+93+92+89+88+85+84+81+80+77+76+73+72+69+68+65+64+61+60+57+56+53+52+49+48+45+44+41+40+37+36+33+32+29+28+25+24+21+20+17+16+13+12+9+8+5+4+1 = \mathbf{2525}.$$

$$Q_B = 99+98+95+94+91+90+87+86+83+82+79+78+75+74+71+70+67+66+63+62+59+58+55+54\ 51+50+47+46+43+42+39+38+35+34+31+30+27+26+23+22+19+18+15+14+11+10+7+6+3+2 = \mathbf{2525}.$$

Conclusion: A receives \$2525 and B receives \$2525, Note the zero error for A and B.

Method 2b: Using tabular form

Step 1: Write the divisors A and B above the numbers (as done in Method 2 of Example 1)

A	B	B	A	A	B	B	A	A	B	B	A	A	B	B	A	A	B	B	A
100	99	98	97	96	95	94	93	92	91	90	89	88	87	86	85	84	83	82	81

A	B	B	A	A	B	B	A	A	B	B	A	A	B	B	A	A	B	B	A
80	79	78	77	76	75	74	73	72	71	70	69	68	67	66	65	64	63	62	61

A	B	B	A	A	B	B	A	A	B	B	A	A	B	B	A	A	B	B	A
60	59	58	57	56	55	54	53	52	51	50	49	48	47	46	45	44	43	42	41

A	B	B	A	A	B	B	A	A	B	B	A	A	B	B	A	A	B	B	A
40	39	38	37	36	35	34	33	32	31	30	29	28	27	26	25	24	23	22	21

A	B	B	A	A	B	B	A	A	B	B	A	A	B	B	A	A	B	B	A
20	19	18	17	16	15	14	13	12	11	10	9	8	7	6	5	4	3	2	1

Step 2: Collect and add the Choices (dividends)

$$Q_A = 100+97+96+93+92+89+88+85+84+81+80+77+76+73+72+69+68+65+64+61+60+57+56+53+52+49+48+45+44+41+40+37+36+33+32+29+28+25+24+21+20+17+16+13+12+9+8+5+4+1 = \mathbf{2525}.$$

$Q_B = 99+98+95+94+91+90+87+86+83+82+79+78+75+74+71+70+67+66+$
$63+62+59+58+55+54+51+50+47+46+43+42+39+38+35+34+31+30+27+$
$26+23+22+19+18+15+14+11+10+7+6+3+2=\mathbf{2525}$.

The above results are pleasantly astonishing. Of the 2^{100} possible ways to divide the above bills, the above technique and consequently the derived formulas divided the above mixture of bills into exactly two equal parts in value. Why has this technique been hiding for nearly 30 years? Note that the ratio $Q_A : Q_B$ is 1:1.

Equations for above: $Q_A = a_1 + \sum_{n=2,4,6,\ldots}^{48} a_{2n} + a_{2n+1} + a_{100}$ and $Q_B = \sum_{n=1,3,5,\ldots}^{49} a_{2n} + a_{2n+1}$

Method 1b: Using term numbers and braces Apply, "AB, BA, AB, BA, AB,..."

$\underbrace{a_1}_{A},\ \underbrace{a_2\ a_3}_{B},\ \underbrace{a_4,\ a_5}_{A},\ \underbrace{a_6,\ a_7}_{B},\ \underbrace{a_8,\ a_9}_{A}\ \underbrace{a_{10},\ a_{11}}_{B},\ \underbrace{a_{12},\ a_{13}}_{A},\ \underbrace{a_{14}\ a_{15}}_{B},\ \underbrace{a_{16},\ a_{17}}_{A},\ \underbrace{a_{18},\ a_{19}}_{B},\ \underbrace{a_{20},\ a_{21}}_{A}\ \underbrace{a_{22},\ a_{23}}_{B},$

$\underbrace{a_{24},\ a_{25}}_{A},\ \underbrace{a_{26},\ a_{27}}_{B}\ \underbrace{a_{28},\ a_{29}}_{A}\ \underbrace{a_{30},\ a_{31}}_{B},\ \underbrace{a_{32},\ a_{33}}_{A}\ \underbrace{a_{34},\ a_{35}}_{B},\ \underbrace{a_{36},a_{37}}_{A}\ \underbrace{a_{38}\ a_{39}}_{B}\ \underbrace{a_{40},a_{41}}_{A},\ \underbrace{a_{42},a_{43}}_{B},\ \underbrace{a_{44},a_{45}}_{A}$

$\underbrace{a_{46},\ a_{47}}_{B},\ \underbrace{a_{48},\ a_{49}}_{A},\ \underbrace{a_{50},\ a_{51}}_{B},\ \underbrace{a_{52}\ a_{53}}_{A},\ \underbrace{a_{54},\ a_{55}}_{B}\ \underbrace{a_{56},\ a_{57}}_{A}\ \underbrace{a_{58},a_{59}}_{B}\ \underbrace{a_{60},\ a_{61}}_{A},\ \underbrace{a_{62},\ a_{63}}_{B}\ \underbrace{a_{64}\ a_{65}}_{A},\ \underbrace{a_{66},\ a_{67}}_{B}$

$\underbrace{a_{68},\ a_{69}}_{A}\ \underbrace{a_{70},a_{71}}_{B}\ \underbrace{a_{72},\ a_{73}}_{A},\ \underbrace{a_{74},\ a_{75}}_{B},\ \underbrace{a_{76},\ a_{77}}_{A}\ \underbrace{a_{78},\ a_{79}}_{B}\ \underbrace{a_{80},\ a_{81}}_{A}\ \underbrace{a_{82},\ a_{83}}_{B},\ \underbrace{a_{84},\ a_{85}}_{A},\ \underbrace{a_{86},\ a_{87}}_{B},\ \underbrace{a_{88},\ a_{89}}_{A},$

$\underbrace{a_{90},\ a_{91}}_{B},\ \underbrace{a_{92},\ a_{93}}_{A},\ \underbrace{a_{94},\ a_{95}}_{B},\ \underbrace{a_{96},\ a_{97}}_{A}\ \underbrace{a_{98},\ a_{99}}_{B},\ \underbrace{a_{100}}_{A},$

Using the term numbers and tabular form

A	B	B	A	A	B	B	A	A	B	B	A	A	B	B	A	A	B	B	A
a_1	a_2	a_3	a_4	a_5	a_6	a_7	a_8	a_9	a_{10}	a_{11}	a_{12}	a_{13}	a_{14}	a_{15}	a_{16}	a_{17}	a_{18}	a_{19}	a_{20}

A	B	B	A	A	B	B	A	A	B	B	A	A	B	B	A	A	B	B	A
a_{21}	a_{22}	a_{23}	a_{24}	a_{25}	a_{26}	a_{27}	a_{28}	a_{29}	a_{30}	a_{31}	a_{32}	a_{33}	a_{34}	a_{35}	a_{36}	a_{37}	a_{38}	a_{39}	a_{40}

A	B	B	A	A	B	B	A	A	B	B	A	A	B	B	A	A	B	B	A
a_{41}	a_{42}	a_{43}	a_{44}	a_{45}	a_{46}	a_{47}	a_{48}	a_{49}	a_{50}	a_{51}	a_{52}	a_{53}	a_{54}	a_{55}	a_{56}	a_{57}	a_{58}	a_{59}	a_{60}

A	B	B	A	A	B	B	A	A	B	B	A	A	B	B	A	A	B	B	A
a_{61}	a_{62}	a_{63}	a_{64}	a_{65}	a_{66}	a_{67}	a_{68}	a_{69}	a_{70}	a_{71}	a_{72}	a_{73}	a_{74}	a_{75}	a_{76}	a_{77}	a_{78}	a_{79}	a_{80}

A	B	B	A	A	B	B	A	A	B	B	A	A	B	B	A	A	B	B	A
a_{81}	a_{82}	a_{83}	a_{84}	a_{85}	a_{86}	a_{87}	a_{88}	a_{89}	a_{90}	a_{91}	a_{92}	a_{93}	a_{94}	a_{95}	a_{96}	a_{97}	a_{98}	a_{99}	a_{100}

Collect the terms for A and add them ; and similarly collect the terms for B and add them.

$Q_A = a_1 + a_4 + a_5 + a_8 + a_9 + a_{12} + a_{13} + a_{16} + a_{17} + a_{20} + a_{21} + a_{24} + a_{25} + a_{28} + a_{29} + a_{32} + a_{33}$
$+ a_{36} + a_{37} + a_{40} + a_{41} + a_{44} + a_{45} + a_{48} + a_{49} + a_{52} + a_{53} + a_{56} + a_{57} + a_{60} + a_{61} + a_{64} + a_{65}$
$+ a_{68} + a_{69} + a_{72} + a_{73} + a_{76} + a_{77} + a_{80} + a_{81} + a_{84} + a_{85} + a_{88} + a_{89} + a_{92} + a_{93} + a_{96} + a_{97} + a_{100}$

$Q_B = a_2 + a_3 + a_6 + a_7 + a_{10} + a_{11} + a_{14} + a_{15} + a_{18} + a_{19} + a_{22} + a_{23} + a_{26} + a_{27} + a_{30} + a_{31} + a_{34}$
$+ a_{35} + a_{38} + a_{39} + a_{42} + a_{43} + a_{46} + a_{47} + a_{50} + a_{51} + a_{54} + a_{55} + a_{58} + a_{59} + a_{62} + a_{63} + a_{66}$
$+ a_{67} + a_{70} + a_{71} + a_{74} + a_{75} + a_{78} + a_{79} + a_{82} + a_{83} + a_{86} + a_{87} + a_{90} + a_{91} + + a_{94} + a_{95} + a_{98} + a_{99}$

Example 2b Consider the existence of dollar bills with denominations \$100, \$99, \$98,...\$2, down to \$1. Suppose the bills are on a table with the \$100 bill at the top, followed by the \$99, \$98, \$97 bills, and so on with the \$1 bill at the bottom of the stack. Now, by mere grabbing in turns always from the top of the stack, the total value of these dollar bills is to be divided equally between A and B.

Question: (a) If a computer costs \$2,000, can A afford to buy this computer?
(b) If a computer costs \$3,000, can A afford to buy this computer?

Answers: (a) From the solution of Example 2a, A received \$2,525, and therefore can afford to buy this computer.
Yes. A can afford to buy this \$2,000 computer.

(b) Since from the solution of Example 2a, A received \$2,525, and the computer costs \$3,000, A cannot afford to buy \$3,000 computer
No. A cannot afford to buy this \$3,000 computer.

Example 3

Let one randomly delete some of the bills in Example 2a, a previous example, and divide as equally as possible the remaining bills between A and B. After the deletion of some of the bills, there are 78 bills remaining.

Solution: Using the numerical values and braces

98.97,96,95,94,93,91,90,89,88,87,86.85,.81,80,79,78,77,76,
A B A B A B A B A B

75,74,73,72,69.68,67,66,65,64,62,61,60,58,56,55,54,53.51.50,
A B A B A B A B A B

49,47,46,45,43.,41,40,39,37,36,35.34,33.32,31,30,29,26,25.24,
A B A B A B A B A B

23,22,21,20,19,18,16,14,13,12,11,10,9,8,7,5,4,2,1
A B A B A B A B A B

$$Q_A = 98+95+94+90+89+86+85+79+78+75+74+69+68+65+64+60+58+54+53$$
$$+49+47+43+41+37+36+33+32+29+26+23+22+19+18+13+12+9+8+4+2$$
$$= \mathbf{1937}$$

$$Q_B = 97+96+93+91+88+87+81+80+77+76+73+72+67+66+62+61+56+55+51$$
$$+50+46+45+40+39+35+34+31+30+25+24+21+20+16+14+11+10+7+5+1$$
$$= \mathbf{1933}$$

Total of Q_A and $Q_B = 3870$. Division by 2 yields 1935.

For Q_A, relative error $= \dfrac{2}{1935} = 0.0010$ or about 0.1%

For Q_B, relative error $= \dfrac{-2}{1935} = -0.0010$

For **equality,** interchange the 47 bill in Q_A and the 45 bill in Q_B. Thus A gives \$2 to B, resulting in equality of **\$1,935** each. Other bills can be interchanged.

Using term numbers

$$Q_A = a_1 + a_4 + a_5 + a_8 + a_9 + a_{12} + a_{13} + a_{16} + a_{17} + a_{20} + a_{21} + a_{24} + a_{25} + a_{28} + a_{29} + a_{32} + a_{33}$$
$$+ a_{36} + a_{37} + a_{40} + a_{41} + a_{44} + a_{45} + a_{48} + a_{49} + a_{52} + a_{53} + a_{56} + a_{57} + a_{60} + a_{61} + a_{64} + a_{65}$$
$$+ a_{68} + a_{69} + a_{72} + a_{73} + a_{76} + a_{77}$$

$$Q_B = a_2 + a_3 + a_6 + a_7 + a_{10} + a_{11} + a_{14} + a_{15} + a_{18} + a_{19} + a_{22} + a_{23} + a_{26} + a_{27} + a_{30} + a_{31} + a_{34}$$
$$+ a_{35} + a_{38} + a_{39} + a_{42} + a_{43} + a_{46} + a_{47} + a_{50} + a_{51} + a_{54} + a_{55} + a_{58} + a_{59} + a_{62} + a_{63} + a_{66}$$
$$+ a_{67} + a_{70} + a_{71} + a_{74} + a_{75} + a_{78}$$

Observe above that the last term for Q_A is a_{77} and the last term for Q_B is a_{78} (there are 78 terms)

Case 2: Three or more divisors

In the previous examples, for communication purposes, A and B were called the "divisors" and the numbers or terms to be divided were called "dividends". The concept of divisors A and B can be extended to three or more divisors such as A, B, C, or A, B, C, D, but in these cases, geometric figures will help keep track of the choices.

Geometric figures to keep track of the order and directions of the divisors (For three or more divisors such as A, B, C; four divisors A, B, C,D)

The arrows are for directions

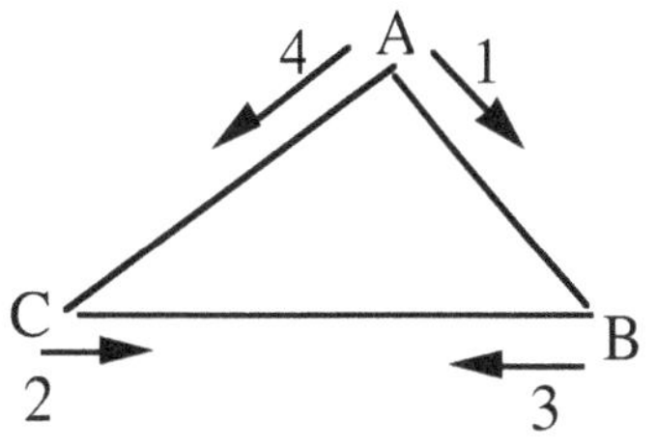

ABC / CBA / BCA / ACB

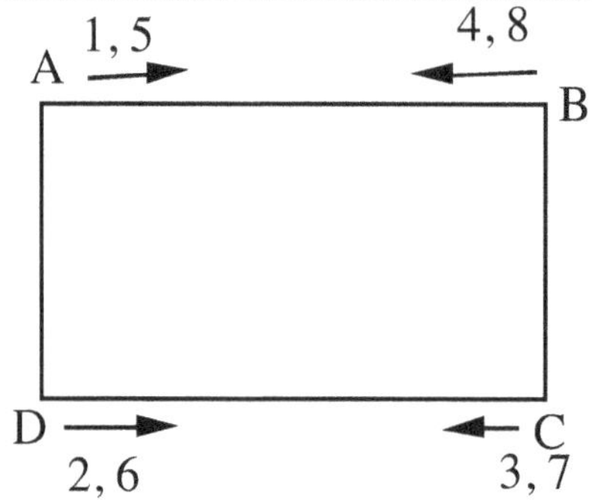

ABCD / DCBA / CDAB / BADC / ABCD

For ABCD

Step 1: Go Clockwise ABCD. (first round)
Step 2: Begin with D, reverse the direction in Step 1 (direction of A) and go DCBA.
Step 3: Begin with C, reverse previous direction (direction of D), and go clockwise CDAB.
Step 4: Begin with B, reverse previous direction (direction of C) and go counterclockwise BADC.
Step 5: Begin with A, reverse pevious direction, but by coincidence go clockwise ABCD. (5th round)

For five divisors A, B, C, D, E

ABCDE, EDCBA, DEABC, CBAED BCDEA

Step 1: Go Clockwise ABCDE
Step 2: Begin with E, reverse the direction in Step 1 and go EDCBA
Step 3: Begin with D, reverse previous direction and go clockwise DEABC
Step 4: Begin with C, reverse previous direction and go counterclockwise CBAED
Step 5: Begin with B, reverse previous direction and go clockwise BCDEA.
Step 6: Beginning again with A, reverse the direction (of B) and go counterclockwise AEDCB.

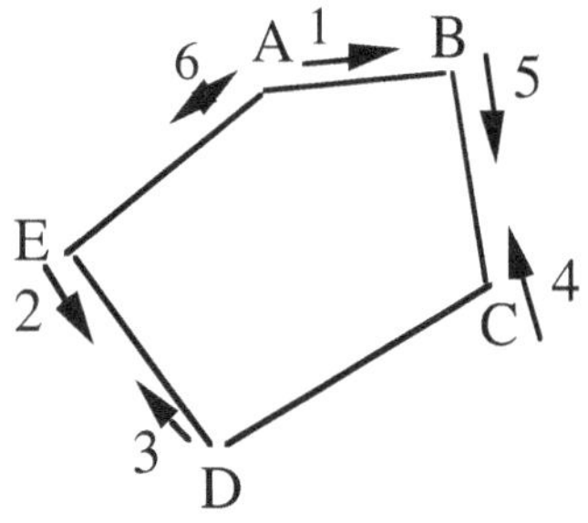

ABCDE, EDCBA DEABC CBAED, BCDEA

Example 4: A businessman wants to take 100 items of different masses to the market. These items are to be packed into boxes. Each box can only hold up to 560 units. The businessman would like to know if 10 boxes would be sufficient to carry all 100 items to the market.

Step 1: Arrange the items in decreasing order of their masses. Let the mass of the first item (largest) be 100 units, and let the masses of the rest of the items be respectively, 99, 98, 97, and so on down to smallest item of mass 1 unit. Let the 10 boxes be labeled A, B, C, D, E, F, G, H, J, and K The ten boxes are to divide the 100 items. **Imitate** Example 2 but with 10 divisors.

Guide1; *ABCDEFGHJK*
Guide 2; *KJHGFEDCBA*
Guide 3 *JKABCDEFGH*
Guide 4; *HGFEDCBAKJ*
Guide 5; *GHJKABCDEF*
Guide 6; *FEDCBAKJHG*
Guide 7: *EFGHJKABCD*
Guide 8; *DCBAKJHGFE*
Guide 9; *CDEFGHJKAB*
Guide10: *BAKJHGFEDC*

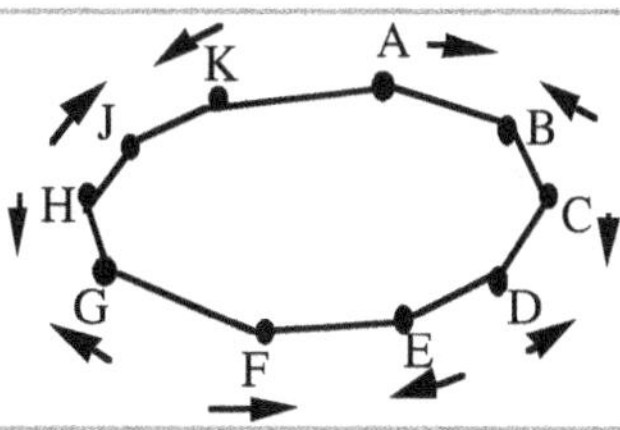

A	B	C	D	E	F	G	H	J	K	K	J	H	G	F	E	D	C	B	A
100	99	98	97	96	95	94	93	92	91	90	89	88	87	86	85	84	83	82	81
a_1	a_2	a_3	a_4	a_5	a_6	a_7	a_8	a_9	a_{10}	a_{11}	a_{12}	a_{13}	a_{14}	a_{15}	a_{16}	a_{17}	a_{18}	a_{19}	a_{20}

J	K	A	B	C	D	E	F	G	H	H	G	F	E	D	C	B	A	K	J
80	79	78	77	76	75	74	73	72	71	70	69	68	67	66	65	64	63	62	61
a_{21}	a_{22}	a_{23}	a_{24}	a_{25}	a_{26}	a_{27}	a_{28}	a_{29}	a_{30}	a_{31}	a_{32}	a_{33}	a_{34}	a_{35}	a_{36}	a_{37}	a_{38}	a_{39}	a_{40}

G	H	J	K	A	B	C	D	E	F	F	E	D	C	B	A	K	J	H	G
60	59	58	57	56	55	54	53	52	51	50	49	48	47	46	45	44	43	42	41
a_{41}	a_{42}	a_{43}	a_{44}	a_{45}	a_{46}	a_{47}	a_{48}	a_{49}	a_{50}	a_{51}	a_{52}	a_{53}	a_{54}	a_{55}	a_{56}	a_{57}	a_{58}	a_{59}	a_{60}

E	F	G	H	J	K	A	B	C	D	D	C	B	A	K	J	H	G	F	E
40	39	38	37	36	35	34	33	32	31	30	29	28	27	26	25	24	23	22	21
a_{61}	a_{62}	a_{63}	a_{64}	a_{65}	a_{66}	a_{67}	a_{68}	a_{69}	a_{70}	a_{71}	a_{72}	a_{73}	a_{74}	a_{75}	a_{76}	a_{77}	a_{78}	a_{79}	a_{80}

C	D	E	F	G	H	J	K	A	B	B	A	K	J	H	G	F	E	D	C
20	19	18	17	16	15	14	13	12	11	10	9	8	7	6	5	4	3	2	1
a_{81}	a_{82}	a_{83}	a_{84}	a_{85}	a_{86}	a_{87}	a_{88}	a_{89}	a_{90}	a_{91}	a_{92}	a_{93}	a_{94}	a_{95}	a_{96}	a_{97}	a_{98}	a_{99}	a_{100}

Step 2: **Collect the choices for A, B, C, D, E, F, G, H, J, K**

Q_A	Q_B	Q_C	Q_D	Q_E	Q_F	Q_G	Q_H	Q_J	Q_K
100 a_1	99, a_2	98 a_3	97 a_4	96 a_5	95 a_6	94 a_7	93 a_8	92 a_9	91 a_{10}
81 a_{20}	82, a_{19}	83 a_{18}	84 a_{17}	85 a_{16}	86 a_{15}	87 a_{14}	88 a_{13}	89 a_{12}	90 a_{11}
78 a_{23}	77, a_{24}	76 a_{25}	75 a_{26}	74 a_{27}	73 a_{28}	72 a_{29}	71 a_{30}	80 a_{21}	79 a_{22}
63 a_{38}	64, a_{37}	65 a_{36}	66 a_{35}	67 a_{34}	68 a_{33}	69 a_{32}	70 a_{31}	61 a_{40}	62 a_{39}
56 a_{45}	55, a_{46}	54 a_{47}	53 a_{48}	52 a_{49}	51 a_{50}	60 a_{41}	59 a_{42}	58 a_{43}	57 a_{44}
45 a_{56}	46 a_{55}	47 a_{54}	48 a_{53}	49 a_{52}	50 a_{51}	41 a_{60}	42 a_{59}	43 a_{58}	44 a_{57}
34 a_{67}	33 a_{68}	32 a_{69}	31 a_{70}	40 a_{61}	39 a_{62}	38 a_{63}	37 a_{64}	36 a_{65}	35 a_{66}
27 a_{74}	28 a_{73}	29 a_{72}	30 a_{71}	21 a_{80}	22 a_{79}	23 a_{78}	24 a_{77}	25 a_{76}	26 a_{75}
12 a_{89}	11 a_{90}	20 a_{81}	19 a_{82}	18 a_{83}	17 a_{84}	16 a_{85}	15 a_{86}	14 a_{87}	13 a_{88}
9 a_{92}	10 a_{91}	1 a_{100}	2 a_{99}	3 a_{98}	4 a_{97}	5 a_{96}	6 a_{95}	7 a_{94}	8 a_{93}
Total: 505	**505**	**505**	**505**	**505**	**505**	**505**	**505**	**505**	**505**

Condition for sufficiency:
The 10 boxes would be sufficient to carry all the 100 items to the market if the mass of the contents of each box is equal to or less than 560 units. Since the mass of the contents in each box is 505 units, which is less than 560 units, each box satisfies this sufficiency condition. Therefore, the 10 boxes would be sufficient to carry the 100 items to the market.
Note above that the ratio

$Q_A : Q_B : Q_C : Q_D : Q_E : Q_F : Q_G : Q_H : Q_{J:} : Q_K = 1:1:1:1:1:1:1:1:1:1$

4b Using the term numbers

A	B	C	D	E	F	G	H	J	K	K	J	H	G	F	E	D	C	B	A
a_1	a_2	a_3	a_4	a_5	a_6	a_7	a_8	a_9	a_{10}	a_{11}	a_{12}	a_{13}	a_{14}	a_{15}	a_{16}	a_{17}	a_{18}	a_{19}	a_{20}
J	K	A	B	C	D	E	F	G	H	H	G	F	E	D	C	B	A	K	J
a_{21}	a_{22}	a_{23}	a_{24}	a_{25}	a_{26}	a_{27}	a_{28}	a_{29}	a_{30}	a_{31}	a_{32}	a_{33}	a_{34}	a_{35}	a_{36}	a_{37}	a_{38}	a_{39}	a_{40}
G	H	J	K	A	B	C	D	E	F	F	E	D	C	B	A	K	J	H	G
a_{41}	a_{42}	a_{43}	a_{44}	a_{45}	a_{46}	a_{47}	a_{48}	a_{49}	a_{50}	a_{51}	a_{52}	a_{53}	a_{54}	a_{55}	a_{56}	a_{57}	a_{58}	a_{59}	a_{60}
E	F	G	H	J	K	A	B	C	D	D	C	B	A	K	J	H	G	F	E
a_{61}	a_{62}	a_{63}	a_{64}	a_{65}	a_{66}	a_{67}	a_{68}	a_{69}	a_{70}	a_{71}	a_{72}	a_{73}	a_{74}	a_{75}	a_{76}	a_{77}	a_{78}	a_{79}	a_{80}
C	D	E	F	G	H	J	K	A	B	B	A	K	J	H	G	F	E	D	C
a_{81}	a_{82}	a_{83}	a_{84}	a_{85}	a_{86}	a_{87}	a_{88}	a_{89}	a_{90}	a_{91}	a_{92}	a_{93}	a_{94}	a_{95}	a_{96}	a_{97}	a_{98}	a_{99}	a_{100}

Collect the terms for A, B, C, D, E, F, G, H, J, K

$Q_A = a_1 + a_{20} + a_{23} + a_{38} + a_{45} + a_{56} + a_{67} + a_{74} + a_{89} + a_{92}$

$Q_B = a_2 + a_{19} + a_{24} + a_{37} + a_{46} + a_{55} + a_{68} + a_{73} + a_{90} + a_{91}$

$Q_C = a_3 + a_{18} + a_{25} + a_{36} + a_{47} + a_{54} + a_{69} + a_{72} + a_{81} + a_{100}$

$Q_D = a_4 + a_{17} + a_{26} + a_{35} + a_{48} + a_{53} + a_{70} + a_{71} + a_{82} + a_{99}$

$Q_E = a_5 + a_{16} + a_{27} + a_{34} + a_{49} + a_{52} + a_{61} + a_{80} + a_{83} + a_{98}$

$Q_F = a_6 + a_{15} + a_{28} + a_{33} + a_{50} + a_{51} + a_{62} + a_{79} + a_{84} + a_{97}$

$Q_G = a_7 + a_{14} + a_{29} + a_{32} + a_{41} + a_{60} + a_{63} + a_{78} + a_{85} + a_{96}$

$Q_H = a_8 + a_{13} + a_{30} + a_{31} + a_{42} + a_{59} + a_{64} + a_{77} + a_{86} + a_{95}$

$Q_J = a_9 + a_{12} + a_{21} + a_{40} + a_{43} + a_{58} + a_{65} + a_{76} + a_{87} + a_{94}$

$Q_K = a_{10} + a_{11} + a_{22} + a_{39} + a_{44} + a_{57} + a_{66} + a_{75} + a_{88} + a_{93}$

Sub-Conclusion
The fairness wisdom method has performed perfectly.
Observe above in Step 2 that the totals for Q_A, Q_B, Q_C, Q_D, Q_E, Q_F, Q_G, Q_H, Q_J, Q_K are all the same. The technique applied picked combinations to produce these equal totals.
Note that in using the technique in this paper, the items involved must be arranged in decreasing order, preferably. Therefore, in programming, the first step should be to arrange the items in decreasing order, a task a computer performs very fast..

In the next eample, Example 5, one will confirm the notion that a method that solves one of the NP problems can be used to solve other similar problems. One will use the results of the above examplc Example 4b to do the next problem.

Example 5: A school offers 100 different courses, and each course requires one hour for the final exam. For each course, all students registered for that course must take the final exam at the same time. Since some students take more than one course, the final exam schedule must be such that students registered for two or more courses will be able to take the exams for all their registered courses. A teacher would like to know if it is possible to schedule all of the exams for the same day so that every student can take the exam for each course registered for.

$$\textbf{Step1: Final Exams 8AM – 6PM}\begin{cases} A=8-9;\ B=9-10;\ C=10-11;\ D=11-12; E=12-1; \\ F=1-2;\ G=2-3;\ H=3-4;\quad J=4-5;\ K=5-6 \end{cases}$$

Let the course numbers be a_1, a_2, a_3,...a_{100} **Using the result of Example 4b**

A	B	C	D	E	F	G	H	J	K	K	J	H	G	F	E	D	C	B	A
a_1	a_2	a_3	a_4	a_5	a_6	a_7	a_8	a_9	a_{10}	a_{11}	a_{12}	a_{13}	a_{14}	a_{15}	a_{16}	a_{17}	a_{18}	a_{19}	a_{20}

J	K	A	B	C	D	E	F	G	H	H	G	F	E	D	C	B	A	K	J
a_{21}	a_{22}	a_{23}	a_{24}	a_{25}	a_{26}	a_{27}	a_{28}	a_{29}	a_{30}	a_{31}	a_{32}	a_{33}	a_{34}	a_{35}	a_{36}	a_{37}	a_{38}	a_{39}	a_{40}

G	H	J	K	A	B	C	D	E	F	F	E	D	C	B	A	K	J	H	G
a_{41}	a_{42}	a_{43}	a_{44}	a_{45}	a_{46}	a_{47}	a_{48}	a_{49}	a_{50}	a_{51}	a_{52}	a_{53}	a_{54}	a_{55}	a_{56}	a_{57}	a_{58}	a_{59}	a_{60}

E	F	G	H	J	K	A	B	C	D	D	C	B	A	K	J	H	G	F	E
a_{61}	a_{62}	a_{63}	a_{64}	a_{65}	a_{66}	a_{67}	a_{68}	a_{69}	a_{70}	a_{71}	a_{72}	a_{73}	a_{74}	a_{75}	a_{76}	a_{77}	a_{78}	a_{79}	a_{80}

C	D	E	F	G	H	J	K	A	B	B	A	K	J	H	G	F	E	D	C
a_{81}	a_{82}	a_{83}	a_{84}	a_{85}	a_{86}	a_{87}	a_{88}	a_{89}	a_{90}	a_{91}	a_{92}	a_{93}	a_{94}	a_{95}	a_{96}	a_{97}	a_{98}	a_{99}	a_{100}

Step 2: Collect the choices for A, B, C, D, E, F, G, H, J, K

Final Exam Schedule: 8 AM-6 PM

$$Q_A=8-9;\ Q_B=9-10;\ Q_C=10-11;\ Q_D=11-12;\ Q_E=12-1;$$
$$Q_F=1-2;\ Q_G=2-3;\ Q_H=3-4;\ Q_J=4-5;\ Q_K=5-6$$

8- 9	9-10	10-11	11-12	12-1	1-2	2-3	3-4	4-5	5-6
Q_A	Q_B	Q_C	Q_D	Q_E	Q_F	Q_G	Q_H	Q_J	Q_K
a_1	a_2	a_3	a_4	a_5	a_6	a_7	a_8	a_9	a_{10}
a_{20}	a_{19}	a_{18}	a_{17}	a_{16}	a_{15}	a_{14}	a_{13}	a_{12}	a_{11}
a_{23}	a_{24}	a_{25}	a_{26}	a_{27}	a_{28}	a_{29}	a_{30}	a_{21}	a_{22}
a_{38}	a_{37}	a_{36}	a_{35}	a_{34}	a_{33}	a_{32}	a_{31}	a_{40}	a_{39}
a_{45}	a_{46}	a_{47}	a_{48}	a_{49}	a_{50}	a_{41}	a_{42}	a_{43}	a_{44}
a_{56}	a_{55}	a_{54}	a_{53}	a_{52}	a_{51}	a_{60}	a_{59}	a_{58}	a_{57}
a_{67}	a_{68}	a_{69}	a_{70}	a_{61}	a_{62}	a_{63}	a_{64}	a_{65}	a_{66}
a_{74}	a_{73}	a_{72}	a_{71}	a_{80}	a_{79}	a_{78}	a_{77}	a_{76}	a_{75}
a_{89}	a_{90}	a_{81}	a_{82}	a_{83}	a_{84}	a_{85}	a_{86}	a_{87}	a_{88}
a_{92}	a_{91}	a_{100}	a_{99}	a_{98}	a_{97}	a_{96}	a_{95}	a_{94}	a_{93}

The final exam for every course has been scheduled. However, if a student takes for example, Course a_1 and course a_{20}, because the duration for the final exams for these two courses is 8-9 AM, the student cannot take the final exams for these two courses simultaneously. Therefore, it is **not** possible to prepare a schedule to allow every student to take the final exams for all registered courses on the same day. However, below is what is possible.

In order for every student to take the final exam for all courses registered for, ten days would be needed as shown below, where the course numbers are $a_1, a_2, a_3, \ldots a_{100}$.

	8- 9	9-10	10-11	11-12	12-1	1-2	2-3	3-4	4-5	5-6
DAY	Q_A	Q_B	Q_C	Q_D	Q_E	Q_F	Q_G	Q_H	Q_J	Q_K
1	a_1	a_2	a_3	a_4	a_5	a_6	a_7	a_8	a_9	a_{10}
2	a_{20}	a_{19}	a_{18}	a_{17}	a_{16}	a_{15}	a_{14}	a_{13}	a_{12}	a_{11}
3	a_{23}	a_{24}	a_{25}	a_{26}	a_{27}	a_{28}	a_{29}	a_{30}	a_{21}	a_{22}
4	a_{38}	a_{37}	a_{36}	a_{35}	a_{34}	a_{33}	a_{32}	a_{31}	a_{40}	a_{39}
5	a_{45}	a_{46}	a_{47}	a_{48}	a_{49}	a_{50}	a_{41}	a_{42}	a_{43}	a_{44}
6	a_{56}	a_{55}	a_{54}	a_{53}	a_{52}	a_{51}	a_{60}	a_{59}	a_{58}	a_{57}
7	a_{67}	a_{68}	a_{69}	a_{70}	a_{61}	a_{62}	a_{63}	a_{64}	a_{65}	a_{66}
8	a_{74}	a_{73}	a_{72}	a_{71}	a_{80}	a_{79}	a_{78}	a_{77}	a_{76}	a_{75}
9	a_{89}	a_{90}	a_{81}	a_{82}	a_{83}	a_{84}	a_{85}	a_{86}	a_{87}	a_{88}
10	a_{92}	a_{91}	a_{100}	a_{99}	a_{98}	a_{97}	a_{96}	a_{95}	a_{94}	a_{93}

Observe how one used the results of the previous example (Example 4b) to solve the above problem, Example 5..

In the next example, one will cover an example involving 1000 items, which will be similar to Example 2a.

Example 6 A builder has 1000 concrete blocks of different masses arranged from 1000 units to one unit. The builder would like to divide the blocks into two piles A and B of equal masses. Prepare a list by masses of all the blocks in pile A, and all the blocks in pile B. Review Example 2a before proceeding

A	B	B	A	A	B	B	A	A	B	B	A	A	B	B	A	A	B	B	A
1000	999	998	997	996	995	994	993	992	991	990	989	988	987	986	985	984	983	982	981
A	B	B	A	A	B	B	A	A	B	B	A	A	B	B	A	A	B	B	A
980	979	978	977	976	975	974	973	972	971	970	969	968	967	966	965	964	963	962	961
A	B	B	A	A	B	B	A	A	B	B	A	A	B	B	A	A	B	B	A
960	959	958	957	956	955	954	953	952	951	950	949	948	947	946	945	944	943	942	941
A	B	B	A	A	B	B	A	A	B	B	A	A	B	B	A	A	B	B	A
940	939	938	937	936	935	934	933	932	931	930	929	928	927	926	925	924	923	922	921
A	B	B	A	A	B	B	A	A	B	B	A	A	B	B	A	A	B	B	A
920	919	918	917	916	915	914	913	912	911	910	909	908	907	906	905	904	903	902	901
A	B	B	A	A	B	B	A	A	B	B	A	A	B	B	A	A	B	B	A
900	899	898	897	896	895	894	893	892	891	890	889	888	887	886	885	884	883	882	881
A	B	B	A	A	B	B	A	A	B	B	A	A	B	B	A	A	B	B	A
880	879	878	877	876	875	874	873	872	871	870	869	868	867	866	865	864	863	862	861
A	B	B	A	A	B	B	A	A	B	B	A	A	B	B	A	A	B	B	A
860	859	858	857	856	855	854	853	852	851	850	849	848	847	846	845	844	843	842	841
A	B	B	A	A	B	B	A	A	B	B	A	A	B	B	A	A	B	B	A
840	839	838	837	836	835	834	833	832	831	830	829	828	827	826	825	824	823	822	821
A	B	B	A	A	B	B	A	A	B	B	A	A	B	B	A	A	B	B	A
820	819	818	817	816	815	814	813	812	811	810	809	808	807	806	805	804	803	802	801
A	B	B	A	A	B	B	A	A	B	B	A	A	B	B	A	A	B	B	A
800	799	798	797	796	795	794	793	792	791	790	789	788	787	786	785	784	783	782	781
A	B	B	A	A	B	B	A	A	B	B	A	A	B	B	A	A	B	B	A
780	779	778	777	776	775	774	773	772	771	770	769	768	767	766	765	764	763	762	761
A	B	B	A	A	B	B	A	A	B	B	A	A	B	B	A	A	B	B	A
760	759	758	757	756	755	754	753	752	751	750	749	748	747	746	745	744	743	742	741
A	B	B	A	A	B	B	A	A	B	B	A	A	B	B	A	A	B	B	A
740	739	738	737	736	735	734	733	732	731	730	729	728	727	726	725	724	723	722	721
A	B	B	A	A	B	B	A	A	B	B	A	A	B	B	A	A	B	B	A
720	719	718	717	716	715	714	713	712	711	710	709	708	707	706	705	704	703	702	701
A	B	B	A	A	B	B	A	A	B	B	A	A	B	B	A	A	B	B	A
700	699	698	697	696	695	694	693	692	691	690	689	688	687	686	685	684	683	682	681
A	B	B	A	A	B	B	A	A	B	B	A	A	B	B	A	A	B	B	A
680	679	678	677	676	675	674	673	672	671	670	669	668	667	666	665	664	663	662	661
A	B	B	A	A	B	B	A	A	B	B	A	A	B	B	A	A	B	B	A
660	659	658	657	656	655	654	653	652	651	650	649	648	647	646	645	644	643	642	641
A	B	B	A	A	B	B	A	A	B	B	A	A	B	B	A	A	B	B	A
640	639	638	637	636	635	634	633	632	631	630	629	628	627	626	625	624	623	622	621
A	B	B	A	A	B	B	A	A	B	B	A	A	B	B	A	A	B	B	A
620	619	618	617	616	615	614	613	612	611	610	609	608	607	606	605	604	603	602	601

P vs NP:

A	B	B	A	A	B	B	A	A	B	B	A	A	B	B	A	A	B	B	A
600	599	598	597	596	595	594	593	592	591	590	589	588	587	586	585	584	583	582	581
A	B	B	A	A	B	B	A	A	B	B	A	A	B	B	A	A	B	B	A
580	579	578	577	576	575	574	573	572	571	570	569	568	567	566	565	564	563	562	561
A	B	B	A	A	B	B	A	A	B	B	A	A	B	B	A	A	B	B	A
560	559	558	557	556	555	554	553	552	551	550	549	548	547	546	545	544	543	542	541
A	B	B	A	A	B	B	A	A	B	B	A	A	B	B	A	A	B	B	A
540	539	538	537	536	535	534	533	532	531	530	529	528	527	526	525	524	523	522	521
A	B	B	A	A	B	B	A	A	B	B	A	A	B	B	A	A	B	B	A
520	519	518	517	516	515	514	513	512	511	510	509	508	507	506	505	504	503	502	501
A	B	B	A	A	B	B	A	A	B	B	A	A	B	B	A	A	B	B	A
500	499	498	497	496	495	494	493	492	491	490	489	488	487	486	485	484	483	482	481
A	B	B	A	A	B	B	A	A	B	B	A	A	B	B	A	A	B	B	A
480	479	478	477	476	475	474	473	472	471	470	469	468	467	466	465	464	463	462	461
A	B	B	A	A	B	B	A	A	B	B	A	A	B	B	A	A	B	B	A
460	459	458	457	456	455	454	453	452	451	450	449	448	447	446	445	444	443	442	441
A	B	B	A	A	B	B	A	A	B	B	A	A	B	B	A	A	B	B	A
440	439	438	437	436	435	434	433	432	431	430	429	428	427	426	425	424	423	422	421
A	B	B	A	A	B	B	A	A	B	B	A	A	B	B	A	A	B	B	A
420	419	418	417	416	415	414	413	412	411	410	409	408	407	406	405	404	403	402	401
A	B	B	A	A	B	B	A	A	B	B	A	A	B	B	A	A	B	B	A
400	399	398	397	396	395	394	393	392	391	390	389	388	387	386	385	384	383	382	381
A	B	B	A	A	B	B	A	A	B	B	A	A	B	B	A	A	B	B	A
380	379	378	377	376	375	374	373	372	371	370	369	368	367	366	365	364	363	362	361
A	B	B	A	A	B	B	A	A	B	B	A	A	B	B	A	A	B	B	A
360	359	358	357	356	355	354	353	352	351	350	349	348	347	346	345	344	343	342	341
A	B	B	A	A	B	B	A	A	B	B	A	A	B	B	A	A	B	B	A
340	339	338	337	336	335	334	333	332	331	330	329	328	327	326	325	324	323	322	321
A	B	B	A	A	B	B	A	A	B	B	A	A	B	B	A	A	B	B	A
320	319	318	317	316	315	314	313	312	311	310	309	308	307	306	305	304	303	302	301
A	B	B	A	A	B	B	A	A	B	B	A	A	B	B	A	A	B	B	A
300	299	298	297	296	295	294	293	292	291	290	289	288	287	286	285	284	283	282	281
A	B	B	A	A	B	B	A	A	B	B	A	A	B	B	A	A	B	B	A
280	279	278	277	276	275	274	273	272	271	270	269	268	267	266	265	264	263	262	261
A	B	B	A	A	B	B	A	A	B	B	A	A	B	B	A	A	B	B	A
260	259	258	257	256	255	254	253	252	251	250	249	248	247	246	245	244	243	242	241
A	B	B	A	A	B	B	A	A	B	B	A	A	B	B	A	A	B	B	A
240	239	238	237	236	235	234	233	232	231	230	229	228	227	226	225	224	223	222	221
A	B	B	A	A	B	B	A	A	B	B	A	A	B	B	A	A	B	B	A
220	219	218	217	216	215	214	213	212	211	210	209	208	207	206	205	204	203	202	201

P vs NP:

A	B	B	A	A	B	B	A	A	B	B	A	A	B	B	A	A	B	B	A
200	199	198	197	196	195	194	193	192	191	190	189	188	187	186	185	184	183	182	181
A	B	B	A	A	B	B	A	A	B	B	A	A	B	B	A	A	B	B	A
180	179	178	177	176	175	174	173	172	171	170	169	168	167	166	165	164	163	162	161
A	B	B	A	A	B	B	A	A	B	B	A	A	B	B	A	A	B	B	A
160	159	158	157	156	155	154	153	152	151	150	149	148	147	146	145	144	143	142	141
A	B	B	A	A	B	B	A	A	B	B	A	A	B	B	A	A	B	B	A
140	139	138	137	136	135	134	133	132	131	130	129	128	127	126	125	124	123	122	121
A	B	B	A	A	B	B	A	A	B	B	A	A	B	B	A	A	B	B	A
120	119	118	117	116	115	114	113	112	111	110	109	108	107	106	105	104	103	102	101

A	B	B	A	A	B	B	A	A	B	B	A	A	B	B	A	A	B	B	A
100	99	98	97	96	95	94	93	92	91	90	89	88	87	86	85	84	83	82	81
A	B	B	A	A	B	B	A	A	B	B	A	A	B	B	A	A	B	B	A
80	79	78	77	76	75	74	73	72	71	70	69	68	67	66	65	64	63	62	61
A	B	B	A	A	B	B	A	A	B	B	A	A	B	B	A	A	B	B	A
60	59	58	57	56	55	54	53	52	51	50	49	48	47	46	45	44	43	42	41
A	B	B	A	A	B	B	A	A	B	B	A	A	B	B	A	A	B	B	A
40	39	38	37	36	35	34	33	32	31	30	29	28	27	26	25	24	23	22	21
A	B	B	A	A	B	B	A	A	B	B	A	A	B	B	A	A	B	B	A
20	19	18	17	16	15	14	13	12	11	10	9	8	7	6	5	4	3	2	1

Concrete masses for Pile A

Step 2: Collect and add the Choices (dividends) :

$Q_{A1} = 1000 + 997 + 996 + 993 + 992 + 989 + 988 + 985 + 984 + 981 + 980 + 977 + 976 + 973$
$+ 972 + 969 + 968 + 965 + 964 + 961 + 960 + 957 + 956 + 953 + 952 + 949 + 948 + 945 +$
$944 + 941 + 940 + 937 + 936 + 933 + 932 + 929 + 928 + 925 + 924 + 921 + 920 + 917 +$
$916 + 913 + 912 + 909 + 908 + 905 + 904 + 901$ **= 47,525**

$Q_A = 900 + 897 + 896 + 893 + 892 + 889 + 888 + 885 + 884 + 881 + 880 + 877 + 876 + 873 + 872 +$
$869 + 868 + 865 + 864 + 861 + 860 + 857 + 856 + 853 + 852 + 849 + 848 + 845 + 844 + 841$
$+840 + 837 + 836 + 833 + 832 + 829 + 828 + 825 + 824 + 821 + 820 + 817 + 816 + 813 + 812$
$+809 + 808 + 805 + 804 + 801$ **= 42,525**

$Q_A = 800 + 797 + 796 + 793 + 792 + 789 + 788 + 785 + 784 + 781 + 780 + 777 + 776 + 773 + 772 +$
$769 + 768 + 765 + 764 + 761 + 760 + 757 + 756 + 753 + 752 + 749 + 748 + 745 + 744 + 741$
$+740 + 737 + 736 + 733 + 732 + 729 + 728 + 725 + 724 + 721 + 720 + 717 + 716 + 713 + 712$
$+709 + 708 + 705 + 704 + 701$ **= 37,525**

$Q_A = 700 + 697 + 696 + 693 + 692 + 689 + 688 + 685 + 684 + 681 + 680 + 677 + 676 + 673 + 672 +$
$669 + 668 + 665 + 664 + 661 + 660 + 657 + 656 + 653 + 652 + 649 + 648 + 645 + 644 + 641$
$+640 + 637 + 636 + 633 + 632 + 629 + 628 + 625 + 624 + 621 + 620 + 617 + 616 + 613 + 612$
$+609 + 608 + 605 + 604 + 601 =$ **32,525**

$Q_A = 600 + 597 + 596 + 593 + 592 + 589 + 588 + 585 + 584 + 581 + 580 + 577 + 576 + 573 + 572 +$
$569 + 568 + 565 + 564 + 561 + 560 + 557 + 556 + 553 + 552 + 549 + 548 + 545 + 544 + 541$
$+540 + 537 + 536 + 533 + 532 + 529 + 528 + 525 + 524 + 521 + 520 + 517 + 516 + 513 + 512$
$+509 + 508 + 505 + 504 + 501$ **= 27,525**

$Q_A = 500 + 497 + 496 + 493 + 492 + 489 + 488 + 485 + 484 + 481 + 480 + 477 + 476 + 473 + 472 +$
$469 + 468 + 465 + 464 + 461 + 460 + 457 + 456 + 453 + 452 + 449 + 448 + 445 + 444 + 441$
$+440 + 437 + 436 + 433 + 432 + 429 + 428 + 425 + 424 + 421 + 420 + 417 + 416 + 413 + 412$
$409 + 408 + 405 + 404 + 401$ **= 22,525**

$Q_A = 400 + 397 + 396 + 393 + 392 + 389 + 388 + 385 + 384 + 381 + 380 + 377 + 376 + 373 + 372 +$
$369 + 368 + 365 + 364 + 361 + 360 + 357 + 356 + 353 + 352 + 349 + 348 + 345 + 344 + 341$
$+340 + 337 + 336 + 333 + 332 + 329 + 328 + 325 + 324 + 321 + 320 + 317 + 316 + 313 + 312$
$309 + 308 + 305 + 304 + 301$ **= 17,525**

$Q_A = 300 + 297 + 296 + 293 + 292 + 289 + 288 + 285 + 284 + 281 + 280 + 277 + 276 + 273 + 272 +$
$269 + 268 + 265 + 264 + 261 + 260 + 257 + 256 + 253 + 252 + 249 + 248 + 245 + 244 + 241$
$+240 + 237 + 236 + 233 + 232 + 229 + 228 + 225 + 224 + 221 + 220 + 217 + 216 + 213 + 212$
$209 + 208 + 205 + 204 + 201$ **= 12,525**

$Q_A = 200 + 197 + 196 + 193 + 192 + 189 + 188 + 185 + 184 + 181 + 180 + 177 + 176 + 173 + 172 +$
$169 + 168 + 165 + 164 + 161 + 160 + 157 + 156 + 153 + 152 + 149 + 148 + 145 + 144 + 141$
$+140 + 137 + 136 + 133 + 132 + 129 + 128 + 125 + 124 + 121 + 120 + 117 + 116 + 113 + 112$
$109 + 108 + 105 + 104 + 101$ **= 7,525**

$Q_A = 100 + 97 + 96 + 93 + 92 + 89 + 88 + 85 + 84 + 81 + 80 + 77 + 76 + 73 + 72 +$
$69 + 68 + 65 + 64 + 61 + 60 + 57 + 56 + 53 + 52 + 49 + 48 + 45 + 44 + 41$
$+40 + 37 + 36 + 33 + 32 + 29 + 28 + 25 + 24 + 21 + 20 + 17 + 16 + 13 + 12$
$9 + 8 + 5 + 4 + 1$ **= 2,525**

Total for Q_A = 250,250 units

Concrete masses for Pile B

$Q_B = 999 + 998 + 995 + 994 + 991 + 990 + 987 + 986 + 983 + 982 + 979 + 978 + 975 + 974 + 971 + 970 + 967 + 966 + 963 + 962 + 959 + 958 + 955 + 954 + 951 + 950 + 947 + 946 + 943 + 942 + 939 + 938 + 935 + 934 + 931 + 930 + 927 + 926 + 923 + 922 + 919 + 918 + 915 + 914 + 911 + 910 + 907 + 906 + 903 + 902 = \mathbf{47{,}525}$

$Q_B = 899 + 898 + 895 + 894 + 891 + 890 + 887 + 886 + 883 + 882 + 879 + 878 + 875 + 874 + 871 + 870 + 867 + 866 + 863 + 862 + 859 + 858 + 855 + 854 + 851 + 850 + 847 + 846 + 843 + 842 + 839 + 838 + 835 + 834 + 831 + 830 + 827 + 826 + 823 + 822 + 819 + 818 + 815 + 814 + 811 + 810 + 807 + 806 + 803 + 802 = \mathbf{42{,}525}$

$Q_B = 799 + 798 + 795 + 794 + 791 + 790 + 787 + 786 + 783 + 782 + 779 + 778 + 775 + 774 + 771 + 770 + 767 + 766 + 763 + 762 + 759 + 758 + 755 + 754 + 751 + 750 + 747 + 746 + 743 + 742 + 739 + 738 + 735 + 734 + 731 + 730 + 727 + 726 + 723 + 722 + 719 + 718 + 715 + 714 + 711 + 710 + 707 + 706 + 703 + 702 = \mathbf{37{,}525}$

$Q_B = 699 + 698 + 695 + 694 + 691 + 690 + 687 + 686 + 683 + 682 + 679 + 678 + 675 + 674 + 671 + 670 + 667 + 666 + 663 + 662 + 659 + 658 + 655 + 654 + 651 + 650 + 647 + 646 + 643 + 642 + 639 + 638 + 635 + 634 + 631 + 630 + 627 + 626 + 623 + 622 + 619 + 618 + 615 + 614 + 611 + 610 + 607 + 606 + 603 + 602 = \mathbf{32{,}525}$

$Q_B = 599 + 598 + 595 + 594 + 591 + 590 + 587 + 586 + 583 + 582 + 579 + 578 + 575 + 574 + 571 + 570 + 567 + 566 + 563 + 562 + 559 + 558 + 555 + 554 + 551 + 550 + 547 + 546 + 543 + 542 + 539 + 538 + 535 + 534 + 531 + 530 + 527 + 526 + 523 + 522 + 519 + 518 + 515 + 514 + 511 + 510 + 507 + 506 + 503 + 502 = \mathbf{27{,}525}$

$Q_B = 499 + 498 + 495 + 494 + 491 + 490 + 487 + 486 + 483 + 482 + 479 + 478 + 475 + 474 + 471 + 470 + 467 + 466 + 463 + 462 + 459 + 458 + 455 + 454 + 451 + 450 + 447 + 446 + 443 + 442 + 439 + 438 + 435 + 434 + 431 + 430 + 427 + 426 + 423 + 422 + 419 + 418 + 415 + 414 + 411 + 410 + 407 + 406 + 403 + 402 = \mathbf{22{,}525}$

$Q_B = 399 + 398 + 395 + 394 + 391 + 390 + 387 + 386 + 383 + 382 + 379 + 378 + 375 + 374 + 371 + 370 + 367 + 366 + 363 + 362 + 359 + 358 + 355 + 354 + 351 + 350 + 347 + 346 + 343 + 342 + 339 + 338 + 335 + 334 + 331 + 330 + 327 + 326 + 323 + 322 + 319 + 318 + 315 + 314 + 311 + 310 + 307 + 306 + 303 + 302 = \mathbf{17{,}525}$

$Q_B = 299 + 298 + 295 + 294 + 291 + 290 + 287 + 286 + 283 + 282 + 279 + 278 + 275 + 274 + 271 + 270 + 267 + 266 + 263 + 262 + 259 + 258 + 255 + 254 + 251 + 250 + 247 + 246 + 243 + 242 + 239 + 238 + 235 + 234 + 231 + 230 + 227 + 226 + 223 + 222 + 219 + 218 + 215 + 214 + 211 + 210 + 207 + 206 + 203 + 202 = \mathbf{12{,}525}$

$Q_B = 199 + 198 + 195 + 194 + 191 + 190 + 187 + 186 + 183 + 182 + 179 + 178 + 175 + 174 + 171 + 170 + 167 + 166 + 163 + 162 + 159 + 158 + 155 + 154 + 151 + 150 + 147 + 146 + 143 + 142 + 139 + 138 + 135 + 134 + 131 + 130 + 127 + 126 + 123 + 122 + 119 + 118 + 115 + 114 + 111 + 110 + 107 + 106 + 103 + 102 = \mathbf{7{,}525}$

$Q_B = 99 + 98 + 95 + 94 + 91 + 90 + 87 + 86 + 83 + 82 + 79 + 78 + 75 + 74 + 71 + 70 + 67 + 66 + 63 + 62 + 59 + 58 + 55 + 54 + 51 + 50 + 47 + 46 + 43 + 42 + 39 + 38 + 35 + 34 + 31 + 30 + 27 + 26 + 23 + 22 + 19 + 18 + 15 + 14 + 11 + 10 + 7 + 6 + 3 + 2 = \mathbf{2{,}525}$

Total for Q_B = 250,250 units

Example 7: Solutions of the Traveling Salesman Problem

Data Ordering and Route Construction Approach

The simplest solution is usually the best solution---Albert Einstein

Abstract

For one more time, yes, P is equal to NP. For the first time in history, the traveling salesman can determine by hand, with zero or negligible error, the shortest route from home base city to visit once, each of three cities, 10 cities, 20 cities, 100 cities, or 1000 cities, and return to the home base city. The formerly NP-hard problem is now NP-easy problem.

The general approach to solving the different types of NP problems are the same, except that sometimes, specific techniques may differ from each other according to the process involved in the problem. The first step is to arrange the data in the problem in increasing or decreasing order. In the salesman problem, the order will be increasing order, since one's interest is in the shortest distances. The main principle here is that the shortest route is the sum of the shortest distances such that the salesman visits each city once and returns to the starting city. The shortest route to visit nine cities and return to the starting city was found in this paper. It was also found out that even though the length of the shortest route is unique, the sequence of the cities involved is not unique.

Since an approach that solves one of these problems can also solve other NP problems. and the traveling salesman problem has been solved, all NP problems can be solved, provided that one has an open mind and continues to think. If all NP problems can be solved, then all NP problems are P problems and therefore, P is equal to NP. The CMI Millennium Prize requirements have been satisfied.

Preliminaries

Given: The distances between each pair of cities.

Required : To find the shortest route to visit each of the cities once and return to the starting city. It is assumed that there is a direct route between each pair of cities.

Note

1. Number of distances required to travel to each city once and return equals the number of cities involved in the problem.

2 The symbol $C_{1,2}$ can mean the distance from City 1 to City 2.

The distance $C_{1,2}$ = the distance $C_{2,1}$.

Used as a sentence, $C_{1,2}$ can mean, from City 1, one visits City 2.

3. C_1 is the home base (starting city) of the traveling salesman.

4. $C_{1,2}(3)$ shows that the numerical value of $C_{1,2}$ is 3.

Determining the Shortest Route

Example From City 1, a traveling salesman would like to visit once each of **nine** other cities, namely, City 2, City 3, City 4, City 5, City 6, City 7, City 8, City 9, City 10; and return to City 1. Determine the shortest route**.**

As it was in the author's previous solutions of NP problems, the first step is to arrange the distances in this problem in increasing order. The main principle in this paper is that the shortest route is the minimum sum of the shortest distances such that the salesman visits each city once and returns to the starting city.

Since there are ten cities, ten distances are needed for the salesman to visit each of nine cities once and return to City 1.

For the departure from City 1, the first subscript of City 1 is 1, and for the return to City 1, the second subscript of the last city visited is 1.

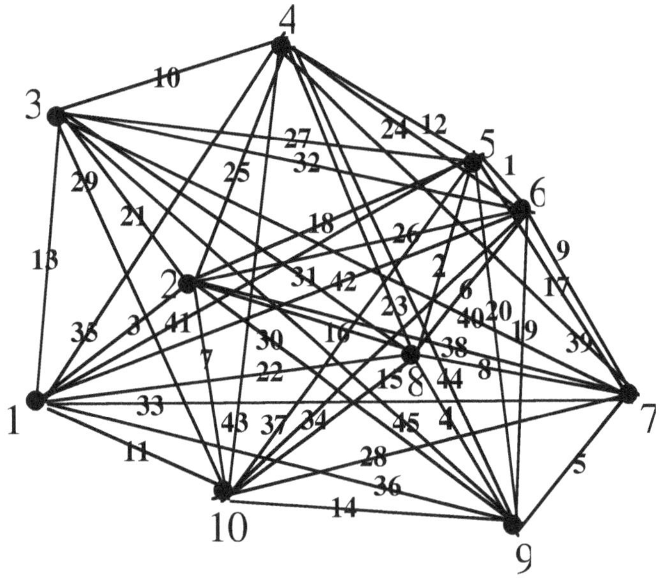

Distances Between Each Pair of Cities

C_1	C_2	C_3	C_4	C_5	C_6	C_7	C_8	C_9
$C_{1,2}$ **3**	$C_{2,3}$ **21**	$C_{3,4}$ **10**	$C_{4,5}$ **12**	$C_{5,6}$ **1**	$C_{6,7}$ **9**	$C_{7,8}$ **8**	$C_{8,9}$ **4**	$C_{9,10}$ **14**
$C_{1,3}$ **13**	$C_{2,4}$ **25**	$C_{3,5}$ **27**	$C_{4,6}$ **24**	$C_{5,7}$ **17**	$C_{6,8}$ **6**	$C_{7,9}$ **5**	$C_{8,10}$ **15**	
$C_{1,4}$ **35**	$C_{2,5}$ **18**	$C_{3,6}$ **32**	$C_{4,7}$ **39**	$C_{5,8}$ **2**	$C_{6,9}$ **19**	$C_{7,10}$ **28**		
$C_{1,5}$ **41**	$C_{2,6}$ **26**	$C_{3,7}$ **40**	$C_{4,8}$ **23**	$C_{5,9}$ **20**	$C_{6,10}$ **34**			
$C_{1,6}$ **42**	$C_{2,7}$ **38**	$C_{3,8}$ **31**	$C_{4,9}$ **44**	$C_{5,10}$ **37**				
$C_{1,7}$ **33**	$C_{2,8}$ **16**	$C_{3,9}$ **45**	$C_{4,10}$ **43**					
$C_{1,8}$ **22**	$C_{2,9}$ **30**	$C_{3,10}$ **29**						
$C_{1,9}$ **36**	$C_{2,10}$ **7**							
$C_{1,10}$ **11**								

Step A: Arrange the numerical values of the distances in increasing order

$C_{5,6}$ **1**	$C_{5,8}$ **2**	$C_{1,2}$ **3**	$C_{8,9}$ **4**	$C_{7,9}$ **5**	$C_{6,8}$ **6**	$C_{2,10}$ **7**	$C_{7,8}$ **8**	$C_{6,7}$ **9**
$C_{3,4}$ **10**	$C_{1,10}$ **11**	$C_{4,5}$ **12**	$C_{1,3}$ **13**	$C_{9,10}$ **14**	$C_{8,10}$ **15**	$C_{2,8}$ **16**	$C_{5,7}$ **17**	$C_{2,5}$ **18**
$C_{6,9}$ **19**	$C_{5,9}$ **20**	$C_{2,3}$ **21**	$C_{1,8}$ **22**	$C_{4,8}$ **23**	$C_{4,6}$ **24**	$C_{2,4}$ **25**	$C_{2,6}$ **26**	$C_{3,5}$ **27**
$C_{7,10}$ **28**	$C_{3,10}$ **29**	$C_{2,9}$ **30**	$C_{3,8}$ **31**	$C_{3,6}$ **32**	$C_{1,7}$ **33**	$C_{6,10}$ **34**	$C_{1,4}$ **35**	$C_{1,9}$ **36**
$C_{5,10}$ **37**	$C_{2,7}$ **38**	$C_{4,7}$ **39**	$C_{3,7}$ **40**	$C_{1,5}$ **41**	$C_{1,6}$ **42**	$C_{4,10}$ **43**	$C_{4,9}$ **44**	$C_{3,9}$ **45**

Step B: Interchange the first and second subscripts of each distance,
Note for example that the distance $C_{1,2}$ = the distance $C_{2,1}$.

$C_{5,6}$ or $C_{6,5}$ **1**	$C_{5,8}$ or $C_{8,5}$ **2**	$C_{1,2}$ or $C_{2,1}$ **3**	$C_{8,9}$ or $C_{9,8}$ **4**
$C_{7,9}$ or $C_{9,7}$ **5**	$C_{6,8}$ or $C_{8,6}$ **6**	$C_{2,10}$ or $C_{10,2}$ **7**	$C_{7,8}$ or $C_{8,7}$ **8**
$C_{6,7}$ o $C_{7,6}$ **9**	$C_{3,4}$ or $C_{4,3}$ **10**	$C_{1,10}$ or $C_{10,1}$ **11**	$C_{4,5}$ or $C_{5,4}$ **12**
$C_{1,3}$ or $C_{3,1}$ **13**	$C_{9,10}$ or $C_{10,9}$ **14**	$C_{8,10}$ or $C_{10,8}$ **15**	$C_{2,8}$ or $C_{8,2}$ **16**
$C_{5,7}$ or $C_{7,5}$ **17**	$C_{2,5}$ or $C_{5,2}$ **18**	$C_{6,9}$ or $C_{9,6}$ **19**	$C_{5,9}$ or $C_{9,5}$ **20**
$C_{2,3}$ or $C_{3,2}$ **21**	$C_{1,8}$ or $C_{8,1}$ **22**	$C_{4,8}$ or $C_{8,4}$ **23**	$C_{4,6}$ or $C_{6,4}$ **24**
$C_{2,4}$ or $C_{4,2}$ **25**	$C_{2,6}$ or $C_{6,2}$ **26**	$C_{3,5}$ or $C_{5,3}$ **27**	$C_{7,10}$ or $C_{10,7}$ **28**
$C_{3,10}$ or $C_{10,3}$ **29**	$C_{2,9}$ or $C_{9,2}$ **30**	$C_{3,8}$ or $C_{8,3}$ **31**	$C_{3,6}$ or $C_{6,3}$ **32**
$C_{1,7}$ or $C_{7,1}$ **33**	$C_{6,10}$ or $C_{10,6}$ **34**	$C_{1,4}$ or $C_{4,1}$ **35**	$C_{1,9}$ or $C_{9,1}$ **36**
$C_{5,10}$ or $C_{10,5}$ **37**	$C_{2,7}$ or $C_{7,2}$ **38**	$C_{4,7}$ or $C_{7,4}$ **39**	$C_{3,7}$ or $C_{7,3}$ **40**
$C_{1,5}$ or $C_{5,1}$ **41**	$C_{1,6}$ or $C_{6,1}$ **42**	$C_{4,10}$ or $C_{10,4}$ **43**	$C_{4,9}$ or $C_{9,4}$ **44**
$C_{3,9}$ or $C_{9,3}$ **45**			

Main Principle

The shortest route is the minimum sum of the shortest distances such that the salesman visits each city once, and returns to the starting city. Since there are ten cities, ten distances are needed to allow the salesman to visit once each of nine cities and return to the starting city.
One will select ten distances, one at a time, to obtain ten well-connected distances to allow the salesman to visit each city once and return to City 1.
Since one is looking for short distances, for the moment, one will work with the ten numbers (distances) up to the value, 14 units in the above table. See the box with thicker lines in the table, below. If necessary, one will move up the table to add some higher numbers and continue.

A $C_{5,6}$ or $C_{6,5}$ **1**	**G** $C_{2,10}$ or $C_{10,2}$ **7**	**N** $C_{1,3}$ or $C_{3,1}$ **13**	**U** $C_{6,9}$ or $C_{9,6}$ **19**
B $C_{5,8}$ or $C_{8,5}$ **2**	**H** $C_{7,8}$ or $C_{8,7}$ **8**	**P** $C_{9,10}$ or $C_{10,9}$ **14**	**V** $C_{5,9}$ or $C_{9,5}$ **20**
C $C_{1,2}$ or $C_{2,1}$ **3**	**J** $C_{6,7}$ or $C_{7,6}$ **9**	**Q** $C_{8,10}$ or $C_{10,8}$ **15**	**W** $C_{2,3}$ or $C_{3,2}$ **21**
D $C_{8,9}$ or $C_{9,8}$ **4**	**K** $C_{3,4}$ or $C_{4,3}$ **10**	**R** $C_{2,8}$ or $C_{8,2}$ **16**	**X** $C_{1,8}$ or $C_{8,1}$ **22**
E $C_{7,9}$ or $C_{9,7}$ **5**	**L** $C_{1,10}$ or $C_{10,1}$ **11**	**S** $C_{5,7}$ or $C_{7,5}$ **17**	**Y** $C_{4,8}$ or $C_{8,4}$ **23**
F $C_{6,8}$ or $C_{8,6}$ **6**	**M** $C_{4,5}$ or $C_{5,4}$ **12**	**T** $C_{2,5}$ or $C_{5,2}$ **18**	**Z** $C_{4,6}$ or $C_{6,4}$ **24**

Solution

Step C: One will now try to construct a ten-distance route using the entries from A to K. If successful, one would surely have constructed the shortest route, since only the least ten numerical distances would have been used. That is, one would have found the sum of the least ten distances.

Note for example that $C_{1,2}(3)$ shows that the numerical value of $C_{1,2}$ is 3. Such notation makes one become aware of a distance size during a route construction.
Below is an attempt to construct a ten-distance route.

$C_{1,2}(3)C_{2,10}(7)--C_{3,4}(10)--C_{5,8}(2)C_{8,6}(6)C_{6,7}(9)C_{7,9}(5)--(A)$

$C_{1,2}(3)\ C_{2,10}(7)--C_{3,4}(10)--C_{5,6}(1)C_{6,7}(9)C_{7,9}(5)C_{9,8}(4)--(B)$

In trying to construct routes in (A), or (B), above, one is unable to complete a ten-distance route, since all the distances needed are not available within entries in boxes A-K. For example, after $C_{2,10}$, the first subscript of the next distance should be 10, (the second subscript of $C_{2,10}$); and there is no distance with this subscript within A-K. Similarly, after $C_{3,4}$, the first subscript of the next distance should be 4; but there is no distance with this subscript within boxes A-K. However, if boxes L, M, N and P are added, the needed distances would be available. One will therefore construct a ten-distance route using boxes A-P. Within boxes A-K, there are only two possible first distances, namely, $C_{1,2}$ and $C_{1,10}$. One of these distances with subscript 1 will be the starting (departure) distance, and the other distance with its subscripts interchanged would be the return distance. After the above expansion to boxes A-P, another possible additional departure or return distance would be $C_{1,3}$. Since there would now be three distances with the subscript 1, one of these distances would be redundant, since one of them is the departure distance, and another with its subscripts interchanged would be the return distance. The additional availability of distances would still allow for the construction of the shortest route, since the addition of distances is very minimum.

Step D:: The dashes above indicate missing distances. After including the entries in boxes L, M, N and P, one obtains the entries in the boxes A -P as shown , above, by the with thick lines. After this minimum addition, one successfully constructed the shortest route to visit nine cities and return to City 1. The shortest route from City 1 to visit nine cities and return to City 1 is given by

$$\boxed{C_{1,3}(13)C_{3,4}(10)C_{4,5}(12)C_{5,6}(1)C_{6,7}(9)C_{7,8}(8)C_{8,9}(4)C_{9,10}(14)C_{10,2}(7)C_{2,1}(3)=81}$$

The details of how the above route was obtained is shown below in Steps 1-11. One is interested in applying the entries in boxes A-P:

Begin from City 1 with $C_{1,2}$ or $C_{1,3}$ or $C_{1,10}$ and return to City 1 with $C_{2,1}$ or $C_{3,1}$ or $C_{10,1}$

Step 1: Begin with first city distance $C_{1,3}(13)$ (from box N, above)

Note: $C_{1,3}$means distance from City 1 to City 3. (From City 1, salesman visits City 3.)

Step: 2: Since the second subscript of $C_{1,3}(13)$ is 3, the first subscript of the next distance will be 3. Inspect each of the above boxes to pick a distance whose first subscript is 3. Box K contains a distance with 3 as a first subscript. We choose the distance in box K, with the numerical value, 10. Connect the chosen distance with the distance in Step 1 to obtain the connected distance $C_{1,3}(13)C_{3,4}(10)$.

Step 3: Since the second subscript of the last distance is 4, the first subscript of the next distance should be 4. Note that the next distance should not contain any of the subscripts already used (i.e., no 1, 3), except that the first subscript of the next distance should be 4. Box M contains a distance with 4 as a first subscript. One chooses the distance $C_{4,5}(12)$ in box M, and attach to obtain the connected distances, $C_{1,3}(13)C_{3,4}(10)C_{4,5}(12)$.

The excluded subscript numbers , except 1, represent the cities already visited.

Step 4: Since the second subscript of the last distance is 5, the first subscript of the next distance should be 5. Note that the next distance should not contain any of the subscripts already used (i.e., no 1, 3, 4), except that the first subscript of the next distance should be 5. One chooses the distance $C_{5,6}(1)$ in box A (with small numerical value, 1) to obtain the connected distances $C_{1,3}(13)C_{3,4}(10)C_{4,5}(12)C_{5,6}(1)$.

Step 5: Since the second subscript of the last distance is 6, the first subscript of the next distance should be 6. Note that the next distance should not contain any of the subscripts already used (i.e., no 1, 3, 4, 5) except that the first subscript of the next distance should be 6, One chooses the distance $C_{6,7}(9)$ in box J to obtain the connected distances

$C_{1,3}(13)C_{3,4}(10)C_{4,5}(12)C_{5,6}(1)C_{6,7}(9)$.

Step 6: Since the second subscript of the last distance is 7, the first subscript of the next distance should be 7. Note that the next distance should not contain any of the subscripts already used (i.e., no 1, 3, 4, 5, 6) except that the first subscript of the next distance should be 7. One chooses the distance $C_{7,8}(8)$ in box H to obtain the connected distances

$C_{1,3}(13)C_{3,4}(10)C_{4,5}(12)C_{5,6}(1)C_{6,7}(9)C_{7,8}(8)$.

Step 7: Since the second subscript of the last distance is 8, the first subscript of the next distance should be 8. Note that the next distance should not contain any of the subscripts already used (i.e., no 1, 3, 4, 5, 6, 7) except that the first subscript of the next distance should be 8. One chooses the distance $C_{8,9}(4)$ in Box D to obtain the connected distances

$C_{1,3}(13)C_{3,4}(10)C_{4,5}(12)C_{5,6}(1)C_{6,7}(9)C_{7,8}(8)C_{8,9}(4)$.

Step 8: Since the second subscript of the last distance 9, the first subscript of the next distance should be 9. Note that the next distance should not contain any of the subscripts already used (i.e., no 1, 3, 4, 5, 6, 7, 8), except that the first subscript of the next distance should be 9. One chooses the distance $C_{9,10}(14)$ in Box P, to obtain the connected distances $C_{1,3}(13)C_{3,4}(10)C_{4,5}(12)C_{5,6}(1)C_{6,7}(9)C_{7,8}(8)C_{8,9}(4)C_{9,10}(14)$

Step 9: Since the second subscript of the last distance is 10, the first subscript of the next distance should be 10. Note that the next distance should not contain any of the subscripts already used (i.e., no 1, 3, 4, 5, 6, 7, 8, 9) except that the first subscript of the next distance should be 10. One chooses the distance $C_{10,2}(7)$ in box G to obtain the connected distances

$C_{1,3}(13)C_{3,4}(10)C_{4,5}(12)C_{5,6}(1)C_{6,7}(9)C_{7,8}(8)C_{8,9}(4)C_{9,10}(14)C_{10,2}(7)$

Step 10: Since the second subscript of the last distance is 2, the first subscript of the next and last distance should be 2. Note that the next distance should not contain any of the subscripts already used (i.e., no 1, 3, 4, 5, 6, 7, 8, 9,10), except that the first subscript of the next distance should be 2 and the second subscript should be 1 (an exception) in order to return to City 1, the starting city. One chooses the distance $C_{2,1}(3)$ in box C to obtain the connected distances

$C_{1,3}(13)C_{3,4}(10)C_{4,5}(12)C_{5,6}(1)C_{6,7}(9)C_{7,8}(8)C_{8,9}(4)C_{9,10}(14)C_{10,2}(7)C_{2,1}(3)$ (Ten distances)

Step 11: Add the distances in parentheses: 13 + 10 + 12 + 1 + 9 + 8 + 4 + 14 + 7 + 3 = 81 and obtain $\boxed{C_{1,3}(13)C_{3,4}(10)C_{4,5}(12)C_{5,6}(1)C_{6,7}(9)C_{7,8}(8)C_{8,9}(4)C_{9,10}(14)C_{10,2}(7)C_{2,1}(3) = 81}$

The above in Step 11 is the shortest route of length 81 units.

EXTRA EXAMPLE (not the shortest route): Using $C_{1,2}$ as the first distance

Step 1 Begin with first city distance $C_{1,2}(3)$ (from box C, above)

Note: $C_{1,2}$ means distance from City 1 to City 2.

Step 2: Since the second subscript of $C_{1,2}(3)$ is 2, the first subscript of the next distance will be 2. Inspect each of the above boxes to pick a distance whose first subscript is 2. Box G contains, a distance with 2 as a first subscript. One chooses the distance in box G, with numerical value 7,

$\boxed{C_{1,2}(3)C_{2,10}(7)}$ Also Do: $C_{1,2}(3)C_{2,3}(21)$; $C_{1,2}(3)C_{2,5}(18)$ $C_{1,2}(3)C_{2,8}(16)$

However, since these connected distances contain values greater than 14, there is no need to continue their construction

Step 3: Since the second subscript of the last distance is 10, the first subscript of the next distance should be 10. Note that the next distance should not contain any of the subscripts already used (i.e., no 1, 2), except that the first subscript of the next distance should be 10. Boxes G and L contain distances with excluded subscripts., One chooses the distance $C_{10,9}(14)$ in box P to obtain the connected distances, $\boxed{C_{1,2}(3)C_{2,10}(7)C_{10,9}(14)}$

The excluded subscript numbers , except 1, represent the cities already visited.

Step 4: Since the second subscript of the last distance is 9, the first subscript of the next distance should be 9. Note that the next distance should not contain any of the subscripts already used (i.e., no 1, 2, 10), except that the first subscript of the next distance should be 9. One chooses the distance $C_{9,8}(4)$ in box D (with numerical value, 4) to obtain the connected distances $\boxed{C_{1,2}(3)C_{2,10}(7)C_{10,9}(14)C_{9,8}(4)}$ Also $C_{1,2}(3)C_{2,10}(7)C_{10,9}(14)C_{9,7}(5)$

$C_{1,2}(3)C_{2,10}(7)C_{10,9}(14)C_{9,5}(20)$ $C_{1,2}(3)C_{2,10}(7)C_{10,9}(14)C_{9,6}(19)$

Step 5 Since the second subscript of the last distance is 8, the first subscript of the next distance should be 8. Note that the next distance should not contain any of the subscripts already used (i.e., no 1, 2, 10, 9), except that the first subscript of the next distance should be 8, One chooses the distance $C_{8,5}(2)$ in box B to obtain the connected distances

$\boxed{C_{1,2}(3)C_{2,10}(7)C_{10,9}(14)C_{9,8}(4)C_{8,5}(2)}$.

Also $C_{1,2}(3)C_{2,10}(7)C_{10,9}(14)C_{9,8}(4)C_{7,5}(17)$; $C_{1,2}(3)C_{2,10}(7)C_{10,9}(14)C_{9,8}(4)C_{7,8}(8)$; $C_{1,2}(3)C_{2,10}(7)C_{10,9}(14)C_{9,8}(4)C_{7,9}(5)$

Step 6: Since the second subscript of the last distance is 5, the first subscript of the next distance should be 5. Note that the next distance should not contain any of the subscripts already used (i.e., no 1, 2, 10, 9, 8) except that the first subscript of the next distance should be 5. One chooses the distance $C_{5,6}(1)$ in box A to obtain the connected distances

$\boxed{C_{1,2}(3)C_{2,10}(7)C_{10,9}(14)C_{9,8}(4)C_{8,5}(2)C_{5,6}(1)}$; Also

$C_{1,2}(3)C_{2,10}(7)C_{10,9}(14)C_{9,8}(4)C_{7,6}(9)C_{6,4}(24)$

Note that $C_{5,6}(1)$ has the least numerical value, 1, among the eligible distances.

Step 7: Since the second subscript of the last distance is 6, the first subscript of the next distance should be 6. Note that the next distance should not contain any of the subscripts already used (i.e., no 1, 2, 10, 9, 8, 5) except that the first subscript of the next distance should be 6. One chooses the distance $C_{6,7}(9)$ in box J to obtain the connected distances

$\boxed{C_{1,2}(3)C_{2,10}(7)C_{10,9}(14)C_{9,8}(4)C_{8,5}(2)C_{5,6}(1)C_{6,7}(9)}$

. Also, $C_{1,2}(3)C_{2,10}(7)C_{10,9}(14)C_{9,8}(4)C_{7,6}(9)C_{6,5}(1)C_{5,4}(12)$

The excluded subscript numbers , except 1, represent the cities already visited.

Step 8: Since the second subscript of the last distance 7, the first subscript of the next distance should be 7. Note that the next distance should not contain any of the subscripts already used (i.e., no 1, 2, 10, 9, 8, 5, 6), except that the first subscript of the next distance should be 7. One chooses the distance $C_{7,4}(39)$ from the original data table to obtain the connected distances $\boxed{C_{1,2}(3)C_{2,10}(7)C_{10,9}(14)C_{9,8}(4)C_{8,5}(2)C_{5,6}(1)C_{6,7}(9)C_{7,4}(39)}$

Note that $C_{7,4}(39)$ has a relatively large numerical value, 39, among the eligible distances. One went up to a larger range of numbers to accommodate $C_{7,4}$. Because a value greater 14 has been used, upon completion of the route construction, the route found would not be the shortest route.The excluded subscript numbers , except 1, represent the cities already visited.

Step 9: Since the second subscript of the last distance is 4, the first subscript of the next distance should be 4. Note that the next distance should not contain any of the subscripts already used (i.e., no 1, 2, 10, 9, 8, 5, 6, 7), except that the first subscript of the next distance should be 4. One chooses the distance $C_{4,3}(10)$ in box K to obtain the connected distances

$\boxed{C_{1,2}(3)C_{2,10}(7)C_{10,9}(14)C_{9,8}(4)C_{8,5}(2)C_{5,6}(1)C_{6,7}(9)C_{7,4}(39)C_{4,3}(10)}$

Alsso: $C_{1,2}(3)C_{2,10}(7)C_{10,9}(14)C_{9,8}(4)C_{7,6}(9)C_{6,5}(1)C_{5,8}(2)C_{8,4}(23)C_{4,3}(10)$

However, since these connected distances contain values greater than 14, there is no need to continue their construction

Note that $C_{4,3}(10)$ has the least numerical value, 2, among the eligible distances.

Step 10: Since the second subscript of the last distance is 3, the first subscript of the next and last distance should be 3. Note that the next distance should not contain any of the subscripts already used (i.e., no 1, 2, 10, 9, 8, 5, 6, 7, 4) except that the first subscript of the next distance should be 3 and the second subscript should be 1 (an exception) in order to return to City 1, he starting city. One chooses the distance $C_{3,1}(13)$ in box N to obtain the connected distances

$$\boxed{C_{1,2}(3)C_{2,10}(7)C_{10,9}(14)C_{9,8}(4)C_{8,5}(2)C_{5,6}(1)C_{6,7}(9)C_{7,4}(39)C_{4,3}(10)C_{3,1}(13)=102}$$

Also $C_{1,2}(3)C_{2,10}(7)C_{10,9}(14)C_{9,8}(4)C_{7,6}(9)C_{6,5}(1)C_{5,8}(2)C_{8,4}(23)C_{4,3}(10)C_{3,1}(13)$

Step 11: Add the distances in parentheses: 3 + 7 + 1 4 + 4 + 2 + 1 + 9 + 39 + 10 + 13 and obtain 102.

For comparison purposes, before proceeding to the discussion and conclusion of the material covered already, one will next summarize the shortcomings of some previous methods for solving the traveling salesman problem,

Shortcomings of the Nearest Neighbor Approach and Grouping of Cities Approach

Shortcoming of the Nearest Neighbor Approach

Consider four cities at A, B, C, D. Let the home base of the salesman be at A.

Case 1: Applying the nearest neighbor approach, one would depart from City A along AD of length 6 units (Note: $6 < 9 < 10$). To visit each of the three cities once and return to A, one would either travel the distances AD + DB + BC+ CA (6 + 4 +12+10 = 32 units) or the distances AD + DC + CB + BA (6 + 9 +12 + 7 = 34 units.

Case 2: If one departs along AB, one would either travel the distances
AB +BD + DC + CA (7 + 4 + 9 +10 = 30 units) or
AB + BC + CD + DA (7 + 12 + 9 + 6 = 34 units)

Case 3: If one departs along AC, one would travel either the distances
AC + CD + DB + BA (10 + 9 + 4 +7 = 30 units) or
AC + CB + BD + DA (10 + 12 + 4 + 6 = 32 units)

Observe above that the shortest route is **not** in Case 1, (of total distance 32 or 34 units) the nearest neighbor approach; but **is** in either **Case 2** or **Case 3 ,** of distance 30 units. Note that the totals in the first parts of Cases 2 and 3 are the same, the same individual distances, except for the order of the addition of the distances.

It is to be observed that departing to the nearest city at D, 6 units away, did not produce the shortest total distance. However, departing to either the city at B, or the city at C produced the shortest route of length 30 units, even though B or C is not the nearest neighbor.

The "culprit" is BC or CB of distance 12 units. If one departs to city at D, one is compelled to travel the longest distance of 12 units, since the options to visit the cities at B and D cannot avoid the 12 units distance. The error for Case 1 is about either 7% or 13%, respectively . As the number of cities increases, the errors will multiply.

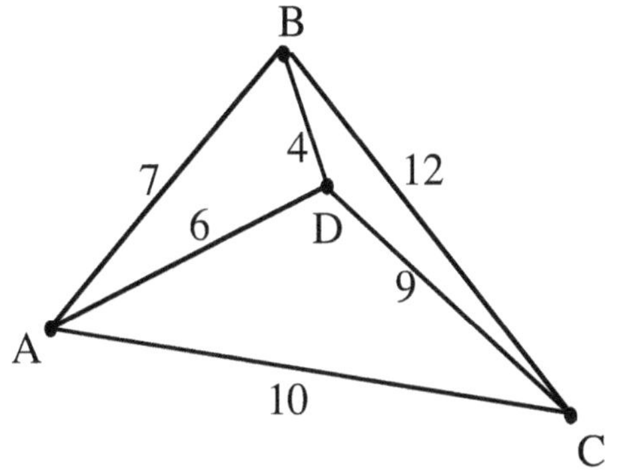

$C_{A,B}$ = distance between City A and City B.

Possible routes for a salesman to visit each of the Cities, B, C, and D without returning to A.

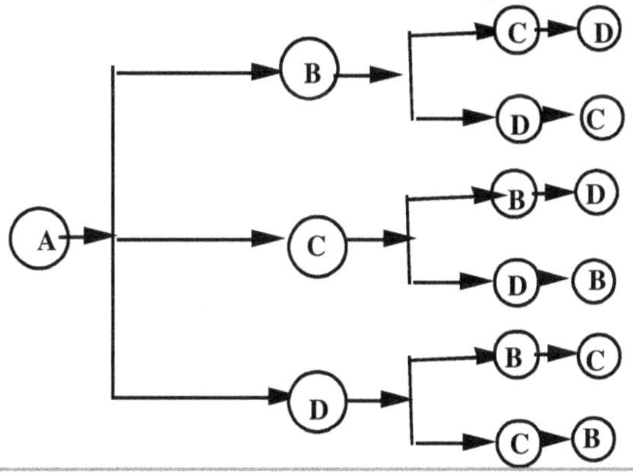

$C_{A,B}$ 7	$C_{A,B}$ 7	$C_{A,C}$ 10	$C_{A,C}$ 10	$C_{A,D}$ 6	$C_{A,D}$ 6
$C_{B,C}$ 12	$C_{B,D}$ 4	$C_{C,B}$ 12	$C_{C,D}$ 9	$C_{D,B}$ 4	$C_{D,C}$ 9
$C_{C,D}$ 9	$C_{D,C}$ 9	$C_{B,D}$ 4	$C_{D,B}$ 4	$C_{B,C}$ 12	$C_{C,B}$ 12
28	20	26	23	22	27

Shortcoming of the Grouping of Cities Approach

Mini-Brute Force plus "Divide and Conquer" Approach

Example : Using a three-distance template to determine the shortest route

From City 1, a traveling salesman would like to visit once each of **nine** other cities, namely, City 2, City 3, City 4, City 5, City 6, City 7, City 8, City 9, City 10; and return to City 1.

Guidelines

Step 1: From City 1 (home base of salesman), consider the possible sub-routes for visiting three other cities (say, Cities 2, 3, and 4) without returning to City 1, and determine the shortest route from City 1 to visit once each of these cities.

Step 2: I From the last city visited by the shortest route, one will next determine the shortest sub-route for visiting three other cities, say, Cities 5, 6 and 7.

Step 3: From the last city visited according to the shortest route for visiting Cities 5, 6 and 7, one will determine the shortest route for visiting Cities 8, 9, and 10. The sums of the distances of the above shortest routes will added, and the distance from City 10 to City 1 will also be added the shortest routes sum. (Review example on previous page, and imitate)

Step 4: The results of Steps 1-3 can be combined into a single table as below (18 columns)

$C_{1,2}$ 3	$C_{1,2}$ 3	$C_{1,3}$ 13	$C_{1,3}$ 13	$C_{1,4}$ 35	$C_{1,4}$ 35	$C_{4,5}$ 12	$C_{4,5}$ 12	$C_{4,6}$ 24	$C_{4,6}$ 24
$C_{2,3}$ 21	$C_{2,4}$ 25	$C_{3,4}$ 10	$C_{3,2}$ 21	$C_{4,3}$ 10	$C_{4,2}$ 25	$C_{5,6}$ 1	$C_{5,7}$ 17	$C_{6,5}$ 1	$C_{6,7}$ 9
$C_{3,4}$ 10	$C_{4,3}$ 10	$C_{4,2}$ 25	$C_{2,4}$ 25	$C_{3,2}$ 21	$C_{2,3}$ 21	$C_{6,7}$ 9	$C_{7,6}$ 9	$C_{5,7}$ 17	$C_{7,5}$ 17
34	38	48	59	66	81	22	38	42	50

$C_{4,7}$ 39	$C_{4,7}$ 39	$C_{7,8}$ 8	$C_{7,8}$ 8	$C_{7,9}$ 5	$C_{7,9}$ 5	$C_{7,10}$ 28	$C_{7,10}$ 28
$C_{7,5}$ 17	$C_{7,6}$ 9	$C_{8,9}$ 4	$C_{8,10}$ 15	$C_{9,8}$ 4	$C_{9,10}$ 14	$C_{10,9}$ 14	$C_{10,8}$ 15
$C_{5,6}$ 1	$C_{6,5}$ 1	$C_{9,10}$ 14	$C_{10,9}$ 14	$C_{8,10}$ 15	$C_{10,8}$ 15	$C_{9,8}$ 4	$C_{8,9}$ 4
57	49	26	37	24	34	46	47

Step 5: Combine the boxed columns (shortest sub-routes) above, and add the distance $C_{10,1}$ ($C_{10,1}$ = 11, is the distance from the last city, City 10, to the home base city of the salesman.

$C_{1,2}$ 3	$C_{4,5}$ 12	$C_{7,9}$ 5
$C_{2,3}$ 21	$C_{5,6}$ 1	$C_{9,8}$ 4
$C_{3,4}$ 10	$C_{6,7}$ 9	$C_{8,10}$ 15
34	22	24

Total = 34 + 22 + 24 + 11 = 91

Shortest route to visit each of the nine cities once and return =

$C_{1,2} + C_{2,3} + C_{3,4} + C_{4,5} + C_{5,6} + C_{6,7} + C_{7,9} + C_{9,8} + C_{8,10} + C_{10,1} = 91$ units.

Observe above that Cities, 2, 3, 4, 5, 6, 7, 8, 9, and 10 have been visited; and by $C_{8,10}$, the salesman is at City 10; and to return to City 1, one adds $C_{10,1}$.

Grouping of cities approach
$C_{1,2}(3)C_{2,3}(21)C_{3,4}(10)C_{4,5}(12)C_{5,6}(1)C_{6,7}(9)C_{7,9}(5)C_{9,8}(4)C_{8,10}(15)C_{10,1}(11) = \mathbf{91}$

Comparison of Approaches for Finding Shortest Routes

Case 1: For the **Nearest Neighbor approach**, the error lies in being compelled to travel an avoidable longer distance as illustrated on page 229.

Case 2: For the **Grouping of Cities approach**, the error emanates from ignoring some of the shortest distances in determining the shortest route. The length of the shortest route by the grouping of cities approach was found to be 91 units, (for sample problem in this paper)

Case 3: For the **Data Ordering and Route Construction approach,** the length of the shortest route determined was 81 units. The error in Case 2 relative to Case 3 is about 12%, In observing the numerical values of the distances for the shortest routes in Cases 2 and 3 as well as the entries in the table used in the construction of the shortest route for Case 3, below, note that Case 3 used numerical values from the table in boxes A-P. (minimum boxes). Even though Case 2 was obtained by a different approach, one can observe that values 15 and 21 in Case 2 are from boxes beyond boxes A-P.

Case 3 **Shortest route**

Data ordering and route construction approach

$C_{1,3}(13)C_{3,4}(10)C_{4,5}(12)C_{5,6}(1)C_{6,7}(9)C_{7,8}(8)C_{8,9}(4)C_{9,10}(14)C_{10,2}(7)C_{2,1}(3) = \mathbf{81}$ <--- **R1**

Numerical distances: 1, 3, 4, 7, 8, 9, 10, 12, 13, 14

Case 2

Grouping of cities approach

$C_{1,2}(3)C_{2,3}(21)C_{3,4}(10)C_{4,5}(12)C_{5,6}(1)C_{6,7}(9)C_{7,9}(5)C_{9,8}(4)C_{8,10}(15)C_{10,1}(11) = \mathbf{91}$

Numerical distances: 1, 3, 4, 5, 9, 10, 11, 12, **15, 21**

A	$C_{5,6}$ or $C_{6,5}$	**1**	**G**	$C_{2,10}$ or $C_{10,2}$	**7**	**N**	$C_{1,3}$ or $C_{3,1}$	**13**	**U**	$C_{6,9}$ or $C_{9,6}$	**19**
B	$C_{5,8}$ or $C_{8,5}$	**2**	**H**	$C_{7,8}$ or $C_{8,7}$	**8**	**P**	$C_{9,10}$ or $C_{10,9}$	**14**	**V**	$C_{5,9}$ or $C_{9,5}$	**20**
C	$C_{1,2}$ or $C_{2,1}$	**3**	**J**	$C_{6,7}$ or $C_{7,6}$	**9**	**Q**	$C_{8,10}$ or $C_{10,8}$	**15**	**W**	$C_{2,3}$ or $C_{3,2}$	**21**
D	$C_{8,9}$ or $C_{9,8}$	**4**	**K**	$C_{3,4}$ or $C_{4,3}$	**10**	**R**	$C_{2,8}$ or $C_{8,2}$	**16**	**X**	$C_{1,8}$ or $C_{8,1}$	**22**
E	$C_{7,9}$ or $C_{9,7}$	**5**	**L**	$C_{1,10}$ or $C_{10,1}$	**11**	**S**	$C_{5,7}$ or $C_{7,5}$	**17**	**Y**	$C_{4,8}$ or $C_{8,4}$	**23**
F	$C_{6,8}$ or $C_{8,6}$	**6**	**M**	$C_{4,5}$ or $C_{5,4}$	**12**	**T**	$C_{2,5}$ or $C_{5,2}$	**18**	**Z**	$C_{4,6}$ or $C_{6,4}$	**24**

Note the following:

$C_{1,3}(13)C_{3,4}(10)C_{4,5}(12)C_{5,6}(1)C_{6,7}(9)C_{7,8}(8)C_{8,9}(4)C_{9,10}(14)C_{10,2}(7)C_{2,1}(3) = 81$ is equivalent to

$C_{1,3}(13) + C_{3,4}(10) + C_{4,5}(12) + C_{5,6}(1) + C_{6,7}(9) + C_{7,8}(8) + C_{8,9}(4) + C_{9,10}(14) + C_{10,2}(7) + C_{2,1}(3) = 81$

(From City 1 to City 3; from City 3 to City 4; from City 4 to City 5; from City 5 to City 6; from City 6 to City 7; from City 7 to City 8; from City 8 to City 9; from City 9 to City 10; .from City 10 to City 2; and finally, from City 2 to City 1.)

Discussion and Conclusion

The length of the shortest route was found to be 81 units; but the sequence of cities of the shortest route is not unique. One sequence of the cities of the shortest route is given by $C_{1,3}(13)C_{3,4}(10)C_{4,5}(12)C_{5,6}(1)C_{6,7}(9)C_{7,8}(8)C_{8,9}(4)C_{9,10}(14)C_{10,2}(7)C_{2,1}(3)$, say R1 If the direction of travel of this route is reversed, one obtains the route given by $C_{1,2}(3)C_{2,10}(7)C_{10,9}(14)C_{9,8}(4)C_{8,7}(8)C_{7,6}(9)C_{6,5}(1)C_{5,4}(12)C_{4,3}(10)C_{3,1}(13)$. Another route of length 81 units is $C_{1,3}(13)C_{3,4}(10)\ C_{4,5}(12)\ C_{5,8}(2)C_{8,6}(6)C_{6,7}(9)C_{7,9}(5)C_{9,10}(14)C_{10,2}(7)C_{2,1}(3)$, and whose reversed travel direction yields another route given by $C_{1,2}(3)C_{2,10}(7)C_{10,9}(14)C_{9,7}(5)C_{7,6}(9)C_{6,8}(6)C_{8,5}(2)C_{5,4}(12)C_{4,3}(10)C_{3,1}(13) = 81$
Therefore, the sequence of cities of the shortest route is not unique, but the length of the route is unique.

Justification of the shortest route.

From City 1, ten distances are needed to visit nine cities and return to City 1.
If each of the distances, $C_{m,n}$, in the ten-distance route were from the least ten distances (i.e., box A-K) in the table, one could immediately conclude that such a ten-distance route is the shortest route. In observing the possible shortest route, $C_{1,3}(13)C_{3,4}(10)C_{4,5}(12)C_{5,6}(1)C_{6,7}(9)C_{7,8}(8)C_{8,9}(4)C_{9,10}(14)C_{10,2}(7)C_{2,1}(3)$,R1, not all the distances are from the least ten distances in the table, and one cannot immediately conclude that R1 is the shortest route. However, the next three distances (except 11 which is not applicable here), 12, 13, and 14 are included in R1.
These additions are minimum additions, and therefore, the shortest route of length 81 units is given by $C_{1,3}(13)C_{3,4}(10)C_{4,5}(12)C_{5,6}(1)C_{6,7}(9)C_{7,8}(8)C_{8,9}(4)C_{9,10}(14)C_{10,2}(7)C_{2,1}(3)$. Perhaps, one should say a shortest route, since the sequence of cities in this paper is not unique.
Observe below that any ten-distance route which contains a distance greater than 14 (largest distance in R1) is at least 6 units greater than that of R1.

$C_{1,3}(13)C_{3,4}(10)C_{4,5}(12)C_{5,6}(1)C_{6,7}(9)C_{7,8}(8)C_{8,9}(4)C_{9,10}(14)C_{10,2}(7)C_{2,1}(3) = \mathbf{81}$	**R1**
Numerical distances: 1, 3, 4, 7, 8, 9, 10, 12, 13, 14	

$C_{1,2}(3)C_{2,10}(7)C_{10,9}(14)C_{9,7}(5)C_{7,6}(9)C_{6,5}(1)C_{5,8}(2)C_{8,4}(23)C_{4,3}(10)C_{3,1}(13) = \mathbf{87}$	**R2**
Numerical distances: 1, 2, 3, 5, 7, 9, 10, 13, 14, **23**	

1

$C_{1,3}(13)C_{3,4}(10)C_{4,5}(12)C_{5,6}(1)C_{6,7}(9)C_{7,9}(5)C_{9,8}(4)C_{8,2}(16)C_{2,10}(7)C_{10,1}(11) = \mathbf{88}$	**R4**
Numerical distances 1, 4, 5, 7, 9, 10, 11, 12, 13, **16**	

A	$C_{5,6}$ or $C_{6,5}$	**1**	**G**	$C_{2,10}$ or $C_{10,2}$	**7**	**N**	$C_{1,3}$ or $C_{3,1}$	**13**	**U**	$C_{6,9}$ or $C_{9,6}$	**19**
B	$C_{5,8}$ or $C_{8,5}$	**2**	**H**	$C_{7,8}$ or $C_{8,7}$	**8**	**P**	$C_{9,10}$ or $C_{10,9}$	**14**	**V**	$C_{5,9}$ or $C_{9,5}$	**20**
C	$C_{1,2}$ or $C_{2,1}$	**3**	**J**	$C_{6,7}$ or $C_{7,6}$	**9**	**Q**	$C_{8,10}$ or $C_{10,8}$	**15**	**W**	$C_{2,3}$ or $C_{3,2}$	**21**
D	$C_{8,9}$ or $C_{9,8}$	**4**	**K**	$C_{3,4}$ or $C_{4,3}$	**10**	**R**	$C_{2,8}$ or $C_{8,2}$	**16**	**X**	$C_{1,8}$ or $C_{8,1}$	**22**
E	$C_{7,9}$ or $C_{9,7}$	**5**	**L**	$C_{1,10}$ or $C_{10,1}$	**11**	**S**	$C_{5,7}$ or $C_{7,5}$	**17**	**Y**	$C_{4,8}$ or $C_{8,4}$	**23**
F	$C_{6,8}$ or $C_{8,6}$	**6**	**M**	$C_{4,5}$ or $C_{5,4}$	**12**	**T**	$C_{2,5}$ or $C_{5,2}$	**18**	**Z**	$C_{4,6}$ or $C_{6,4}$	**24**

The future in the approach for solving the traveling salesman problem lies in the approach (data ordering and route construction)) whereby one concentrates on the smallest distances, and by judicious selection, construct the shortest route. Such an approach reduces the redundant use of brute force. For the nine cities visit, using brute-force, one would have to consider about 362,880 possibilities. Each possibility would be a column of nine distances. One of these 362,880 columns would be the shortest route to visit the nine cities without returning to City1.

Bye-bye: nearest neighbor approach. You compelled the salesman to travel a longer distance. Bye-bye: grouping of cities approach. You ignored some of the shortest distances. Welcome: Data Ordering and Route Construction. Continue to refine and you would always be welcome

The error in the shortest route of length 81 units determined is zero or negligible.

Now, by moving the tip of a pencil, enjoy the following travel:

$C_{1,3}(13)C_{3,4}(10)C_{4,5}(12)C_{5,6}(1)C_{6,7}(9)C_{7,8}(8)C_{8,9}(4)C_{9,10}(14)C_{10,2}(7)C_{2,1}(3) = 81$ is equivalent to

$C_{1,3}(13) + C_{3,4}(10) + C_{4,5}(12) + C_{5,6}(1) + C_{6,7}(9) + C_{7,8}(8) + C_{8,9}(4) + C_{9,10}(14) + C_{10,2}(7) + C_{2,1}(3) = 81$

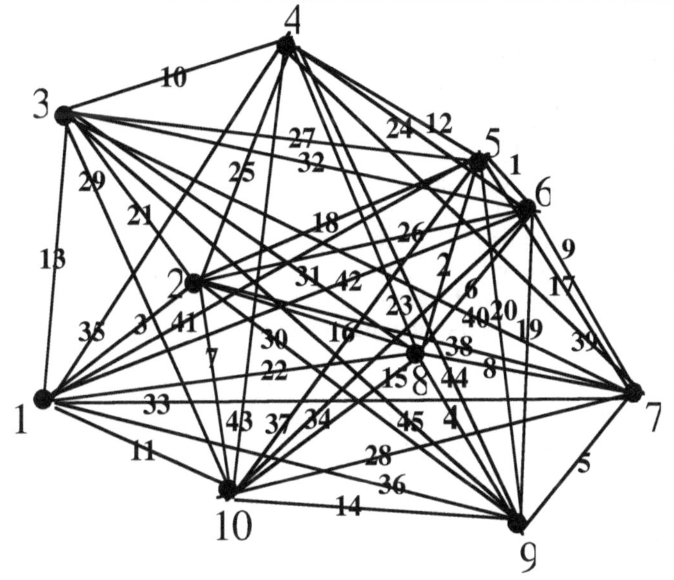

Travel Route

From City 1 to City 3;
from City 3 to City 4;
from City 4 to City 5;
from City 5 to City 6;
from City 6 to City 7;
from City 7 to City 8;
from City 8 to City 9;
from City 9 to City 10;
.from City 10 to City 2;
and finally, from City 2 to City 1.

The above paper has been added to the author's previous solutions of NP problems in the paper, P vs NP: Solutions of NP problems (Example 7) at viXra:, 1408.0204

Adonten

Overall Conclusion

Three different types of NP problems were solved, The first type involved the division of items of different sizes, lengths, masses, volumes, or values into equal parts by combinations only. The second type covered possible final exam schedules for schools. The third type was the traveling salesman problem. The general approach to solving the different types of NP problems were the same, except that sometimes, specific techniques may differ from each other according to the process involved in the problem. The first step in these problems is to arrange the given data in either increasing or decreasing order. For the solutions of the first type of NP problems, an extended Ashanti fairness wisdom technique was applied to a set of 100 items of different values or masses. Two people A and B were able to divide items equally by merely choosing in turns from a set of ordered items. The total value or mass of A's items was found to be equal total value or mass of B's items, and these results are combinations of the items of different values or masses. It is very pleasing that such a simple technique can produce desired combinations. High school and middle school graduates could be taught the technique involved. From the solutions, formulas or simple equations were produced to help programmers apply the techniques. Note that in using the technique in this paper, the items involved must be arranged in decreasing order, preferably. Therefore, in programming, the first step should be to arrange the items in decreasing order. The technique was also applied to 1000 items; and the results were perfect, just like the results for the 100 items. Therefore, the technique covered does not care whether there are 2^{100} or 2^{1000} possibilities. There are social consequences of the method and principles used to divide the set of items into equal totals. The results can be applied by government agencies in the distribution of goods and services. Management personnel should be aware of the principles involved in the above technique. From the elementary school, through high school, and perhaps college, students should be taught the principles in the above wisdom technique, since throughout life, one is going to encounter situations in which two or more people are asked to choose in turns, from items of different values or sizes, and in this case, the sequence by which the choices are made matters; one may be either a participant or one may be in charge of the distribution process.
By hand, the techniques can be used to prepare final exam schedules for 100 or 1000 courses. School secretaries and office assistants can learn and apply the techniques covered.
A new approach to solving the traveling salesman was used to determine the shortest route to visit nine cities and return to the starting city. The distances involved were arranged in increasing order and by inspection, ten distances were selected from a set of the shortest 14 distances, instead of the overall set of 45 distances. The selected distances were used to construct the shortest route. Finally, if an approach can solve one NP problem, that approach can also solve other NP problems. Since three different types NP problems (seven problems) were solved using the same approach outlined above, all NP problems can be solved. The formerly NP problems are now P problems, and therefore, it is concluded that P is equal to NP.

Perhaps, one may make the following statements.
1. NP plus human ability equals P. **2.** NP plus human inability is **not** equal to P.
3. NP minus human inability equals P.

Appendix 7
Navier-Stokes Equations Solutions Completed
Abstract

"5% of the people think; 10% of the people think that they think; and the other 85% would rather die than think."----Thomas Edison

"The simplest solution is usually the best solution"---Albert Einstein

Over nearly a year and half ago, the Navier-Stokes equations in 3-D for incompressible fluid low were analytically solved by the author. However, some of the solutions contained implicit terms. In this paper, the implicit terms have been expressed explicitly in terms of x, y, z and t. The author proposed and applied a new law, the law of definite ratio for incompressible fluid flow. This law states that in incompressible fluid flow, the other terms of the fluid flow equation divide the gravity term in a definite ratio, and each term utilizes gravity to function. The sum of the terms of the ratio is always unity. It was mathematically shown that without gravity forces on earth, there would be no incompressible fluid flow on earth as is known, and also, there would be no magnetohydrodynamics. In addition to the usual method of solving these equations, the N-S equations were also solved by a second method in which the three equations in the system were added to produce a single equation which was then integrated. The solutions by the two methods were identical, except for the constants involved. Ratios were used to split-up the equations; and the resulting sub-equations were readily integrable, and even, the nonlinear sub-equations were readily integrated. The examples in the preliminaries show everyday examples on using ratios to divide a quantity into parts, as well as possible applications of the solution method in mathematics, science, engineering, business, economics, finance, nvestment and personnel management decisions. The x–direction Navier-Stokes equation was linearized, solved, and the solution analyzed. This solution was followed by the solution of the Euler equation of fluid flow. The Euler equation represents the nonlinear part of the Navier-Stokes equation. Following the Euler solution, the Navier-Stokes equation was solved essentially by combining the solutions of the linearized equation and the Euler solution. For the Navier-Stokes equation, the linear part of the relation obtained from the integration of the linear part of the equation satisfied the linear part of the equation; and the relation from the integration of the non-linear part satisfied the non-linear part of the equation. The solutions and relations revealed the role of each term of the Navier-Stokes equations in fluid flow. The gravity term is the indispensable term in fluid flow, and it is involved in the parabolic and forward motion of fluids. The pressure gradient term is also involved in the parabolic motion. The viscosity terms are involved in the parabolic, periodic and decreasingly exponential motion. Periodicity increases with viscosity. The variable acceleration term is also involved in the periodic and decreasingly exponential motion. The fluid flow in the Navier-Stokes solution may be characterized as follows. The x–direction solution consists of linear, parabolic and hyperbolic terms. The first three terms characterize parabolas. If one assumes that in laminar flow, the axis of symmetry of the parabola for horizontal velocity flow profile is in the direction of fluid flow, then in turbulent flow, some of the axes of symmetry of the parabolas would be at right angles to that of laminar flow. The characteristic curve for the integral of the x–nonlinear term is such a parabola whose axis of symmetry is at right angles to that of laminar flow.

The integral of the y–nonlinear term is similar parabolically to that of the x–nonlinear term.

The integral of the z–nonlinear term is a combination of two similar parabolas and a hyperbola.

If the above x–direction flow is repeated simultaneously in the y– and z– directions, th e flow is chaotic and consequently turbulent.

For a spin-off, the smooth solutions from above are specialized and extended to satisfy the requirements of the CMI Millennium Prize Problems, and thus prove the existence of smooth solutions of the Navier-Stokes equations.

Introduction

Solutions of the Navier-Stokes Equations

Case 1: Solutions of the Linearized Navier-Stokes Equations (x–direction)

Equation $$-\mu(\frac{\partial^2 v_x}{\partial x^2}+\frac{\partial^2 v_x}{\partial y^2}+\frac{\partial^2 v_x}{\partial z^2})+\frac{\partial p}{\partial x}+4\rho(\frac{\partial v_x}{\partial t})=\rho g_x$$

Solutions $$V_x(x,y,z,t)=-\frac{\rho g_x}{2\mu}(ax^2+by^2+cz^2)+C_1x+C_3y+C_5z+\frac{fg_x}{4}t+C_9;\quad P(x)=d\rho g_x x$$

Case 2: Solutions of the Euler Equations for Incompressible Fluid Flow ()

Equation $$\rho(\frac{\partial v_x}{\partial t}+V_x\frac{\partial V_x}{\partial x}+V_y\frac{\partial v_x}{\partial y}+V_z\frac{\partial v_x}{\partial z})+\frac{\partial p_x}{\partial x}=\rho g_x$$ x–direction

Solutions

$$V_x(x,y,z,t)=fg_xt\pm\sqrt{2hg_xx}+\frac{ng_xy}{V_y}+\frac{qg_xz}{V_z}+\underbrace{\frac{\psi_y(V_y)}{V_y}+\frac{\psi_z(V_z)}{V_z}}_{\text{arbitrary functions}};V_y\neq0,V_z\neq0;P(x)=d\rho g_xx;$$

Case 3: Solutions of the Navier-Stokes Equations (Original) : x–direction

Equation: $$-\mu\frac{\partial^2 V_x}{\partial x^2}-\mu\frac{\partial^2 V_x}{\partial y^2}-\mu\frac{\partial^2 V_x}{\partial z^2}+\frac{\partial p}{\partial x}+\rho\frac{\partial V_x}{\partial t}+\rho V_x\frac{\partial V_x}{\partial x}+\rho V_y\frac{\partial V_x}{\partial y}+\rho V_z\frac{\partial V_x}{\partial z}=\rho g_x$$

Solutions

$$V_x=-\frac{\rho g_x}{2\mu}(ax^2+by^2+cz^2)+C_1x+C_3y+C_5z+fg_xt\pm\sqrt{2hg_xx}+\frac{ng_xy}{V_y}+\frac{qg_xz}{V_z}+\frac{\psi_y(V_y)}{V_y}+\frac{\psi_z(V_z)}{V_z}+C_9$$
$$P(x)=d\rho g_xx;\quad (a+b+c+d+h+n+q=1)\quad V_y\neq0,\ V_z\neq0$$

Summary for the fractional terms of the x–direction

$\frac{ng_xy}{V_y}$ and $\frac{qg_xz}{V_z}$ in terms of x, y, z and t (for Case 3)

$$\frac{ng_xy}{V_y}=\frac{-(ng_x)(-\frac{\rho g_z}{2\mu}(\beta_1x^2+\beta_2y^2+\beta_3z^2)+C_1x+C_3y+C_5z+\beta_5g_zt\pm\sqrt{2\beta_8g_zz})}{\beta_7g_z}$$

$$\frac{qg_xz}{V_z}=\frac{-(qg_xz)\{[(\beta_7g_zy)(-\frac{\rho g_x}{2\mu}(ax^2+by^2+cz^2)+C_1x+C_3y+C_5z+fg_xt\pm\sqrt{2hg_xx}]-[CE]\}}{(\beta_7g_zy)(qg_xz-\beta_6g_zx)}$$

$$(CE=-(ng_xy)(-\frac{\rho g_z}{2\mu}(\beta_1x^2+\beta_2y^2+\beta_3z^2)+C_{14}x+C_{15}y+C_{16}z+\beta_5g_zt\pm\sqrt{2\beta_8g_zz})$$

One observes above that the most important insight of the above solutions is the indispensability of the gravity term in incompressible fluid flow. Observe that if gravity, g_x, were zero, for Case 1, the first three terms, the seventh, and $P(x)$ would all be zero; for Case 2, the first four terms and $P(x)$ would all be zero; and for Case 3, the first three terms, the seventh, the eighth, the ninth, the tenth terms and $P(x)$ would all be zero. These results can be stated emphatically that without gravity forces on earth, there would be no incompressible fluid flow on earth as is known. It would not therefore be meaningful to write a Navier-Stokes equation for incompressible fluid flow without the gravity term, since there would be no fluid flow.

More Observations Comparison of the N-S solutions with equations of motion under gravity and liquid pressure of elementary physics

Motion equations of elementary physics:

(B): $V_f = V_0 + gt$; (C): $V_f^2 = V_0^2 + 2gx$; (D): $V = \sqrt{2gx}$; (E): $x = V_0 t + \frac{1}{2}gt^2$

The **liquid pressure**, P at the bottom of a liquid of depth h units is given by $P = \rho gh$

Observe the following about the Navier-Stokes Solutions (Case 3)

1. The first three terms are parabolic in x, y, and z; the minus sign shows the usual inverted parabola when a projectile is fired upwards at an acute angle to the horizontal; also note the " gt" in $V = gt$ of (B) of the motion equations and the $fg_x t$ in the Navier-Stokes solution.

2. The pressure, $P = \rho gh$ of the liquid pressure and the $P(x) = d\rho g_x x$ of the Navier-Stokes solution.

Note that, only the approach in this paper could yield $P(x) = d\rho g_x x$ by integrating $dp/dx = d\rho g_x$

3. Observe the " $\sqrt{2gx}$ " in $V = \sqrt{2gx}$ of (D) and the $\sqrt{2hg_x x}$ in the Navier-Stokes solution.

In fact, the N-S solution term $\sqrt{2hg_x x}$ could have been obtained from $V_f^2 = V_0^2 + 2gx$,(C) , of the equation of motion by letting $V_0 = 0$ (for the convective term) ignoriiiing the ratio term "h" of the N-S radicand. There are eight main terms (ignoring the arbitrary functions) in the N-S solution. Of these eight terms, six terms, namely, $-\frac{a\rho g_x}{2\mu}x^2$, $-\frac{b\rho g_x}{2\mu}y^2$, $-\frac{c\rho g_x}{2\mu}z^2$, $fg_x t$, $\sqrt{2hg_x x}$ and $d\rho g_x x$ are similar (except for the constants involved) to the terms in the equations of motion and fluid pressure. This similarity means that the approach used in solving the Navier-Stokes equation is sound. One should also note that to obtain these six terms simultaneously on integration, only the equation with the gravity term as the subject of the equation will yield these six terms. The author suggests that this form of the equation with the gravity term as the subject of the equation be called the standard form of the Navier-Stokes equation, since in this form, one can immediately split-up the equations using ratios, and integrate.

4. With regards to the variables x, y, and z, the parabolicity of the first three terms and the parabolicity of the eighth, ninth and tenth terms hint at inverse relations.. For examples, $V_x = x^2$ and $V_x = \pm\sqrt{x}$ are inverse relations of each other, $V_x = y^2$ and $V_x = \pm\sqrt{y}$ are inverse relations of each other, $V_x = z^2$ and $V_x = \pm\sqrt{z}$ are inverse relations of each other. The implications of knowing these relationships is that if one knows the steps, rules or formulas for designing for laminar flow, one can deduce the steps, rules or formulas for designing for turbulent flow by reversing the steps and using opposite operations in each step of the corresponding laminar flow design. Thus for every method, or formula for laminar flow, there is a corresponding method, formula for turbulent flow design (see also, "Power of Ratios" book by A. A. Frempong, p. 28).

For the **velocity profile,** the x–direction solution consists of linear, parabolic, and hyperbolic terms. The first three terms characterize inverted parabolas. Flow distribution for laminar flow is parabolic with the axis of symmetry of the parabola in direction of the fluid flow. If one assumes that in laminar flow, the axis of symmetry of the parabola for horizontal velocity flow profile is in the direction of fluid flow, then in turbulent flow, the axes of symmetry of some of the parabolas would have been rotated 90 degrees from that for laminar flow. The characteristic curve for the integral of the x–nonlinear term is such a parabola whose axis of symmetry is at right angles to that of laminar flow. The integral of the y–nonlinear term is similar parabolically to the integral of the x–nonlinear term. The characteristic curve for the integral of the z–nonlinear term is a combination of two similar parabolas and a hyperbola. If the above x–direction flow is repeated simultaneously in the y– and z– directions, the flow is chaotic and consequently turbulent.

Options

The Navier-Stokes equations in three dimensions are three simultaneous equations in Cartesian coordinates for the flow of incompressible fluids. The equations are presented below:

$$\begin{cases} \mu(\frac{\partial^2 V_x}{\partial x^2}+\frac{\partial^2 V_x}{\partial y^2}+\frac{\partial^2 V_x}{\partial z^2})-\frac{\partial p}{\partial x}+\rho g_x=\rho(\frac{\partial V_x}{\partial t}+V_x\frac{\partial V_x}{\partial x}+V_y\frac{\partial V_x}{\partial y}+V_z\frac{\partial V_x}{\partial z}) & (N_x) \\ \mu(\frac{\partial^2 V_y}{\partial x^2}+\frac{\partial^2 V_y}{\partial y^2}+\frac{\partial^2 V_y}{\partial z^2})-\frac{\partial p}{\partial y}+\rho g_y=\rho(\frac{\partial V_y}{\partial t}+V_x\frac{\partial V_y}{\partial x}+V_y\frac{\partial V_y}{\partial y}+V_z\frac{\partial V_y}{\partial z}) & (N_y) \\ \mu(\frac{\partial^2 V_z}{\partial x^2}+\frac{\partial^2 V_z}{\partial y^2}+\frac{\partial^2 V_z}{\partial z^2})-\frac{\partial p}{\partial z}+\rho g_z=\rho(\frac{\partial V_z}{\partial t}+V_x\frac{\partial V_z}{\partial x}+V_y\frac{\partial V_z}{\partial y}+V_z\frac{\partial V_z}{\partial z}) & (N_z) \end{cases}$$

Equation (N_x) will be the first equation to be solved; and based on its solution, one will be able to write down the solutions for the other two equations, (N_y), and (N_z).

Dimensional Consistency

The Navier-Stokes equations are dimensionally consistent as shown below:

$$\mu(\frac{\partial^2 V_x}{\partial x^2}+\frac{\partial^2 V_x}{\partial y^2}+\frac{\partial^2 V_x}{\partial z^2})-\frac{\partial p}{\partial x}+\rho g_x=\rho(\frac{\partial V_x}{\partial t}+V_x\frac{\partial V_x}{\partial x}+V_y\frac{\partial V_x}{\partial y}+V_z\frac{\partial V_x}{\partial z})$$

Using *MLT*

$$\boxed{M(L^{-2}T^{-2}+L^{-2}T^{-2}+L^{-2}T^{-2}-L^{-2}T^{-2}+L^{-2}T^{-2})=M(L^{-2}T^{-2}+L^{-2}T^{-2}+L^{-2}T^{-2}+L^{-2}T^{-2})}$$

Using *kg–m–s*

$$\boxed{kg(m^{-2}s^{-2}+m^{-2}s^{-2}+m^{-2}s^{-2}-m^{-2}s^{-2}+m^{-2}s^{-2}=kg(m^{-2}s^{-2}+m^{-2}s^{-2}+m^{-2}s^{-2}+m^{-2}s^{-2}}$$

$$\overbrace{\mu(\frac{\partial^2 V_x}{\partial x^2}+\frac{\partial^2 V_x}{\partial y^2}+\frac{\partial^2 V_x}{\partial z^2})}^{\text{viscosity}}-\overbrace{\frac{\partial p_x}{\partial x}}^{\text{pressure gradient}}+\overbrace{\rho g_x}=\rho(\underbrace{\overbrace{\frac{\partial V_x}{\partial t}}^{\substack{\text{variable acceleration}\\ \text{(local rate of change of } V_x)}}+\overbrace{V_x\frac{\partial V_x}{\partial x}+V_y\frac{\partial V_x}{\partial y}+V_z\frac{\partial V_x}{\partial z}}^{\substack{\text{convective acceleration}\\ \text{(rate of change in } V_x \text{ due to motion)}}}}_{\text{inertia per volume}})$$

Option 1

Solution of 3-D Linearized Navier-Stokes Equation in the *x*-direction

The equation will be linearized by redefinition. The nine-term equation will be reduced to six terms.

Given: $$\mu(\frac{\partial^2 v_x}{\partial x^2}+\frac{\partial^2 v_x}{\partial y^2}+\frac{\partial^2 v_x}{\partial z^2})-\frac{\partial p}{\partial x}+\rho g_x=\rho(\frac{\partial v_x}{\partial t}+V_x\frac{\partial v_x}{\partial x}+V_y\frac{\partial v_x}{\partial y}+V_z\frac{\partial v_x}{\partial z}) \quad \text{(A)}$$

$$-\mu\frac{\partial^2 v_x}{\partial x^2}-\mu\frac{\partial^2 v_x}{\partial y^2}-\mu\frac{\partial^2 v_x}{\partial z^2}+\frac{\partial p}{\partial x}+\rho\frac{\partial v_x}{\partial t}+\rho V_x\frac{\partial v_x}{\partial x}+\rho V_y\frac{\partial v_x}{\partial y}+\rho V_z\frac{\partial v_x}{\partial z}=\rho g_x \quad \text{(B)}$$

$$-\mu(\frac{\partial^2 v_x}{\partial x^2}+\frac{\partial^2 v_x}{\partial y^2}+\frac{\partial^2 v_x}{\partial z^2})+\frac{\partial p}{\partial x}+4\rho(\frac{\partial v_x}{\partial t})=\rho g_x \quad \text{(C)}$$

Plan: One will split-up equation (C) into five sub-equations, solve them, and combine the solutions. On splitting-up the equations and proceeding to solve them, the non linear terms could be redefined and made linear. This linearization is possible if the gravitational force term is the subject of the equation as in equation (B). After converting the non-linear terms to linear terms by redefinition, one will have only six terms as in equation (C). One will show logically how equation (C) was obtained from equation (B), using a ratio method.

Three main steps are covered.
In main Step 1, one shows how equation (C) was obtained from equation (B)
In main Step 2, equation (C) will be split-up into five equations.
In main Step 3, each equation will be solved.
In main Step 4, the solutions from the five equations will be combined.
In main Step 5, the combined relation will be checked in equation (C). for identity.

Preliminaries

Requirements and procedure for solving a partial differential equation

1. Integrate the partial differential equation.
2. Find the partial derivatives from the integration relation from Step 1
3. Substitute the derivatives from Step 2 in the original partial differential equation and simplify. both sides of the equation.
4. If the left-hand side of the equation is equal to the right-hand side of the equation, then the integration relation from Step 1 is a solution to the partial differential equation.
   (Steps 2-4 can be summarized as checking for identity, or determining if the integration relation satisfies the original partial differential equation.)

Note: If one does not successfully check for identity, one cannot claim a solution.

A ratio method will be used to split-up the partial differential equations into sub-equation which are then integrated.

Example 1: A grandmother left \$45,000 in her will to be divided between eight grandchildren, Betsy, Comfort, Elaine, Ingrid, Elizabeth, Maureen, Ramona, Marilyn, in the ratio $\frac{1}{36}:\frac{1}{18}:\frac{1}{12}:\frac{1}{9}:\frac{5}{36}:\frac{1}{6}:\frac{7}{36}:\frac{2}{9}$. (**Note**: $\frac{1}{36}+\frac{1}{18}+\frac{1}{12}+\frac{1}{9}+\frac{5}{36}+\frac{1}{6}+\frac{7}{36}+\frac{2}{9}=1$)
How much does each receive?

Solution:

Betsy's share of \$45,000 $=\frac{1}{36}\times \$45{,}000=\$1{,}250$

Comfort's share of \$45,000 $=\frac{1}{18}\times \$45{,}000=\$2{,}500$

Elaine's share of \$45,000 $=\frac{1}{12}\times \$45{,}000=\$3{,}750$

Ingrid's share of \$45,000 $=\frac{1}{9}\times \$45{,}000=\$5{,}000$

Elizabeth's share of \$45,000 $=\frac{5}{36}\times \$45{,}000=\$6{,}250$

Maureen's share of \$45,000 $=\frac{1}{6}\times \$45{,}000=\$7{,}500$

Ramona's share of \$45,000 $=\frac{7}{36}\times \$45{,}000=\$8{,}750$

Marilyn's share of \$45,000 $=\frac{2}{9}\times \$45{,}000=\$10{,}000$

Check; Sum of shares $\boxed{=\$45{,}000}$
Sum of the fractions = 1

Example 2: Sir Isaac Newton left ρg_x units in his will to be divided between $-\mu\frac{\partial^2 V_x}{\partial x^2}$, $-\mu\frac{\partial^2 V_x}{\partial y^2}$, $-\mu\frac{\partial^2 V_x}{\partial z^2}$, $\frac{\partial p}{\partial x}$, $\rho\frac{\partial V_x}{\partial t}$, $\rho V_x\frac{\partial V_x}{\partial x}$, $\rho V_y\frac{\partial V_x}{\partial y}$, $\rho V_z\frac{\partial V_x}{\partial z}$ in the ratio $a:b:c:d:f:h:m:n$.
where $a+b+c+d+f+h+m+n=1$. How much does each receive?

Solution $-\mu\frac{\partial^2 V_x}{\partial x^2}$'s share of ρg_x units $= a\rho g_x$ units

$-\mu\frac{\partial^2 V_x}{\partial y^2}$'s share of ρg_x units $= b\rho g_x$ units

$-\mu\frac{\partial^2 V_x}{\partial z^2}$'s share of ρg_x units $= c\rho g_x$ units

$\frac{\partial p}{\partial x}$'s share of ρg_x units $= d\rho g_x$ units

$\rho\frac{\partial V_x}{\partial t}$'s share of ρg_x units $= f\rho g_x$ units

$\rho V_x\frac{\partial V_x}{\partial x}$'s share of ρg_x units $= h\rho g_x$ units

$\rho V_y\frac{\partial V_x}{\partial y}$'s share of ρg_x units $= m\rho g_x$ units

$\rho V_z\frac{\partial V_x}{\partial z}$'s share of ρg_x units $= n\rho g_x$ units

Sum of shares = $\boxed{\rho g_x \text{ units}}$ **Note**: $a+b+c+d+f+h+m+n=1$

Example 3:
The returns on investments A, B, C, D are in the ratio $a:b:c:d$. If the total return on these four investments is P dollars, what is the return on each of these investments?

$$(a+b+c+d=1)$$

Solution Return on investment $A = aP$ dollars
Return on investment $B = bP$ dollars
Return on investment $C = cP$ dollars
Return on investment $D = dP$ dollars

Check

$aP + bP + cP + dP = P$
$P(a+b+c+d) = P$
$a+b+c+d = 1$ (dividing both sides by P)

Example 4: Solve the quadratic equation;

$$6x^2 + 11x - 10 = 0$$

Method 1 (a common method)
By factoring,

$6x^2 + 11x - 10 = 0$
$(3x-2)(2x+5) = 0$ and solving,
$(3x-2) = 0$ or $(2x+5) = 0$
$x = \frac{2}{3}, x = -\frac{5}{2}.$
Solution set: $\{-\frac{5}{2}, \frac{2}{3}\}$

Example 4, Method 2:
One applies the discussion in Example 2
One will call this method the **multiplier method.**

Step 1: From $6x^2 + 11x - 10 = 0$ (1)

$6x^2 + 11x = 10$
$6x^2 = 10a$; (Here, a is a multiplier)
$3x^2 = 5a$ (2)
$11x = 10b$ (Here, b is a multiplier)
$11x = 10(1-a)$ $(a+b) = 1$
$11x = 10 - 10a$
$x = \frac{10-10a}{11}$
$3(\frac{10-10a}{11})^2 = 5a$ (Substituting for x in (2)
$3(\frac{100-200a+100a^2}{121}) = 5a$

Step 2: $300a^2 - 1205a + 300 = 0$

$$60a^2 - 241a + 60 = 0$$

$a = \frac{241 \pm \sqrt{241^2 - 4(60)(60)}}{120}$
$a = \frac{241 \pm \sqrt{43681}}{120}$
$a = \frac{241 \pm 209}{120}$
$a = \frac{241 \pm 209}{120} = \frac{241+209}{120}$ or $\frac{241-209}{120}$
$= \frac{450}{120}$ or $\frac{32}{120}$
$= \frac{15}{4}$ or $\frac{4}{15}$

Step 3: Since $a+b=1$, when $a = \frac{15}{4}$ or $3\frac{3}{4}$

$b = 1 - 3\frac{3}{4} = -2\frac{3}{4}$ or $-\frac{11}{4}$
when $a = \frac{4}{15}$, $b = 1 - \frac{4}{15} = \frac{11}{15}$

Step 4: When $b = -\frac{11}{4}$, $11x = 10(-\frac{11}{4})$

$x = -\frac{5}{2}$
When $b = \frac{11}{15}$, $11x = 10(\frac{11}{15})$
$x = \frac{10}{11}(\frac{11}{15})$; $x = \frac{2}{3}$

Again, one obtains the same solution set $\{-\frac{5}{2}, \frac{2}{3}\}$ as by the factoring method.

The objective of presenting examples 1, 2, 3,, and 4 was to convince the reader that the principles to be used in splitting the Navier-Stokes equations are valid.. In Examples 4, one could have used the quadratic formula directly to solve for x, without finding a and b first. The objective was to show that the introduction of a and b did not change the solution set of the original equation.For the rest of the coverage in this paper, a multiplier is the same as a ratio term
The multiplier method is the same as the ratio method.

Main Step 1
Linearization of the Non-Linear Terms

Step 1: The main principle is to multiply the right side of the equation by the ratio terms
This step is critical to the removal of the non-linearity of the equation.
ρg_x is to be divided by the terms on the left-hand--side of the equation in the ratio
$a:b:c:d:f:h:m:n \quad (a+b+c+d+f+h+m+n=1$

$$-\mu\frac{\partial^2 v_x}{\partial x^2}-\mu\frac{\partial^2 v_x}{\partial y^2}-\mu\frac{\partial^2 v_x}{\partial z^2}+\frac{\partial p}{\partial x}+\underbrace{\rho\frac{\partial V_x}{\partial t}+\overbrace{\rho V_x\frac{\partial V_x}{\partial x}+\rho V_y\frac{\partial V_x}{\partial y}+\rho V_z\frac{\partial V_x}{\partial z}}^{\text{nonlinear terms}}}_{\text{all acceleration terms}}=\rho g_x \quad (1)$$

Apply the principles involved in the ratio method covered in the preliminaries, to the nonlinear terms (the last three terms.)

Then $\rho V_z\frac{\partial V_x}{\partial z}=n\rho g_x$, where n is the ratio term corresponding to $\rho V_z\frac{\partial V_x}{\partial z}$.

$$V_z\frac{\partial v_x}{\partial z}=ng_x \qquad (2)$$

$V_z\frac{dV_x}{dz}=ng_x$ (One drops the partials symbol, since a single independent variable is involved)

$\frac{dz}{dt}\frac{dV_x}{dz}=ng_x \quad (V_z=\frac{dz}{dt}$, by definition)

$$\frac{dv_x}{dt}=ng_x \qquad (3)$$

Therefore, $\boxed{V_z\frac{\partial V_x}{\partial z}=\frac{dV_x}{dt}=ng_x}$ (4)

Step 2: Similarly, Let $\rho V_y\frac{\partial V_x}{\partial y}=m\rho g_x$ (m is the ratio term corresponding to $\rho V_y\frac{\partial V_x}{\partial y}$) (5)

$V_y\frac{dV_x}{dy}=mg_x$ (One drops the partials symbol, since a single independent variable is involved)

$\frac{dy}{dt}\frac{dV_x}{dy}=mg_x \quad (V_y=\frac{dy}{dt})$

$$\frac{dv_x}{dt}=mg_x \qquad (6)$$

Therefore, $\boxed{V_y\frac{dV_x}{dy}=\frac{dV_x}{dt}=mg_x}$ (7)

Step 3: Let $\rho V_x\frac{\partial v_x}{\partial x}=h\rho g_x$ where h is the ratio term corresponding to $\rho V_x\frac{\partial v_x}{\partial x}$.

$$V_x\frac{\partial V_x}{\partial x}=hg_x \qquad (8)$$

$V_x\frac{dV_x}{dx}=hg_x$ (One drops the partials symbol, since a single independent variable is involved)

$\frac{dx}{dt}\frac{dV_x}{dx}=hg_x \quad (V_x=\frac{dx}{dt})$

$\frac{dv_x}{dt}=hg_x$ (9) Therefore, $\boxed{V_x\frac{\partial V_x}{\partial x}=\frac{dV_x}{dt}=hg_x}$ (10)

From equations (4), (7), (10), $V_x \frac{\partial v_x}{\partial x} = V_y \frac{\partial v_x}{\partial y} = V_z \frac{\partial v_x}{\partial z} = \frac{dv_x}{dt}$ and

$$V_x \frac{\partial v_x}{\partial x} + V_y \frac{\partial v_x}{\partial y} + V_z \frac{\partial v_x}{\partial z} = \boxed{3\frac{dv_x}{dt}} \quad (11)$$

Thus, the ratio of the linear term $\frac{\partial V_x}{\partial t}$ to the nonlinear sum $V_x \frac{\partial v_x}{\partial x} + V_y \frac{\partial v_x}{\partial y} + V_z \frac{\partial v_x}{\partial z}$ in equation (1) is 1 to 3. Unquestionably, there is a ratio between the sum of the nonlinear terms and the linear term $\frac{\partial v_x}{\partial t}$. This ratio must be verified experimentally.

Note: One could have obtained equation (C) from equation (A) by redefining the nonlinear terms by **carelessly** disregarding the partial derivatives of the nonlinear terms in equation (1). However, the author did not do that, but logically, the terms became linearized.

Note also that the above linearization is possible only if ρg_x is the subject of the equation, and it will later be learned that a solution to the logically linearized Navier-Stokes equation is obtained only if ρg_x is the subject of the equation.

Step 4: Substitute the right side of equation (11) for the nonlinear terms on the left- side of

$$-\mu\frac{\partial^2 v_x}{\partial x^2} - \mu\frac{\partial^2 v_x}{\partial y^2} - \mu\frac{\partial^2 v_x}{\partial z^2} + \frac{\partial p}{\partial x} + \underbrace{\rho\frac{\partial V_x}{\partial t} + \overbrace{\rho V_x\frac{\partial V_x}{\partial x} + \rho V_y\frac{\partial V_x}{\partial y} + \rho V_z\frac{\partial V_x}{\partial z}}^{\text{nonlinear terms}}}_{\text{all acceleration terms}} = \rho g_x \quad (12)$$

Then one obtains $-\mu\frac{\partial^2 v_x}{\partial x^2} - \mu\frac{\partial^2 v_x}{\partial y^2} - \mu\frac{\partial^2 v_x}{\partial z^2} + \frac{\partial p}{\partial x} + \underbrace{\rho\frac{\partial V_x}{\partial t} + 3\rho\frac{\partial V_x}{\partial x}}_{\text{all acceleration terms}} = \rho g_x$

$$\boxed{-\mu\frac{\partial^2 v_x}{\partial x^2} - \mu\frac{\partial^2 v_x}{\partial y^2} - \mu\frac{\partial^2 v_x}{\partial z^2} + \frac{\partial p}{\partial x} + 4\rho\frac{\partial v_x}{\partial t} = \rho g_x} \quad \text{(simplifying)} \quad (13)$$

Now, instead of solving equation (1), previous page, one will solve the following equation

$$\boxed{-K\frac{\partial^2 V_x}{\partial x^2} - K\frac{\partial^2 V_x}{\partial y^2} - K\frac{\partial^2 V_x}{\partial z^2} + \frac{1}{\rho}\frac{\partial p}{\partial x} + 4\frac{\partial V_x}{\partial t} = g_x} \quad (k = \frac{\mu}{\rho}) \quad (14)$$

Main Step 2

Step 5: In equation (14) divide g_x by the terms on the left side in the ratio $a:b:c:d:f$.

$$\boxed{-K\frac{\partial^2 V_x}{\partial x^2} = ag_x;\quad -K\frac{\partial^2 V_x}{\partial y^2} = bg_x;\quad -K\frac{\partial^2 V_x}{\partial z^2} = cg_x;\quad \frac{1}{\rho}\frac{\partial p}{\partial x} = dg_x;\quad 4\frac{\partial V_x}{\partial t} = fg_x}$$

(a, b, c, d, f are the ratio terms and $a + b + c + d + f = 1$).

As proportions: $\frac{-K\frac{\partial^2 V_x}{\partial x^2}}{a} = \frac{g_x}{1}$; $\frac{-K\frac{\partial^2 V_x}{\partial y^2}}{b} = \frac{g_x}{1}$; $\frac{-K\frac{\partial^2 V_x}{\partial z^2}}{c} = \frac{g_x}{1}$; $\frac{\frac{1}{\rho}\frac{\partial p}{\partial x}}{d} = \frac{g_x}{1}$; $\frac{4\frac{\partial V_x}{\partial t}}{f} = \frac{g_x}{1}$

One can view each of the ratio terms a, b, c, d, f as a fraction (a real number) of $\boxed{g_x}$ contributed by each expression on the left-hand side of equation (14) above.

Main Step 3

Step 6: Solve the differential equations in Step 5.

Solutions of the five sub-equations

$$\boxed{-K\frac{\partial^2 V_x}{\partial x^2} = ag_x}$$

$$k\frac{\partial^2 V_x}{\partial x^2} = -ag$$

$$\frac{\partial^2 V_x}{\partial x^2} = -\frac{a}{k}g$$

$$\frac{\partial V_x}{\partial x} = -\frac{ag}{k}x + C_1$$

$$V_{x1} = -\frac{ag}{2k}x^2 + C_1 x + C_2$$

$$\boxed{-K\frac{\partial^2 V_x}{\partial y^2} = bg_x}$$

$$K\frac{\partial^2 V_x}{\partial y^2} = -bg$$

$$\frac{\partial^2 V_x}{\partial y^2} = -\frac{b}{k}g$$

$$\frac{\partial V_x}{\partial y} = -\frac{bg}{k}y + C_3$$

$$V_{x2} = -\frac{bg}{2k}y^2 + C_3 y + C_4$$

$$\boxed{-K\frac{\partial^2 V_x}{\partial z^2} = cg_x}$$

$$K\frac{\partial^2 V_x}{\partial z^2} = -cg$$

$$\frac{\partial^2 V_x}{\partial z^2} = -\frac{c}{k}g$$

$$\frac{\partial V_x}{\partial z} = -\frac{cg}{k}z + C_5$$

$$V_{x3} = -\frac{cg}{2k}z^2 + C_5 z + C_6$$

$$\boxed{\frac{1}{\rho}\frac{\partial p}{\partial x} = dg_x}$$

$$\frac{1}{\rho}\frac{\partial p}{\partial x} = dg$$

$$\frac{\partial p}{\partial x} = d\rho g$$

$$p = d\rho g x + C_7$$

$$\boxed{4\frac{\partial V_x}{\partial t} = fg_x}$$

$$\frac{\partial V_x}{\partial t} = \frac{f}{4}g_x$$

$$V_{x4} = \frac{fg_x}{4}t$$

Main Step 4

Step 7: One combines the above solutions

$$\begin{aligned}
V_x &= V_{x1} + V_{x2} + V_{x3} + V_{x4} \\
&= -\frac{ag_x}{2k}x^2 + C_1 x + C_2 - \frac{bg_x}{2k}y^2 + C_3 y + C_4 - \frac{cg_x}{2k}z^2 + C_5 z + C_6 + \frac{fg_x}{4}t + C_7 \\
&= -\frac{ag_x}{2k}x^2 + C_1 x - \frac{bg_x}{2k}y^2 + C_3 y - \frac{cg_x}{2k}z^2 + C_5 z + \frac{fg_x}{4}t + C_9 \\
&= -\frac{ag_x}{2k}x^2 - \frac{bg_x}{2k}y^2 - \frac{cg_x}{2k}z^2 + C_1 x + C_3 y + C_5 z + \frac{fg_x}{4}t + C_9 \\
&= -\frac{ag_x}{2k}x^2 - \frac{bg_x}{2k}y^2 - \frac{cg_x}{2k}z^2 + C_1 x + C_3 y + C_5 z + \frac{fg_x}{4}t + C_9 \\
&= -\frac{g_x}{2k}(ax^2 + by^2 + cz^2) + C_1 x + C_3 y + C_5 z + \frac{fg_x}{4}t + C_9 \\
V_x &= -\frac{\rho g_x}{2\mu}(ax^2 + by^2 + cz^2) + C_1 x + C_3 y + C_5 z + \frac{fg_x}{4}t + C_9
\end{aligned}$$

$$P(x) = d\rho g_x x$$

$$\boxed{\begin{aligned}
&V_x = V_{x1} + V_{x2} + V_{x3} + V_{x4} \\
&V_x(x,y,z,t) = -\frac{\rho g_x}{2\mu}(ax^2 + by^2 + cz^2) + C_1 x + C_3 y + C_5 z + \frac{fg_x}{4}t + C_9 \\
&P(x) = d\rho g_x x
\end{aligned}}$$

Main Step 5
Checking in equation (C)

Step 8: Find the derivatives, using

$$V_x = -\frac{\rho g_x}{2\mu}(ax^2 + by^2 + cz^2) + C_1x + C_3y + C_5z + \frac{fg_x}{4}t + C_9$$

$$\boxed{P(x) = d\rho g_x x}$$

$$\frac{\partial V_x}{\partial x} = -\frac{\rho g_x}{2\mu}(2ax) + C_1$$

1. $\boxed{\frac{\partial^2 V_x}{\partial x^2} = -\frac{a\rho g_x}{\mu}}$

$$\frac{\partial V_x}{\partial y} = -\frac{\rho g_x}{\mu}(by) + C_3$$

2. $\boxed{\frac{\partial^2 V_x}{\partial y^2} = -\frac{b\rho g_x}{\mu}}$

$$\frac{\partial V_x}{\partial z} = -\frac{\rho g_x}{\mu}(cz)$$

3. $\boxed{\frac{\partial^2 V_x}{\partial z^2} = -\frac{c\rho g_x}{\mu};}$

4. $\boxed{\frac{\partial p}{\partial x} = d\rho g_x;}$

5. $\boxed{\frac{\partial V_x}{\partial t} = \frac{fg_x}{4}}$

Step 9: Substitute the derivatives from Step 8 in $-\mu(\frac{\partial^2 v_x}{\partial x^2} + \frac{\partial^2 v_x}{\partial y^2} + \frac{\partial^2 v_x}{\partial z^2}) + \frac{\partial p_x}{\partial x} + 4\rho\frac{\partial v_x}{\partial t} = \rho g_x$ to check for identity (to determine if the relation obtained satisfies the original equation).

$$-\mu(\frac{\partial^2 V_x}{\partial x^2} + \frac{\partial^2 V_x}{\partial y^2} + \frac{\partial^2 V_x}{\partial z^2}) + \frac{\partial p}{\partial x} + 4\rho\frac{\partial V_x}{\partial t} = \rho g_x$$

$$-\mu(-\frac{a\rho g_x}{\mu} - \frac{b\rho g_x}{\mu} - \frac{c\rho g_x}{\mu}) + d\rho g_x + 4\rho\frac{f}{4}g_x \stackrel{?}{=} \rho g_x$$

$$a\rho g_x + b\rho g_x + c\rho g_x + d\rho g_x + \rho f g_x \stackrel{?}{=} \rho g_x$$

$$ag_x + bg_x + cg_x + dg_x + fg_x \stackrel{?}{=} g_x$$

$$g_x(a + b + c + d + f) \stackrel{?}{=} g_x$$

$$g_x(1) \stackrel{?}{=} g_x \quad (a + b + c + d + f = 1)$$

$$g_x \stackrel{?}{=} g_x \quad \text{Yes}$$

Scrapwork

$$\boxed{\begin{aligned}\frac{\partial^2 V_x}{\partial x^2} &= -\frac{a\rho g_x}{\mu};\\ \frac{\partial^2 V_x}{\partial y^2} &= -\frac{b\rho g_x}{\mu};\\ \frac{\partial^2 V_x}{\partial z^2} &= -\frac{c\rho g_x}{\mu};\end{aligned}}$$

$\boxed{\frac{\partial p}{\partial x} = d\rho g_x;}$ $\boxed{\frac{\partial V_x}{\partial t} = \frac{fg_x}{4}}$

An identity is obtained and therefore, the solution of equation (C), p.96, is given by

$$\boxed{V_x(x,y,z,t) = -\frac{\rho g_x}{2\mu}(ax^2 + by^2 + cz^2) + C_1x + C_3y + C_5z + \frac{fg_x}{4}t + C_9;\ P(x) = d\rho g_x x}$$

The above solution is unique, because all possible equations were integrated but only a single equation, the equation with the gravity term as the subject of the equation produced the solution.

Solution Summary for v_x, v_y and v_z

For v_x $\quad a+b+c+d+f=1$

$$\mu(\frac{\partial^2 v_x}{\partial x^2}+\frac{\partial^2 v_x}{\partial y^2}+\frac{\partial^2 v_x}{\partial z^2})-\frac{\partial p}{\partial x}+\rho g_x=\rho(\frac{\partial v_x}{\partial t}+V_x\frac{\partial v_x}{\partial x}+V_y\frac{\partial v_x}{\partial y}+V_z\frac{\partial v_x}{\partial z})$$

$$-K\frac{\partial^2 v_x}{\partial x^2}-K\frac{\partial^2 v_x}{\partial y^2}-K\frac{\partial^2 v_x}{\partial z^2}+\frac{1}{\rho}\frac{\partial p}{\partial x}+4\frac{\partial v_x}{\partial t}=g_x$$

$$V_x=V_{x1}+V_{x2}+V_{x3}+V_{x4}$$

$$=-\frac{ag_x}{2k}x^2+C_1x+C_2-\frac{bg_x}{2k}y^2+C_3y+C_4-\frac{cg_x}{2k}z^2+C_5z+C_6+\frac{fg_x}{4}t+C_7+C_8$$

$$=-\frac{ag_x}{2k}x^2+C_1x-\frac{bg_x}{2k}y^2+C_3y-\frac{cg_x}{2k}z^2+C_5z+\frac{fg_x}{4}t+C_9$$

$$=-\frac{ag_x}{2k}x^2-\frac{bg_x}{2k}y^2-\frac{cg_x}{2k}z^2+C_1x+C_3y+C_5z+\frac{fg_x}{4}t+C_9$$

$$V_y(x,y,z,t)=-\frac{\rho g_x}{2\mu}(ax^2+by^2+cz^2)+C_1x+C_3y+C_5z+\frac{fg_x}{4}t+C_9$$

$$P(x)=d\rho g_x x$$

For v_y $\quad h+j+m+n+q=1$

$$\mu(\frac{\partial^2 v_y}{\partial x^2}+\frac{\partial^2 v_y}{\partial y^2}+\frac{\partial^2 v_y}{\partial z^2})-\frac{\partial p}{\partial y}+\rho g_y=\rho(\frac{\partial v_y}{\partial t}+V_x\frac{\partial v_y}{\partial x}+V_y\frac{\partial v_y}{\partial y}+V_z\frac{\partial v_y}{\partial z})$$

$$-K\frac{\partial^2 V_y}{\partial x^2}-K\frac{\partial^2 V_y}{\partial y^2}-K\frac{\partial^2 V_y}{\partial z^2}+\frac{1}{\rho}\frac{\partial p}{\partial y}+4\frac{\partial V_y}{\partial t}=g_y$$

$$V_y=-\frac{hg_y}{2k}x^2+C_1x-\frac{jg_y}{2k}y^2+C_3y-\frac{mg_y}{2k}z^2+C_5z+\frac{ng_y}{4}t$$

$$V_y(x,y,z,t)=-\frac{\rho g_y}{2\mu}(hx^2+jy^2+mz^2)+C_1x+C_3y+C_5z+\frac{qg_y}{4}t+C$$

$$P(y)=n\rho g_y y$$

For v_z $\quad r+s+u+v+w=1$

$$\mu(\frac{\partial^2 v_z}{\partial x^2}+\frac{\partial^2 v_z}{\partial y^2}+\frac{\partial^2 v_z}{\partial z^2})-\frac{\partial p}{\partial z}+\rho g_z=\rho(\frac{\partial v_z}{\partial t}+V_x\frac{\partial v_z}{\partial x}+V_y\frac{\partial v_z}{\partial y}+V_z\frac{\partial v_z}{\partial z})$$

$$-k\frac{\partial^2 v_z}{\partial x^2}-k\frac{\partial^2 v_z}{\partial y^2}-k\frac{\partial^2 v_z}{\partial z^2}+\frac{1}{\rho}\frac{\partial p}{\partial z}+4\frac{\partial v_z}{\partial t}=g_z$$

$$V_z=-\frac{rg_z}{2k}x^2+C_1x-\frac{sg_z}{2k}y^2+C_3y-\frac{ug_z}{2k}z^2+C_5z+\frac{wg_z}{4}t$$

$$V_z(x,y,z,t)=-\frac{\rho g_z}{2\mu}(rx^2+sy^2+uz^2)+C_1x+C_3y+C_5z+\frac{wg_z}{4}t+C$$

$$P(z)=v\rho g_z z$$

Discussion About Linearized N-S Solutions

A solution to equation $-\mu(\frac{\partial^2 v_x}{\partial x^2}+\frac{\partial^2 v_x}{\partial y^2}+\frac{\partial^2 v_x}{\partial z^2})+\frac{\partial p}{\partial x}+4\rho(\frac{\partial v_x}{\partial t})=\rho g_x$ (C) is

$$\boxed{\begin{array}{l} V_x(x,y,z,t)=-\frac{\rho g_x}{2\mu}(ax^2+by^2+cz^2)+C_1x+C_3y+C_5z+\frac{fg_x}{4}t+C_9 \\ P(x)=d\rho g_x x; \quad (a+b+c+d+f=1) \end{array}}.$$

This relation gives an identity when checked in Equation (C) above.

One observes above that the most important insight of the above solution is the indispensability of the gravity term in incompressible fluid flow. Observe that if gravity, g, were zero, the first three terms, the seventh term, and $P(x)$ would all be zero.. This result can be stated emphatically that without gravity forces on earth, there would be no incompressible fluid flow on earth as is known. The above result will be the same when one covers the general case, Option 4.

The above parabolic solution is also encouraging. It reminds one of the parabolic curve obtained when a stone is projected upwards at an acute angle to the horizontal..

More Observations **Comparison of the Navier-Stokes solutions with equations of motion under gravity and liquid pressure of elementary physics**

Motion equations of elementary physics:

(B): $V_f=V_0+gt$; (C): $V_f^2=V_0^2+2gx$; (D): $V=\sqrt{2gx}$; (E): $x=V_0t+\frac{1}{2}gt^2$

Liquid Pressure,

The liquid pressure, P at the bottom of a liquid of depth h units is given by $P=\rho gh$

$x-$ direction linearized Navier–Stokes equation:

$$-\mu(\frac{\partial^2 v_x}{\partial x^2}+\frac{\partial^2 v_x}{\partial y^2}+\frac{\partial^2 v_x}{\partial z^2})+\frac{\partial p_x}{\partial x}+4\rho\frac{\partial v_x}{\partial t}=\rho g_x$$

$x-$ direction Navier–Stokes solution:

$$\boxed{V_x(x,y,z,t)=-\frac{\rho g_x}{2\mu}(ax^2+by^2+cz^2)+C_1x+C_3y+C_5z+\tfrac{1}{4}fg_xt+C_9;\ P(x)=d\rho g_x x}$$

Observe the following above:

1. Observe that the first three terms of the solution are parabolic in x, y, and z; the minus sign showing the inverted parabola when a projectile is fired upwards at an acute angle to the horizontal; Also note the "gt" in $V=gt$ of (B) of the motion equations and the fg_xt in the Navier-Stokes solution.

2. Observe the $P=\rho gh$ of the liquid pressure and the $P(x)=d\rho g_x x$ of the Navier-Stokes solution. Note that d is a ratio term.

There are five main terms in the solution of the linearized Navier-Stokes equation. All of t hese five terms, namely, $-\frac{a\rho g_x}{2\mu}x^2$, $-\frac{b\rho g_x}{2\mu}y^2$, $-\frac{c\rho g_x}{2\mu}z^2$, fg_xt, and $d\rho g_x x$ are similar (except for the constants involved) to the terms in the equations of motion and fluid pressure of elementary physics. This similarity means that the approach used in solving the Navier-Stokes equation is sound. One should also note that to obtain these five terms simultaneously, only the equation with the gravity term as the subject of the equation will yield these six terms. The author suggests that this form of the equation with the gravity term as the subject of the equation be called the standard form of the linearized Navier-Stokes equation, since in this form, one can immediately split-up the equation using ratios, and integrate.

The author also tried the following possible approaches: (D), (E) and (F), but none of the possible solutions completely satisfied the corresponding original equations (D), (E) or (F) .

$$\mu\frac{\partial^2 v_x}{\partial x^2}+\mu\frac{\partial^2 v_x}{\partial y^2}+\mu\frac{\partial^2 v_x}{\partial z^2}+\rho g_x-4\rho\frac{\partial v_x}{\partial t}=\frac{\partial p}{\partial x} \quad \text{(D)} \quad \text{(One uses the subject } \boxed{\frac{\partial p}{\partial x}}$$

$$\frac{K}{4}\frac{\partial^2 V_x}{\partial x^2}+\frac{K}{4}\frac{\partial^2 V_x}{\partial y^2}+\frac{K}{4}\frac{\partial^2 V_x}{\partial z^2}-\frac{1}{4\rho}\frac{\partial p}{\partial x}+\frac{g_x}{4}=\frac{\partial V_x}{\partial t} \quad \text{(E), (One uses the subject } \boxed{\frac{\partial V_x}{\partial t}}$$

$$-\frac{\partial^2 v_x}{\partial y^2}-\frac{\partial^2 v_x}{\partial z^2}-\frac{\rho g_x}{\mu}+\frac{4\rho}{\mu}\frac{\partial v_x}{\partial t}+\frac{1}{\mu}\frac{\partial p}{\partial x}=\frac{\partial^2 v_x}{\partial x^2} \quad \text{(F)} \quad \text{(One uses subject } \boxed{\frac{\partial^2 V_x}{\partial x^2}}$$

Integration Results Summary

Case 1: $-\mu(\frac{\partial^2 V_x}{\partial x^2}+\frac{\partial^2 V_x}{\partial y^2}+\frac{\partial^2 V_x}{\partial z^2})+\frac{\partial p}{\partial x}+4\rho(\frac{\partial V_x}{\partial t})=\rho g_x$ (C)

$$\boxed{\begin{aligned}&V_x(x,y,z,t)=-\frac{\rho g_x}{2\mu}(ax^2+by^2+cz^2)+C_1x+C_3y+C_5z+\frac{fg_x}{4}t+C_9\\&P(x)=d\rho g_x x; \quad (a+b+c+d+f=1)\end{aligned}}$$

<----Solution

Case 2: $\mu\frac{\partial^2 v_x}{\partial x^2}+\mu\frac{\partial^2 v_x}{\partial y^2}+\mu\frac{\partial^2 v_x}{\partial z^2}+\rho g_x-4\rho\frac{\partial v_x}{\partial t}=\frac{\partial p}{\partial x}$ (D). (One uses the subject $\boxed{\frac{\partial p}{\partial x}}$

$$\boxed{\begin{aligned}&V_x(x,y,z.t)=\frac{\lambda_x}{2\mu}(ax^2+by^2+cz^2)+C_1x+\lambda_p x+C_3y+C_5z-\frac{f\lambda}{4\rho}t+C\\&P(x)=\frac{1}{d}\rho g_x x\end{aligned}}$$

Case 3: $\frac{K}{4}\frac{\partial^2 V_x}{\partial x^2}+\frac{K}{4}\frac{\partial^2 V_x}{\partial y^2}+\frac{K}{4}\frac{\partial^2 V_x}{\partial z^2}-\frac{1}{4\rho}\frac{\partial p}{\partial x}+\frac{g_x}{4}=\frac{\partial V_x}{\partial t}$ (E). (One uses the subject $\boxed{\frac{\partial V_x}{\partial t}}$

$$\boxed{\begin{aligned}&V_x(x,y,z,t)=(C_1\cos\lambda_x x+C_2\sin\lambda_x x)e^{-(\lambda^2/\beta)t}+(C_3\cos\lambda_y y+C_4\sin\lambda_y y)e^{-(\lambda_y{}^2/\omega)t}\\&+(C_5\cos\lambda_z z+C_6\sin\lambda_z z)e^{-(\lambda_z{}^2/\varepsilon)t}+\frac{g}{4f}t+\lambda x+C_8\\&P(x)=\lambda x=d\rho g_x x\end{aligned}}$$

Case 4: $-\frac{\partial^2 v_x}{\partial y^2}-\frac{\partial^2 v_x}{\partial z^2}-\frac{\rho g_x}{\mu}+\frac{4\rho}{\mu}\frac{\partial v_x}{\partial t}+\frac{1}{\mu}\frac{\partial p}{\partial x}=\frac{\partial^2 v_x}{\partial x^2}$ (F). (One uses the subject $\boxed{\frac{\partial^2 V_x}{\partial x^2}}$

$$\boxed{\begin{aligned}&V_x(x,y,z,t)=(A\cos\lambda y+B\sin\lambda y)\left(Ce^{(\frac{\lambda\sqrt{a}}{a})x}+De^{-(\frac{\lambda\sqrt{a}}{a})x}\right)\\&+(E\cos\lambda z+F\sin\lambda z)\left(He^{(\frac{\lambda\sqrt{b}}{b})x}+Le^{(-\frac{\lambda\sqrt{b}}{b})x}\right)-\frac{\rho g_x x^2}{2c\mu}+Ax+B+(A_1\cos\lambda x+B_1\sin\lambda x)e^{-(\lambda^2/\alpha)t}\\&+\frac{\lambda}{2\mu f}x^2+C_2x+C_3); \qquad P(x)=d\rho g_x x\end{aligned}}$$

Note: Relations for equations with subjects g_x and $\frac{\partial p}{\partial x}$ are almost identical.

By comparing possible solutions for equations (C) and (D), $\lambda_x=-\rho g_x$ in relation for (D).

$$\boxed{V_x(x,y,z,t)=\frac{\lambda_x}{2\mu}(ax^2+by^2+cz^2)+C_1x+\lambda_p x+C_3y+C_5z-\frac{f\lambda}{4\rho}t+C; \quad P(x)=\frac{1}{d}\rho g_x x}$$

Comparative analysis of the possible solutions when checked in each corresponding equation

Equation	Equation Subject	Number of terms of possible solutions **not** satisfying original equation
Case 1: $-\mu(\frac{\partial^2 V_x}{\partial x^2}+\frac{\partial^2 V_x}{\partial y^2}+\frac{\partial^2 V_x}{\partial z^2})+\frac{\partial p}{\partial x}+4\rho(\frac{\partial V_x}{\partial t})=\rho g_x$	ρg_x	None Case 1 yields the solution
Case 2: $\mu\frac{\partial^2 V_x}{\partial x^2}+\mu\frac{\partial^2 V_x}{\partial y^2}+\mu\frac{\partial^2 V_x}{\partial z^2}+\rho g_x-4\rho\frac{\partial V_x}{\partial t}=\frac{\partial p}{\partial x}$	$\frac{\partial p}{\partial x}$	One term
Case 3: $\frac{K}{4}\frac{\partial^2 V_x}{\partial x^2}+\frac{K}{4}\frac{\partial^2 V_x}{\partial y^2}+\frac{K}{4}\frac{\partial^2 V_x}{\partial z^2}-\frac{1}{4\rho}\frac{\partial p}{\partial x}+\frac{g_x}{4}=\frac{\partial V_x}{\partial t}$	$\frac{\partial V_x}{\partial t}$	At least 2 terms
Case 4: $-\frac{\partial^2 V_x}{\partial y^2}-\frac{\partial^2 V_x}{\partial z^2}-\frac{\rho g_x}{\mu}+\frac{4\rho}{\mu}\frac{\partial V_x}{\partial t}+\frac{1}{\mu}\frac{\partial p}{\partial x}=\frac{\partial^2 V_x}{\partial x^2}$	$\frac{\partial^2 V_x}{\partial x^2}$	At least 2 terms
Case 5: $-\frac{\partial^2 V_x}{\partial x^2}-\frac{\partial^2 V_x}{\partial z^2}-\frac{\rho g_x}{\mu}+\frac{4\rho}{\mu}\frac{\partial V_x}{\partial t}+\frac{1}{\mu}\frac{\partial p}{\partial x}=\frac{\partial^2 V_x}{\partial y^2}$	$\frac{\partial^2 V_x}{\partial y^2}$	At least 2 terms
Case 6: $-\frac{\partial^2 V_x}{\partial y^2}-\frac{\partial^2 V_x}{\partial x^2}-\frac{\rho g_x}{\mu}+\frac{4\rho}{\mu}\frac{\partial V_x}{\partial t}+\frac{1}{\mu}\frac{\partial p}{\partial x}=\frac{\partial^2 V_x}{\partial z^2}$	$\frac{\partial^2 V_x}{\partial z^2}$	At least 2 terms

Note above that only Case 1 is the solution, and this may imply that the solution to the Navier-Stokes equation is unique. Out of six possible subjects, only one subject produced a solution. The above results show that a relation obtained by the integration of a partial differential equation must be checked in the corresponding equation for identity before the relation becomes a solution, Cases 2, 3, 4, 5 and 6, are not solutions but integration relations. For example, it would be incorrect to say that the equation in Case 3 has a periodic solution; but it would be correct to say that the equation in Case 3 has a periodic relation, since the relation obtained by integration does not satisfy its corresponding equation. It would be correct to say that the equation in Case 1 has a parabolic solution or a parabolic relation. Below are detailed explanation of results of the identity checking process.

Outcome 1: With g_x included and with g_x as the subject of the equation. The solution is straightforward and the possible solution checks well in the original equation (C). Also, if g_x or ρg_x is not the subject of the equation, the linearization of the nonlinear terms could not be justified.

Outcome 2: With g_x included but with $\frac{\partial V_x}{\partial t}$ as the subject of the equation.

There are two problems when checking . **1.** For $\frac{\partial V_x}{\partial t}=-\frac{1}{4\rho}\frac{\partial p}{\partial x}\rightarrow-\frac{\lambda t}{4\rho d}$; **2.** $\frac{g_x}{4}=\frac{\partial V_x}{\partial t}\rightarrow\frac{g_x t}{4f}$

With d and f in the denominators, the multipliers sum $a+b+c+d+f=1$ is false.

Outcome 3 : With g_x excluded, and $\frac{\partial V_x}{\partial t}$ as the subject of the equation, there is one problem:

$-\frac{1}{4\rho}\frac{\partial p}{\partial x}=\frac{\partial V_x}{\partial t}\rightarrow-\frac{\lambda t}{4\rho d}$.With d in the denominator $a+b+c+d+f=1$ is false

Outcome 4 : With g_x included, and $\frac{\partial^2 V_x}{\partial x^2}$ as the subject of the equation, there are at least, two problems in the checking with the multipliers c and f in the denominators. Checking for $a+b+c+d+f=1$ is impossible.

Outcomes 5 and 6 are similar to Outcome 4.

Characteristic curves of the integration results

Equations	Equation Subject	Curve characteristics
Case 1: $-\mu(\frac{\partial^2 V_x}{\partial x^2}+\frac{\partial^2 V_x}{\partial y^2}+\frac{\partial^2 V_x}{\partial z^2})+\frac{\partial p}{\partial x}+4\rho(\frac{\partial V_x}{\partial t})=\rho g_x$	ρg_x	Parabolic and Inverted
Case 2: $\mu\frac{\partial^2 v_x}{\partial x^2}+\mu\frac{\partial^2 v_x}{\partial y^2}+\mu\frac{\partial^2 v_x}{\partial z^2}+\rho g_x-4\rho\frac{\partial v_x}{\partial t}=\frac{\partial p}{\partial x}$	$\frac{\partial p}{\partial x}$	Parabolic
Case 3: $\frac{K}{4}\frac{\partial^2 V_x}{\partial x^2}+\frac{K}{4}\frac{\partial^2 V_x}{\partial y^2}+\frac{K}{4}\frac{\partial^2 V_x}{\partial z^2}-\frac{1}{4\rho}\frac{\partial p}{\partial x}+\frac{g_x}{4}=\frac{\partial V_x}{\partial t}$	$\frac{\partial V_x}{\partial t}$	Periodic and decreasingly exponential
Case 4: $-\frac{\partial^2 v_x}{\partial y^2}-\frac{\partial^2 v_x}{\partial z^2}-\frac{\rho g_x}{\mu}+\frac{4\rho}{\mu}\frac{\partial v_x}{\partial t}+\frac{1}{\mu}\frac{\partial p}{\partial x}=\frac{\partial^2 v_x}{\partial x^2}$	$\frac{\partial^2 V_x}{\partial x^2}$	Periodic, parabolic, and decreasingly exponential
Case 5: $-\frac{\partial^2 v_x}{\partial x^2}-\frac{\partial^2 v_x}{\partial z^2}-\frac{\rho g_x}{\mu}+\frac{4\rho}{\mu}\frac{\partial v_x}{\partial t}+\frac{1}{\mu}\frac{\partial p}{\partial x}=\frac{\partial^2 v_x}{\partial y^2}$	$\frac{\partial^2 v_x}{\partial y^2}$	Periodic, parabolic, and decreasingly exponential
Case 6: $-\frac{\partial^2 V_x}{\partial y^2}-\frac{\partial^2 V_x}{\partial x^2}-\frac{\rho g_x}{\mu}+\frac{4\rho}{\mu}\frac{\partial V_x}{\partial t}+\frac{1}{\mu}\frac{\partial p}{\partial x}=\frac{\partial^2 V_x}{\partial z^2}$	$\frac{\partial^2 V_x}{\partial z^2}$	Periodic, parabolic, and decreasingly exponential

The following are possible interpretations of the roles of the terms based on the types of curves produced when using the terms as subjects of the equations.

1. g_x and $\frac{\partial p}{\partial x}$ are involved in the parabolic motion; g_x is responsible for the forward motion.

2. $\frac{\partial V_x}{\partial t}$ is involved in the periodic and decreasingly exponential behavior.

3. $\frac{\partial^2 V_x}{\partial x^2}$, $\frac{\partial^2 V_x}{\partial y^2}$ and $\frac{\partial^2 V_x}{\partial z^2}$ are involved in the parabolic, periodic and decreasingly exponential motion. As μ increases, the periodicity increases

Definitions and Classification of Equations

$$\boxed{-K\frac{\partial^2 V_x}{\partial x^2} - K\frac{\partial^2 V_x}{\partial y^2} - K\frac{\partial^2 V_x}{\partial z^2} + \frac{1}{\rho}\frac{\partial p}{\partial x} + 4\frac{\partial V_x}{\partial t} = g_x} \qquad (k = \frac{\mu}{\rho})$$

One may classify the equations involved in Option 1 according to the following:

Driver Equation: A differential equation whose integration relation satisfies its corresponding equation.

Supporter equation: A differential equation which contains the same terms as the driver equation but whose integration relation does not satisfy its corresponding equation but provides useful information about the driver equation.

Note that the driver equation and a supporter equation differ only in the subject of the equation.

Equation	Equation Subject	Type of equation	# of terms of relation not satisfying original equation
Case 1: $-\mu(\frac{\partial^2 V_x}{\partial x^2} + \frac{\partial^2 V_x}{\partial y^2} + \frac{\partial^2 V_x}{\partial z^2}) + \frac{\partial p}{\partial x} + 4\rho(\frac{\partial V_x}{\partial t}) = \rho g_x$	ρg_x	Driver Equation	None
Case 2: $\mu\frac{\partial^2 v_x}{\partial x^2} + \mu\frac{\partial^2 v_x}{\partial y^2} + \mu\frac{\partial^2 v_x}{\partial z^2} + \rho g_x - 4\rho\frac{\partial v_x}{\partial t} = \frac{\partial p}{\partial x}$	$\frac{\partial p}{\partial x}$	Supporter equation	One term
Case 3: $\frac{K}{4}\frac{\partial^2 V_x}{\partial x^2} + \frac{K}{4}\frac{\partial^2 V_x}{\partial y^2} + \frac{K}{4}\frac{\partial^2 V_x}{\partial z^2} - \frac{1}{4\rho}\frac{\partial p}{\partial x} + \frac{g_x}{4} = \frac{\partial V_x}{\partial t}$	$\frac{\partial V_x}{\partial t}$	Supporter equation	At least 2 terms
Case 4: $-\frac{\partial^2 V_x}{\partial y^2} - \frac{\partial^2 v_x}{\partial z^2} - \frac{\rho g_x}{\mu} + \frac{4\rho}{\mu}\frac{\partial v_x}{\partial t} + \frac{1}{\mu}\frac{\partial p}{\partial x} = \frac{\partial^2 v_x}{\partial x^2}$	$\frac{\partial^2 V_x}{\partial x^2}$	Supporter equation	At least 2 terms
Case 5: $-\frac{\partial^2 V_x}{\partial x^2} - \frac{\partial^2 v_x}{\partial z^2} - \frac{\rho g_x}{\mu} + \frac{4\rho}{\mu}\frac{\partial v_x}{\partial t} + \frac{1}{\mu}\frac{\partial p}{\partial x} = \frac{\partial^2 v_x}{\partial y^2}$	$\frac{\partial^2 v_x}{\partial y^2}$	Supporter equation	At least 2 terms
Case 6: $-\frac{\partial^2 V_x}{\partial y^2} - \frac{\partial^2 V_x}{\partial x^2} - \frac{\rho g_x}{\mu} + \frac{4\rho}{\mu}\frac{\partial V_x}{\partial t} + \frac{1}{\mu}\frac{\partial p}{\partial x} = \frac{\partial^2 V_x}{\partial z^2}$	$\frac{\partial^2 V_x}{\partial z^2}$	Supporter equation	At least 2 terms

The uniqueness of the above solution will guide one to save time and not try to solve some forms of Euler or Navier-Stokes equation which do not produce solutions. That is, one will solve only the equations with the gravity term as the subject. This uniqueness will also guide one to solve the magnetohydrodynamic equations.

Applications of the splitting technique in science, engineering, business fields

The approach used in solving the equations allows for how the terms interact with each other. The author has not seen this technique anywhere, but the results are revealing and promising.

Fluid flow design considerations:

1. Maximize the role of g_x forces, followed by; **2.** $\frac{\partial p}{\partial x}$ forces; then 3. $\frac{\partial V_x}{\partial t}$

Make g_x happy by always providing a workable slope.

For long distance flow design such as for water pipelines, water channels, oil pipelines. whenever possible, the design should facilitate and maximize the role of gravity forces, and if design is impossible to facilitate the role of gravity forces, design for $\frac{\partial p}{\partial x}$ to take over flow.

The performance of $\frac{\partial^2 V_x}{\partial x^2}$ should be studied further, since its role is the most complicated: periodic, parabolic, and decreasingly exponential.

Tornado Effect Relief

Perhaps, machines can be designed and built to chase and neutralize or minimize tornadoes during touch-downs. The energy in the tornado at touch-down can be harnessed for useful purposes.

Business and economics applications.

1. Figuratively, if g_x is the president of a company, it will have good working relationships with all the members of the board of directors, according to the solution of the Navier-Stokes equation. If g_x is present at a meeting g_x must preside over the meeting for the best outcome.

2. If g_x is absent from a meeting, let $\frac{\partial p}{\partial x}$ preside over the meeting, and everything will workout well. However, if g_x is present, g_x must preside over the meeting.

To apply the results of the solutions of the Navier-Stokes equations in other areas or fields, the properties, characteristics and functions of g_x, $\frac{\partial p}{\partial x}$, $\frac{\partial v_x}{\partial t}$ must be studied to determine analogous terms in those areas of possible applications. Other areas of applications include investments choice decisions, financial decisions, personnel management and family relationships.

Option 2
Solutions of 4-D Linearized Navier-Stokes Equations

One advantage of the pairing approach is that the above solution can easily be extended to any number of dimensions.

If one adds $\mu\frac{\partial^2 V_x}{\partial s^2}$ and $\rho V_s\frac{\partial V_x}{\partial s}$ to the 3-D x–direction equation, one obtains the 4-D Navier--Stokes equation $-\mu(\frac{\partial^2 v_x}{\partial x^2}+\frac{\partial^2 v_x}{\partial y^2}+\frac{\partial^2 v_x}{\partial z^2}+\frac{\partial^2 v_x}{\partial s^2})+\frac{\partial p}{\partial x}+4\rho(\frac{\partial v_x}{\partial t})+\rho V_s\frac{\partial V_x}{\partial s}=\rho g_x$

After linearization, $-\mu(\frac{\partial^2 v_x}{\partial x^2}+\frac{\partial^2 v_x}{\partial y^2}+\frac{\partial^2 v_x}{\partial z^2}+\frac{\partial^2 v_x}{\partial s^2})+\frac{\partial p}{\partial x}+5\rho(\frac{\partial v_x}{\partial t})=\rho g_x$ and its solution is

$$\boxed{\begin{array}{l} V_x(x,y,z,s,t)=-\frac{\rho g_x}{2\mu}(ax^2+by^2+cz^2+es^2)+C_1x+C_3y+C_5z+C_7s+\frac{fg_x}{5}t+C_9 \\ P(x)=d\rho g_x x \qquad (a+b+c+d+e+f=1) \end{array}}$$

For n–dimensions one can repeat the above as many times as one wishes.

Option 3
Solutions of the Euler Equations of Fluid flow

In the Navier-Stokes equation, if $\mu = 0$, one obtains the Euler equation. From

$$\mu(\frac{\partial^2 v_x}{\partial x^2}+\frac{\partial^2 v_x}{\partial y^2}+\frac{\partial^2 v_x}{\partial z^2})-\frac{\partial p}{\partial x}+\rho g_x=\rho(\frac{\partial v_x}{\partial t}+V_x\frac{\partial v_x}{\partial x}+V_y\frac{\partial v_x}{\partial y}+V_z\frac{\partial v_x}{\partial z})$$, one obtains

Euler equation : $(\mu = 0)$ $-\frac{\partial p}{\partial x}+\rho g_x=\rho(\frac{\partial v_x}{\partial t}+V_x\frac{\partial v_x}{\partial x}+V_y\frac{\partial v_x}{\partial y}+V_z\frac{\partial v_x}{\partial z})$ or

$$\boxed{\rho(\frac{\partial v_x}{\partial t}+V_x\frac{\partial V_x}{\partial x}+V_y\frac{\partial v_x}{\partial y}+V_z\frac{\partial v_x}{\partial z})+\frac{\partial p_x}{\partial x}=\rho g_x}$$ <---driver equation.

Euler equation $(\mu = 0)$: $\frac{\partial v_x}{\partial t}+V_x\frac{\partial v_x}{\partial x}+V_y\frac{\partial v_x}{\partial y}+V_z\frac{\partial v_x}{\partial z}+\frac{1}{\rho}\frac{\partial p}{\partial x}=g_x$ <---driver equation

Split the equation using the ratio terms f, h, n, q, d,, and solve. $(f+h+n+q+d=1)$

1. $\frac{\partial V_x}{\partial t}=fg_x$ $V_{x4}=fg_x t$ $V_{x4}=fg_x t$	**2.** $V_x\frac{\partial V_x}{\partial x}=hg_x$ $V_x\frac{dV_x}{dx}=hg_x$ $V_x dV_x=hg_x dx$ $\frac{V_x^2}{2}=hg_x x$ or $V_x^2=2hg_x x$ $V_x=\pm\sqrt{2hg_x x}$	**3.** $V_y\frac{\partial V_x}{\partial y}=ng_x$ $V_y\frac{dV_x}{dy}=ng_x$ $V_y dV_x=ng_x \mathrm{dy}$ $V_y V_x=ng_x \mathrm{y}+\psi_y(V_y)$ $V_{x6}=\frac{ng_x \mathrm{y}}{V_y}+\frac{\psi_y(V_y)}{V_y}$ $V_y\neq 0$	**4.** $V_z\frac{\partial V_x}{\partial z}=qg_x$ $V_z\frac{dV_x}{dz}=qg_x$ $V_z dV_x=qg_x \mathrm{dz}$; $V_z V_x=qg_x \mathrm{z}+\psi_z(V_z)$ $V_{x7}=\frac{qg_x \mathrm{z}}{V_z}+\frac{\psi_z(V_z)}{V_z}$ $V_z\neq 0$	**5.** $\frac{1}{\rho}\frac{\partial p}{\partial x}=dg_x$ $\frac{1}{\rho}\frac{\partial p}{\partial x}=dg_x$ $\frac{\partial p}{\partial x}=d\rho g_x$ $p=d\rho g_x x+C_7$

$$\boxed{\begin{array}{l} V_x(x,y,z,t)=fg_x t\pm\sqrt{2hg_x x}+\frac{ng_x \mathrm{y}}{V_y}+\frac{qg_x \mathrm{z}}{V_z}+\frac{\psi_y(V_y)}{V_y}+\frac{\psi_z(V_z)}{V_z}+C \\ P(x)=d\rho g_x x \qquad (f+h+n+q+d=1)\ V_y\neq 0,\ V_z\neq 0 \end{array}}$$

Find the test derivatives to check in the original equation.

1. $\frac{\partial V_x}{\partial t}=fg_x$	**2.** $V_x^2=2hg_x x$; $2V_x\frac{\partial V_x}{\partial x}=2hg_x$; $\frac{\partial V_x}{\partial x}=\frac{hg_x}{V_x}$, $V_x\neq 0$	3. $\frac{\partial V_x}{\partial y}=\frac{ng_x}{V_y}$ $V_y\neq 0$	4. $\frac{\partial V_x}{\partial z}=\frac{qg_x}{V_z}$ $V_z\neq 0$	5. $\frac{\partial p}{\partial x}=d\rho g_x$

$\frac{\partial v_x}{\partial t}+V_x\frac{\partial v_x}{\partial x}+V_y\frac{\partial v_x}{\partial y}+V_z\frac{\partial v_x}{\partial z}+\frac{1}{\rho}\frac{\partial p}{\partial x}=g_x$ (Above, $\psi_y(V_y)$ and $\psi_z(V_z)$ are arbitrary functions)

$$fg_x+V_x\frac{hg_x}{V_x}+V_y\frac{ng_x}{V_y}+V_z\frac{qg_x}{V_z}+\frac{1}{\rho}d\rho g_x\overset{?}{=}g_x$$

$$fg_x+hg_x+ng_x+qg_x+dg_x\overset{?}{=}g_x$$

$$g_x(f+h+n+q+d)\overset{?}{=}g_x$$

$$g_x(1)\overset{?}{=}g_x \qquad (f+h+n+q+d=1)$$

$$g_x\overset{?}{=}g_x$$ Yes

The relation obtained satisfies the Euler equation. Therefore the solution to the Euler equation

$$:\frac{\partial p}{\partial x}+\rho\frac{\partial V_x}{\partial t}+\rho V_x\frac{\partial V_x}{\partial x}+\rho V_y\frac{\partial V_x}{\partial y}+\rho V_z\frac{\partial V_x}{\partial z})=\rho g_x \text{ is}$$

$$\boxed{\begin{array}{l} V_x(x,y,z,t)=fg_xt\pm\sqrt{2hg_xx}+\dfrac{ng_xy}{V_y}+\dfrac{qg_xz}{V_z}+\underbrace{\dfrac{\psi_y(V_y)}{V_y}+\dfrac{\psi_z(V_z)}{V_z}}_{\text{arbitrary functions}};\ P(x)=d\rho g_xx \\ V_y\neq 0,\ V_z\neq 0;\ \ (d+f+h+n+q=1)\end{array}} \quad x\text{-direction}$$

Similarly, the equations and solutions for the other two directions are respectively

$$\text{For } V_y,\ \frac{\partial p}{\partial y}+\rho\frac{\partial V_y}{\partial t}+\rho V_x\frac{\partial V_y}{\partial x}+\rho V_y\frac{\partial V_y}{\partial y}+\rho V_z\frac{\partial V_y}{\partial z}=\rho g_y$$

$$\boxed{\begin{array}{l} V_y(x,y,z,t)=\lambda_5g_yt\pm\sqrt{2\lambda_7g_yy}+\dfrac{\lambda_6g_yx}{V_x}+\dfrac{\lambda_8g_yz}{V_z}+\dfrac{\psi_x(V_x)}{V_x}+\dfrac{\psi_z(V_z)}{V_z};\ P(y)=\lambda_4\rho g_yy \\ V_x\neq 0,\ V_z\neq 0;\ \ (\lambda_4+\lambda_5+\lambda_6+\lambda_7+\lambda_8=1)\end{array}} \quad y\text{-direction}$$

$$\text{For } V_z:\ \frac{\partial p}{\partial z}+\rho\frac{\partial V_z}{\partial t}+\rho V_x\frac{\partial V_z}{\partial x}+\rho V_y\frac{\partial V_z}{\partial y}+\rho V_z\frac{\partial V_z}{\partial z}=\rho g_z$$

$$\boxed{\begin{array}{l} V_z(x,y,z,t)=\beta_5g_zt\pm\sqrt{2\beta_8g_zz}+\dfrac{\beta_6g_zx}{V_x}+\dfrac{\beta_7g_zy}{V_y}+\dfrac{\psi_x(V_x)}{V_x}+\dfrac{\psi_y(V_y)}{V_y};\ P(z)=\beta_4\rho g_zz \\ V_x\neq 0,\ V_y\neq 0;\ \ (\beta_4+\beta_5+\beta_6+\beta_7+\beta_8=1)\end{array}} \quad z\text{-direction}$$

One will next solve the above system of solutions for V_x, V_y, V_z in order to express $\frac{ng_xy}{V_y}$ and $\frac{q_eg_xz}{V_z}$ in terms of x, y, z, and t.

Solving for V_x, V_y, V_z $\frac{ng_xy}{V_y}$, and $\frac{q_eg_xz}{V_z}$

$$\begin{cases} V_x=fg_xt\pm\sqrt{2hg_xx}+\dfrac{ng_xy}{V_y}+\dfrac{qg_xz}{V_z}+\dfrac{\psi_y(V_y)}{V_y}+\dfrac{\psi_z(V_z)}{V_z}. & \text{(A)} \\ V_y=\lambda_5g_yt\pm\sqrt{2\lambda_7g_yy}+\dfrac{\lambda_6g_yx}{V_x}+\dfrac{\lambda_8g_yz}{V_z}+\dfrac{\psi_x(V_x)}{V_x}+\dfrac{\psi_z(V_z)}{V_z}. & \text{(B)} \\ V_z=\beta_5g_zt\pm\sqrt{2\beta_8g_zz}+\dfrac{\beta_6g_zx}{V_x}+\dfrac{\beta_7g_zy}{V_y}+\dfrac{\psi_x(V_x)}{V_x}+\dfrac{\psi_y(V_y)}{V_y}. & \text{(C)}\end{cases}$$

Let $V_x=x$, $V_y=y$ and $V_z=z$. (x, y and z are being used for simplicity. They will be changed back to V_x, V_y, and V_z later, and they do not represent the variables x, y and z in the system of solutions)

Step 1 From the above system of solutions, let

$A = (fgt + \sqrt{2hg_x x})$; $D = (qg_x z)$; $E = (ng_x y)$

$B = (\lambda_5 g_y t + \sqrt{2\lambda_7 g_y y})$; $F = (\lambda_6 g_y x)$; $G = (\lambda_8 g_y z)$

$C = (\beta_5 g_z t + \sqrt{2\beta_8 g_z z})$; $J = (\beta_6 g_z x)$; $L = (\beta_7 g_z y)$

Step 2: Then the solutions to the Euler system of equations become (ignoring the arbitrary functions)

$$\left.\begin{aligned} x &= A + \frac{D}{z} + \frac{E}{y} \\ y &= B + \frac{F}{x} + \frac{G}{z} \\ z &= C + \frac{J}{x} + \frac{L}{y} \end{aligned}\right\} \text{M}$$

Step 3

$$\left.\begin{aligned} xyz &= Ayz + Dy + Ez && (1) \\ xyz &= Bxz + Fz + Gx && (2) \\ xyz &= Cxy + Jy + Lx && (3)) \end{aligned}\right\} \text{N}$$

Step 4

$$\left.\begin{aligned} 0 &= Ayz + Dy + Ez - Bxz - Fz - Gx && (4) \\ 0 &= Ayz + Dy + Ez - Cxy - Jy - Lx && (5) \\ 0 &= Bxz + Fz + Gx - Cxy - Jy + -Lx && (6) \end{aligned}\right\} \text{P}$$

Maples software was used to solve system P to obtain

Step 5

$$x = \frac{L(FCD - FCJ - JLA + JCE)}{C(-BLD + BLJ + GLA - GCE)}$$

$$V_x = \frac{L(FCD - FCJ - JLA + JCE)}{C(-BLD + BLJ + GLA - GCE)} \quad \text{(back to } V_x\text{)}$$

$$y = -\frac{L}{C};$$

$$V_y = -\frac{L}{C} \quad \text{(changing back to } V_y \text{ as agreed to)}$$

$$z = -\frac{L(D-J)}{LA - CE};$$

$$V_z = -\frac{L(D-J)}{LA - CE} \quad \text{(changing back to } V_z \text{ as agrred to)}$$

Note:
None of the popular academic programs could solve the system in M.
Maples solved system P (step 4 above) for x, y, and z in terms of A, B, C, D. E. F, G. J. and L.
Note also that x, y and z are not the same as the x, y and z in the system of equations.. They were used for convenience and simplicity .

Step 5: Apply and substitute from in steps 6-8 below

$A = (fgt \pm \sqrt{2hg_x x})$; $B = (\lambda_5 g_y t \pm \sqrt{2\lambda_7 g_y y})$; $C = (\beta_5 g_z t \pm \sqrt{2\beta_8 g_z z})$; $D = (qg_x z)$;
$E = (ng_x y)$; $F = (\lambda_6 g_y x)$; $G = (\lambda_8 g_y z) J = (\beta_6 g_z x)$; $L = (\beta_7 g_z y)$

Step 6

$$V_y = -\frac{L}{C} = -\frac{(\beta_7 g_z y)}{(\beta_5 g_z t \pm \sqrt{2\beta_8 g_z z})}$$

$$\frac{ng_x y}{V_y} = ng_x y \div -\left(\frac{(\beta_7 g_z y)}{(\beta_5 g_z t \pm \sqrt{2\beta_8 g_z z})}\right)$$

$$\frac{ng_x y}{V_y} = -\frac{(ng_x y)[\beta_5 g_z t \pm \sqrt{2\beta_8 g_z z}]}{\beta_7 g_z y}; \quad y \neq 0$$

Step 7

$$\frac{ng_x y}{V_y} = -\frac{[\beta_5 g_z t](ng_x y) \pm [\sqrt{2\beta_8 g_z z}](ng_x y)}{\beta_7 g_z y}$$

$$\frac{ng_x y}{V_y} = -\frac{n\beta_5 g_z t}{\beta_7} \pm \frac{(\sqrt{2\beta_8 g_z z})(ng_x)}{\beta_7 g_z}$$

$$\boxed{\frac{ng_x y}{V_y} = -\frac{n\beta_5 g_z t}{\beta_7} \pm \frac{(\sqrt{2\beta_8 g_z z}(ng_x)}{\beta_7 g_z}}$$

Step 8:

$$V_z = -\frac{L(D-J)}{LA-CE}$$

$$V_z = -\frac{(\beta_7 g_z y)[qg_x z - \beta_6 g_z x]}{(\beta_7 g_z y)(fg_x t \pm \sqrt{2hg_x x}) - (\beta_5 g_z t \pm \sqrt{2\beta_8 g_z z})(ng_x y)}$$

$$\frac{qg_x z}{V_z} = (qg_x z) \div -\left(\frac{(\beta_7 g_z y)[qg_x z - \beta_6 g_z x]}{(\beta_7 g_z y)(fg_x t \pm \sqrt{2hg_x x}) - (\beta_5 g_z t \pm \sqrt{2\beta_8 g_z z})(ng_x y)}\right)$$

$$= -\frac{(qg_x z) \bullet (\beta_7 g_z y)(fg_x t \pm \sqrt{2hg_x x}) - (\beta_5 g_z t \pm \sqrt{2\beta_8 g_z z})(ng_x y)}{(\beta_7 g_z y)[qg_x z - \beta_6 g_z x]}$$

$$= -\frac{(qg_x z) \bullet (\beta_7 g_z y fg_x t \pm \sqrt{2hg_x x}\beta_7 g_z y - \beta_5 g_z t ng_x y \pm \sqrt{2\beta_8 g_z z}\ ng_x y}{(\beta_7 g_z y)[qg_x z - \beta_6 g_z x]}$$

$$= -\frac{qg_x g_x z\beta_7 fg_z yt \pm \sqrt{2hg_x x}\beta_7 g_x g_z qyz - \beta_5 g_x g_z nqtyz \pm \sqrt{2\beta_8 g_z z}\ g_x g_x nqyz}{(\beta_7 g_z y)[qg_x z - \beta_6 g_z x]}$$

$$\frac{qg_x z}{V_z} = -\frac{qg_x g_x z\beta_7 fg_z t \pm \sqrt{2hg_x x}\beta_7 g_x g_z qz - \beta_5 g_x g_z nqtz \pm \sqrt{2\beta_8 g_z z}\ g_x g_x nqz}{(\beta_7 g_z)[qg_x z - \beta_6 g_z x]}$$

((Dividing out the "y" in the numerator and the denominator)

$$\frac{qg_x z}{V_z} = -\frac{qg_x g_x z\beta_7 fg_z t \pm \sqrt{2hg_x x}\beta_7 g_x g_z qz - \beta_5 g_x g_z nqtz \pm \sqrt{2\beta_8 g_z z}\ g_x g_x nqz}{(\beta_7 g_z)[qg_x z - \beta_6 g_z x]}$$ or

$$\frac{qg_x z}{V_z} = \frac{qg_x g_x z\beta_7 fg_z t \pm \sqrt{2hg_x x}\beta_7 g_x g_z qz - \beta_5 g_x g_z nqtz \pm \sqrt{2\beta_8 g_z z}\ g_x g_x nqz}{\beta_7 \beta_6 g_z g_z x - \beta_7 qg_z g_x z}$$

$$\boxed{\frac{qg_x z}{V_z} = \frac{(\beta_7 fg_z g_x g_x q - \beta_5 g_x g_z nq)tz \pm \sqrt{2hg_x x}\beta_7 g_x g_z qz \pm \sqrt{2\beta_8 g_z z}\ g_x g_x nqz}{\beta_7 \beta_6 g_z g_z x - \beta_7 qg_z g_x z}}$$

Summary for the fractional terms of the *x*-direction solution

$\frac{ng_x y}{V_y}$ and $\frac{qg_x z}{V_z}$ in terms of x, y, z and t

$\left.\frac{ng_x y}{V_y} = -\frac{n\beta_5 g_z t}{\beta_7} \pm \frac{(\sqrt{2\beta_8 g_z z})(ng_x)}{\beta_7 g_z}\right\}$ B	$\frac{ng_x y}{V_y} = -k_1 g_z t \pm \frac{\sqrt{2k_2 g_z z} \bullet g_x k_3}{g_z}$ $k_1 = \frac{n\beta_5}{\beta_7}$; $k_2 = \beta_8$; $k_3 = \frac{n}{\beta_7}$;

$$\left.\frac{qg_x z}{V_z} = \frac{(\beta_7 f_e g_z g_x g_x q - \beta_5 g_x g_z n_e q)tz \pm \sqrt{2hg_x x}\beta_7 g_x g_z qz \pm \sqrt{2\beta_8 g_z z}\ g_x g_x nqz}{\beta_7 \beta_6 g_z g_z x - \beta_7 qg_z g_x z}\right\} C$$

$$\frac{qg_x z}{V_z} = \frac{(g_x^2 g_z k_4 - g_x g_z k_5)tz \pm \sqrt{2g_x k_6 x} \bullet g_x g_z k_7 z \pm \sqrt{2g_z k_8 z} \bullet g_x^2 k_9 z}{g_z^2 k_{10} x - g_x g_z k_{11} z}$$

$k_4 = \beta_7 fq$; $k_5 = \beta_5 nq$;; $k_6 = h$; $k_8 = \beta_8$; $k_9 = nq$ $k_{10} = \beta_7 \beta_6$ $k_{11} = \beta_7 q$

Analysis of the Euler Solutions

$$V_x(x,y,z,t) = fg_xt \pm \sqrt{2hg_xx} + \frac{ng_xy}{V_y} + \frac{qg_xz}{V_z} + \underbrace{\frac{\psi_y(V_y)}{V_y} + \frac{\psi_z(V_z)}{V_z}}_{\text{arbitrary functions}};\ P(x) = d\rho g_xx \quad x\text{-direction}$$

$$V_y \neq 0,\ V_z \neq 0;\ (d+f+h+n+q=1)$$

$$\left.\frac{ng_xy}{V_y} = -\frac{n\beta_5g_zt}{\beta_7} \pm \frac{(\sqrt{2\beta_8g_zz})(ng_x)}{\beta_7g_z}\right\} \text{B}$$

$$\left.\frac{qg_xz}{V_z} = \frac{(\beta_7fg_zg_x^2q - \beta_5g_xg_znq)tz \pm \sqrt{2hg_xx}\beta_7g_xg_zqz \pm \sqrt{2\beta_8g_zz}\ g_x^2nqz}{\beta_7\beta_6g_z^2x - \beta_7qg_zg_xz}\right\} \text{C}$$

$$d+f+h+n+q=1;\quad \lambda_4+\lambda_5+\lambda_6+\lambda_7+\lambda_8=1;\quad \beta_4+\beta_5+\beta_6+\beta_7+\beta_8=1$$

One observes above that the most important insight of the above solution is the indispensability of the gravity term in incompressible fluid flow. Observe that if gravity, g_x were zero, the first four terms of the velocity solution and $P(x)$ would all be zero. This result can be stated emphatically that without gravity forces on earth, there would be no incompressible fluid flow on earth as is known.

More Observations: Comparison of the Euler solutions with equations of motion under gravity and liquid pressure of elementary physics

Motion under gravity equations: (B): $V = gt$; (C): $V = \sqrt{2gx}$;

Liquid Pressure, P at the bottom of a liquid of depth h units is given by $P = \rho gh$

Observe the following similarities above:

1. Observe the "gt" in $V = gt$ of (B) of the motion equations and the fg_xt in the Euler solution.
2. Observe the "$\sqrt{2gx}$" in $V = \sqrt{2gx}$ of (C) and the $\sqrt{2hg_xx}$ in the Euler solution.
3. Observe the $P = \rho gh$ of the liquid pressure and the $P(x) = d\rho g_xx$ of the Euler solution.

There are five main terms (ignoring the arbitrary functions) in the Euler solution. Of these five terms, three terms, namely, fg_xt, $\sqrt{2hg_xx}$, $d\rho g_xx$ are the same (except for the constants involved) as the terms in the equations of motion under gravity. This similarity means that the approach used in solving the Euler equation is sound. One should also note that to obtain these three terms simultaneously, only the equation with the gravity term as the subject of the equation will yield these three terms. The author suggests that this form of the equation with the gravity term as the subject of the equation be called the standard form of the Euler equation, since in this form, one can immediately split-up the equations using ratios, and integrate.

The **velocity profile** of the x–direction solution consists of linear, parabolic, and hyperbolic terms. If one assumes that in laminar flow, the axis of symmetry of the parabola for horizontal velocity flow profile is in the direction of fluid flow, then in turbulent flow, the axis of symmetry of the parabola would be at right angles to that for laminar flow. The characteristic curve for the integral of the x–nonlinear term is such a parabola whose axis of symmetry is at right angles to that of laminar flow. The integral of the y–nonlinear term is similar parabolically to that of the x–nonlinear term. The characteristic curve for the integral of the z–nonlinear term is a combination of two similar parabolas and a hyperbola. If the above x–direction flow is repeated simultaneously in the y– and z– directions, the flow is chaotic and consequently turbulent

Standard form of the x-direction Euler equation for incompressible fluid flow
One will call the Euler equation with the gravity term as the subject of equation in (A) , the standard form of the Euler equation for the ratio method of solving these equations, since this form produces a solution on integration. None of the other forms in (B), (C), (D), (E) , or (F), produces a solution. That is, the integration results of each of the other five equations do not satisfy the corresponding equation.

$$\boxed{\rho\frac{\partial V_x}{\partial t}+\rho V_x\frac{\partial V_x}{\partial x}+\rho V_y\frac{\partial V_x}{\partial y}+\rho V_z\frac{\partial V_x}{\partial z}+\frac{\partial p_x}{\partial x}=\rho g_x \quad \text{(A)}} \quad < \text{standard form}$$

$$-\rho\frac{\partial V_x}{\partial t}-\rho V_x\frac{\partial V_x}{\partial x}-\rho V_y\frac{\partial V_x}{\partial y}-\rho V_z\frac{\partial V_x}{\partial z}+\rho g_x=\frac{\partial p_x}{\partial x} \quad \text{(B)}$$

$$-\rho V_x\frac{\partial V_x}{\partial x}-\rho V_y\frac{\partial V_x}{\partial y}-\rho V_z\frac{\partial V_x}{\partial z}+\frac{\partial p_x}{\partial x}+\rho g_x=\rho\frac{\partial V_x}{\partial t} \quad \text{(C)}$$

$$-\rho\frac{\partial V_x}{\partial t}-\rho V_y\frac{\partial V_x}{\partial y}-\rho V_z\frac{\partial V_x}{\partial z}-\frac{\partial p_x}{\partial x}+\rho g_x=\rho V_x\frac{\partial V_x}{\partial x} \quad \text{(D)}$$

$$-\rho\frac{\partial V_x}{\partial t}-\rho V_x\frac{\partial V_x}{\partial x}-\rho V_z\frac{\partial V_x}{\partial z}-\frac{\partial p_x}{\partial x}+\rho g_x=\rho V_y\frac{\partial V_x}{\partial y} \quad \text{(E)}$$

$$-\rho\frac{\partial V_x}{\partial t}-\rho V_x\frac{\partial V_x}{\partial x}-\rho V_y\frac{\partial V_x}{\partial y}-\frac{\partial p_x}{\partial x}+\rho g_x=\rho V_z\frac{\partial V_x}{\partial z} \quad \text{(F)}$$

Uniqueness of the solution of the Euler equation
When each term of the linearized Navier-Stokes equation was made subject of the N-S equation , only the equation with the gravity term as the subject of the equation produced a solution. (vixra:1405.0251 of 2014). Similarly. the solution of the Euler solution is unique.

Extra:

Linearized Euler Equation: If one linearizes the Euler equation as was done in the linearization of the Navier-Stokes equation, one obtains $4\frac{\partial V_x}{\partial t}+\frac{1}{\rho}\frac{\partial p}{\partial x}=g_x$; whose solution is $V_x=\frac{fg_x}{4}t+C$; $P(x)=d\rho g_x x$.

One May also solve the following for V_x , V_y, V_z in terms of x, y and z.

A **Euler solution system to solve for V_x, V_y, V_z**

$$V_x=\frac{(fg_x t\pm\sqrt{2hg_x x})V_yV_z+[qg_x z+\psi_z(V_z)]V_y+[ng_x y+\psi_y(V_y)]V_z}{V_yV_z}$$

$$V_y=\frac{(\lambda_5 g_y t\pm\sqrt{2\lambda_7 g_y y})V_xV_z+[\lambda_8 g_y z+\psi_z(V_z)]V_x+[\lambda_6 g_y x+\psi_x(V_x)]V_z}{V_xV_z}$$

$$V_z=\frac{(\beta_5 g_z t\pm\sqrt{2\beta_8 g_z z})V_xV_y+[\beta_6 g_z x+\psi_x(V_x)]V_y+[\beta_7 g_z y+\psi_y(V_y)]V_x}{V_xV_y}$$

Note: By comparison with Navier-Stokes equation and its relation, a relation to Euler equation can be found by deleting the Navier-Stokes relation resulting from the μ-terms.

Option 4
Solutions of 3-D Navier-Stokes Equations (Original)
Method 1

As in Option 1 for solving these equations, the first step here, is to split-up the equation into eight sub-equations using the ratio method. One will solve **only** the driver equation, based on the experience gained in solving the linearized equation. There are 8 supporter equations.

$$-\mu\frac{\partial^2 V_x}{\partial x^2}-\mu\frac{\partial^2 V_x}{\partial y^2}-\mu\frac{\partial^2 V_x}{\partial z^2}+\frac{\partial p}{\partial x}+\rho\frac{\partial V_x}{\partial t}+\overbrace{\rho V_x\frac{\partial V_x}{\partial x}+\rho V_y\frac{\partial V_x}{\partial y}+\rho V_z\frac{\partial V_x}{\partial z}}^{\text{nonlinear terms}}=\rho g_x \qquad \text{(A)}$$

$$-K\frac{\partial^2 V_x}{\partial x^2}-K\frac{\partial^2 V_x}{\partial y^2}-K\frac{\partial^2 V_x}{\partial z^2}+\frac{1}{\rho}\frac{\partial p}{\partial x}+\frac{\partial V_x}{\partial t}+V_x\frac{\partial V_x}{\partial x}+V_y\frac{\partial V_x}{\partial y}+V_z\frac{\partial V_x}{\partial z}=g_x \qquad (K=\frac{\mu}{\rho}) \quad \text{(B)}$$

Step 1: Apply the ratio method to equation (B) to obtain the following equations:

$$1.\ -K\frac{\partial^2 V_x}{\partial x^2}=ag_x;\ 2.\ -K\frac{\partial^2 V_x}{\partial y^2}=bg_x;\ 3.\ -K\frac{\partial^2 V_x}{\partial z^2}=cg_x;\ 4.\ \frac{1}{\rho}\frac{\partial p}{\partial x}=dg_x;\ 5.\ \frac{\partial V_x}{\partial t}=fg_x$$

$$6.\ V_x\frac{\partial V_x}{\partial x}=hg_x;\ 7.\ V_y\frac{\partial V_x}{\partial y}=qg_x;\ 8.\ V_z\frac{\partial V_x}{\partial z}=ng_x$$

where $a,\ b,\ c,\ d, f,\ h,\ n,\ q$ are the ratio terms and $a+b+c+d+f+h+\ n+q=1$

Step 2: Solve the differential equations in Step 1.

Note that after splitting the equations, the equations can be solved using techniques of ordinary differential equations.

One can view each of the ratio terms $a,\ b,\ c,\ d, f,\ h,\ n,\ q$ as a fraction (a real number) of $\boxed{g_x}$ contributed by each expression on the left-hand side of equation (B) above.

Solutions of the eight sub-equations

$$\boxed{1.\ -k\frac{\partial^2 V_x}{\partial x^2}=ag_x}$$

$$k\frac{\partial^2 V_x}{\partial x^2}=-ag_x$$

$$\frac{\partial^2 V_x}{\partial x^2}=-\frac{a}{k}g_x$$

$$\frac{\partial V_x}{\partial x}=-\frac{ag}{k}x+C_1$$

$$V_{x1}=-\frac{ag_x}{2k}x^2+C_1x+C_2$$

$$\boxed{2.\ -K\frac{\partial^2 V_x}{\partial y^2}=bg_x}$$

$$K\frac{\partial^2 V_x}{\partial y^2}=-bg_x$$

$$\frac{\partial^2 V_x}{\partial y^2}=-\frac{b}{k}g_x$$

$$\frac{\partial V_x}{\partial y}=-\frac{bg_x}{k}y+C_3$$

$$V_{x2}=-\frac{bg_x}{2k}y^2+C_3y+C_4$$

$$\boxed{3.\ -K\frac{\partial^2 V_x}{\partial z^2}=cg_x}$$

$$K\frac{\partial^2 V_x}{\partial z^2}=-cg_x$$

$$\frac{\partial^2 V_x}{\partial z^2}=-\frac{c}{k}g_x$$

$$\frac{\partial V_x}{\partial z}=-\frac{cg_x}{k}z+C_5$$

$$V_{x3}=-\frac{cg_x}{2k}z^2+C_5z+C_6$$

$$\boxed{4.\ \frac{1}{\rho}\frac{\partial p}{\partial x}=dg_x}$$

$$\frac{1}{\rho}\frac{\partial p}{\partial x}=dg_x$$

$$\frac{\partial p}{\partial x}=d\rho g_x$$

$$p=d\rho g_x x+C_7$$

$$\boxed{5.\ \frac{\partial V_x}{\partial t}=fg_x}$$

$$V_{x4}=fgt$$

$$\boxed{6.\ V_x\frac{\partial V_x}{\partial x}=hg_x}$$

$$V_x\frac{dV_x}{dx}=hg_x$$

$$V_xdV_x=hg_xdx$$

$$\frac{V_x^2}{2}=hg_xx$$

$$V_{x5}=\pm\sqrt{2hg_xx+C_7}$$

$$\boxed{7.\ V_y\frac{\partial V_x}{\partial y}=ng_x}$$

$$V_y\frac{dV_x}{dy}=ng_x$$

$$V_ydV_x=ng_x\text{dy}$$

$$V_yV_x=ng_x\text{y}+\psi_y(V_y)$$

$$V_{x6}=\frac{ng_x\text{y}}{V_y}+\frac{\psi_y(V_y)}{V_y}$$

$$\boxed{8.\ V_z\frac{\partial V_x}{\partial z}=qg_x}$$

$$V_z\frac{dV_x}{dz}=qg_x$$

$$V_zdV_x=qg_x\text{dz};$$

$$V_zV_x=qg_x\text{z}+\psi_z(V_z)$$

$$V_{x7}=\frac{qg_x\text{z}}{V_z}+\frac{\psi_z(V_z)}{V_z}$$

Note:
$\psi_y(V_y),\ \psi_z(V_z)$ are arbitrary functions, (integration constants)

$V_y\neq 0$

$V_z\neq 0$

Step 3: One combines the above solutions

$$V_x(x,y,z,t) = V_{x1} + V_{x2} + V_{x3} + V_{x4} + V_{x5} + V_{x6} + V_{x7}$$

$$= -\frac{ag_x}{2k}x^2 + C_1x - \frac{bg_x}{2k}y^2 + C_3y - \frac{cg_x}{2k}z^2 + C_5z + fg_xt \pm \sqrt{2hg_xx} + \frac{ng_xy}{V_y} + \frac{qg_xz}{V_z} + \frac{\psi_y(V_y)}{V_y} + \frac{\psi_z(V_z)}{V_z}$$

relation for linear terms — relation for non - linear terms

$$\overbrace{-\frac{\rho g_x}{2\mu}(ax^2 + by^2 + cz^2) + C_1x + C_3y + C_5z + fg_xt}^{\text{relation for linear terms}} \pm \overbrace{\sqrt{2hg_xx} + \frac{ng_xy}{V_y} + \frac{qg_xz}{V_z} + \underbrace{\frac{\psi_y(V_y)}{V_y} + \frac{\psi_z(V_z)}{V_z}}_{\text{arbitrary functions}}}^{\text{relation for non-linear terms}} + C_9$$

$$P(x) = d\rho g_xx; \quad (a+b+c+d+f+h+n+q=1) \quad V_y \neq 0,\ V_z \neq 0$$

Step 4: Find the test derivatives

Test derivatives for the linear part					Test derivatives for the non-linear part		
$\frac{\partial^2 V_x}{\partial x^2} = -\frac{a\rho g_x}{\mu}$	$\frac{\partial^2 V_x}{\partial y^2} = -\frac{b\rho g_x}{\mu}$	$\frac{\partial^2 V_x}{\partial z^2} = -\frac{c\rho g_x}{\mu}$	$\frac{\partial p}{\partial x} = d\rho g_x$	$\frac{\partial V_x}{\partial t} = fg_x$	$V_x^2 = 2hg_xx$; $2V_x\frac{\partial V_x}{\partial x} = 2hg_x$; $\frac{\partial V_x}{\partial x} = \frac{hg_x}{V_x}, V_x \neq 0$	$\frac{\partial V_x}{\partial y} = \frac{ng_x}{V_y}$	$\frac{\partial V_x}{\partial z} = \frac{qg_x}{V_z}$

Step 5: **S**ubstitute the derivatives from Step 4 in equation (A) for the checking.

$$-\mu\frac{\partial^2 V_x}{\partial x^2} - \mu\frac{\partial^2 V_x}{\partial y^2} - \mu\frac{\partial^2 V_x}{\partial z^2} + \frac{\partial p}{\partial x} + \rho\frac{\partial V_x}{\partial t} + \rho V_x\frac{\partial V_x}{\partial x} + \rho V_y\frac{\partial V_x}{\partial y} + \rho V_z\frac{\partial V_x}{\partial z} = \rho g_x \quad \textbf{(A)}$$

$$-\mu\left(-\frac{a\rho g_x}{\mu} - \frac{b\rho g_x}{\mu} - \frac{c\rho g_x}{\mu}\right) + d\rho g_x + f\rho g_x + \rho\left(V_x\frac{hg_x}{V_x}\right) + \rho V_y\left(\frac{ng_x}{V_y}\right) + \rho V_z\left(\frac{qg_x}{V_z}\right) \stackrel{?}{=} \rho g_x$$

$$a\rho g_x + b\rho g_x + c\rho g_x + d\rho g_x + f\rho g_x + h\rho g_x + n\rho g_x + q\rho g_x \stackrel{?}{=} \rho g_x$$

$$ag_x + bg_x + cg_x + dg_x + fg_x + hg_x + ng_x + qg_x \stackrel{?}{=} g_x$$

$$g_x(a+b+c+d+f+h+n+q) \stackrel{?}{=} g_x$$

$$g_x(1) \stackrel{?}{=} g_x \quad \text{Yes} \quad (a+b+c+d+f+h+n+q=1)$$

Step 6: The linear part of the relation satisfies the linear part of the equation; and the non-linear part of the relation satisfies the non-linear part of the equation.(B) below is the solution.

Analogy for the Identity Checking Method: If one goes shopping with American dollars and Japanese yens (without any currency conversion) and after shopping, if one wants to check the cost of the items purchased, one would check the cost of the items purchased with dollars against the receipts for the dollars; and one would also check the cost of the items purchased with yens against the receipts for the yens purchase. However, if one converts one currency to the other, one would only have to check the receipts for only a single currency, dollars or yens. This conversion case is similar to the linearized equations, where there was no partitioning in identity checking. Note that for the Euler equations, there was no partitioning in taking derivatives for identity checking.

Note: After expressing $\frac{ng_xy}{V_y}$ and $\frac{q_eg_xz}{V_z}$ in terms of x, y, z, and t, there would be no partitioning in identity checking.

Summary of solutions for $V_x \quad V_y, V_z \quad (P(x) = d\rho g_x x;\ P(y) = \lambda_4 \rho g_y y\ ,\ P(z) = \beta_4 \rho g_z z)$

$$V_x = -\frac{\rho g_x}{2\mu}(ax^2 + by^2 + cz^2) + C_1 x + C_3 y + C_5 z + f g_x t \pm \sqrt{2hg_x x} + \frac{ng_x y}{V_y} + \frac{qg_x z}{V_z} + \frac{\psi_y(V_y)}{V_y} + \frac{\psi_z(V_z)}{V_z} + C_9$$

$$P(x) = d\rho g_x x; \quad (a+b+c+d+h+n+q=1) \quad V_y \neq 0,\ V_z \neq 0$$

$$V_y = -\frac{\rho g_y}{2\mu}(\lambda_1 x^2 + \lambda_2 y^2 + \lambda_3 z^2) + C_{10} x + C_{11} y + C_{12} z + \lambda_5 g_y t \pm \sqrt{2\lambda_7 g_y y} + \frac{\lambda_6 g_y x}{V_x} + \frac{\lambda_8 g_y z}{V_z} + \frac{\psi_x(V_x)}{V_x} + \frac{\psi_z(V_z)}{V_z}$$

$$V_z = -\frac{\rho g_z}{2\mu}(\beta_1 x^2 + \beta_2 y^2 + \beta_3 z^2) + C_{14} x + C_{15} y + C_{16} z + \beta_5 g_z t \pm \sqrt{2\beta_8 g_z z} + \frac{\beta_6 g_z x}{V_x} + \frac{\beta_7 g_z y}{V_y} + \frac{\psi_x(V_x)}{V_x} + \frac{\psi_y(V_y)}{V_y}$$

The above solutions are unique, because from the experience in Option 1, only the equations with the gravity terms as the subjects of the equations produced the solutions.

$$\begin{cases} V_x = -\frac{\rho g_x}{2\mu}(ax^2 + by^2 + cz^2) + C_1 x + C_3 y + C_5 z + f g_x t \pm \sqrt{2hg_x x} + \frac{ng_x y}{V_y} + \frac{qg_x z}{V_z} + \frac{\psi_y(V_y)}{V_y} + \frac{\psi_z(V_z)}{V_z} \\ V_y = -\frac{\rho g_y}{2\mu}(\lambda_1 x^2 + \lambda_2 y^2 + \lambda_3 z^2) + C_{10} x + C_{11} y + C_{12} z + \lambda_5 g_y t \pm \sqrt{2\lambda_7 g_y y} + \frac{\lambda_6 g_y x}{V_x} + \frac{\lambda_8 g_y z}{V_z} + \frac{\psi_x(V_x)}{V_x} + \frac{\psi_z(V_z)}{V_z} \\ V_z = -\frac{\rho g_z}{2\mu}(\beta_1 x^2 + \beta_2 y^2 + \beta_3 z^2) + C_{14} x + C_{15} y + C_{16} z + \beta_5 g_z t \pm \sqrt{2\beta_8 g_z z} + \frac{\beta_6 g_z x}{V_x} + \frac{\beta_7 g_z y}{V_y} + \frac{\psi_x(V_x)}{V_x} + \frac{\psi_y(V_y)}{V_y} \end{cases}$$

One will next solve the above system of solutions for V_x , V_y, V_z in order to express $\frac{ng_x y}{V_y}$ and $\frac{q_e g_x z}{V_z}$ in terms of x, y, z, and t The author used the help of the Maples software for the simultaneous algebraic solutions for $V_y \quad V_z$. The basic expressions are of the forms $-\frac{\rho g_x}{2\mu}ax^2$, $-\frac{\rho g_x}{2\mu}by^2$, $-\frac{\rho g_x}{2\mu}cz^2$, $fg_x t$, $\sqrt{2hg_x x}$, and $d\rho g_x x$; These expressions are similar to the terms of the equations of motion under gravity and liquid pressure of elementary physics. Note that the explicit solutions will be the results of the basic operations (addition, subtraction, multiplication, division, power finding and root extraction) on the expressions in Step 1 below.

Solving for V_x , V_y, V_z $\frac{ng_x y}{V_y}$, and $\frac{q_e g_x z}{V_z}$

Let $V_x = x$, $V_y = y$ and $V_z = z$. (x, y and z are being used for simplicity. They will be change d back to V_x , V_y, and V_z later, and they do not represent the variables x, y and z in the solutions.

Step 1 From the above system of solutions, let

$$A = -\frac{\rho g_x}{2\mu}(ax^2 + by^2 + cz^2) + C_1 x + C_3 y + C_5 z + f g_x t \pm \sqrt{2hg_x x}$$

$$B = -\frac{\rho g_y}{2\mu}(\lambda_1 x^2 + \lambda_2 y^2 + \lambda_3 z^2) + C_{10} x + C_{11} y + C_{12} z + \lambda_5 g_y t \pm \sqrt{2\lambda_7 g_y y}$$

$$C = -\frac{\rho g_z}{2\mu}(\beta_1 x^2 + \beta_2 y^2 + \beta_3 z^2) + C_{14} x + C_{15} y + C_{16} z + \beta_5 g_z t \pm \sqrt{2\beta_8 g_z z}$$

$$D = qg_x z;\ E = ng_x y;\ F = \lambda_6 g_y x$$

$$G = \lambda_8 g_y z;\ J = \beta_6 g_z x;\ L = \beta_7 g_z y$$

Step 2 Then the solutions to the N-S system of equations become (ignoring the arbitrary functions)

$$\left.\begin{aligned} x &= A + \frac{D}{z} + \frac{E}{y} \\ y &= B + \frac{F}{x} + \frac{G}{z} \\ z &= C + \frac{J}{x} + \frac{L}{y} \end{aligned}\right\} \text{M}$$

Step 3	**Step 4**
$xyz = Ayz + Dy + Ez$ (1); $xyz = Bxz + Fz + Gx$ (2); $xyz = Cxy + Jy + Lx$ (3)) } N	$0 = Ayz + Dy + Ez - Bxz - Fz - Gx$ (4); $0 = Ayz + Dy + Ez - Cxy - Jy - Lx$ (5); $0 = Bxz + Fz + Gx - Cxy - Jy + -Lx$ (6) } P

Maples software was used to solve system P to obtain

Step 5

$$x = \frac{L(FCD - FCJ - JLA + JCE)}{C(-BLD + BLJ + GLA - GCE)}$$

$$V_x = \frac{L(FCD - FCJ - JLA + JCE)}{C(-BLD + BLJ + GLA - GCE)} \quad \text{(back to } V_x \text{)}$$

$$y = -\frac{L}{C};$$

$$V_y = -\frac{L}{C} \quad \text{(changing back to } V_y \text{ as agreed to)}$$

$$z = -\frac{L(D - J)}{LA - CE};$$

$$V_z = -\frac{L(D - J)}{LA - CE} \quad \text{(changing back to } V_z \text{ as agrred to)}$$

Note:
None of the popular academic programs could solve the system in M.
Maples solved system P (step 4 above) for x, y,and z in terms of A, B, C, D. E. F, G. J. and L.
Note also that x, y and z are not the same as the x, y and z in the system of equations..
They were used for convenience and simplicity .

Step 5: Apply the following and substitute for A, B, C, D. E. F, G., J. and L in steps 6-8 below

$$A = -\frac{\rho g_x}{2\mu}(ax^2 + by^2 + cz^2) + C_1x + C_3y + C_5z + fg_xt \pm \sqrt{2hg_xx}$$

$$B = -\frac{\rho g_y}{2\mu}(\lambda_1x^2 + \lambda_2y^2 + \lambda_3z^2) + C_{10}x + C_{11}y + C_{12}z + \lambda_5g_yt \pm \sqrt{2\lambda_7g_yy}$$

$$C = -\frac{\rho g_z}{2\mu}(\beta_1x^2 + \beta_2y^2 + \beta_3z^2) + C_{14}x + C_{15}y + C_{16}z + \beta_5g_zt \pm \sqrt{2\beta_8g_zz}$$

$$D = qg_xz\ ;\ E = ng_xy;\ \ F = \lambda_6g_yx$$

$$G = \lambda_8g_yz;\ \ J = \beta_6g_zx;\ \ L = \beta_7g_zy$$

Step 6

$$V_y = -\frac{L}{C} = -\frac{(\beta_7g_zy)}{(-\frac{\rho g_z}{2\mu}(\beta_1x^2 + \beta_2y^2 + \beta_3z^2) + C_1x + C_3y + C_5z + \beta_5g_zt \pm \sqrt{2\beta_8g_zz}}$$

$$\frac{ng_xy}{V_y} = ng_xy \div -\left(\frac{(\beta_7g_zy)}{(-\frac{\rho g_z}{2\mu}(\beta_1x^2 + \beta_2y^2 + \beta_3z^2) + C_1x + C_3y + C_5z + \beta_5g_zt \pm \sqrt{2\beta_8g_zz}}\right)$$

$$\frac{ng_xy}{V_y} = -\frac{(ng_xy)[(-\frac{\rho g_z}{2\mu}(\beta_1x^2 + \beta_2y^2 + \beta_3z^2) + C_1x + C_3y + C_5z + \beta_5g_zt \pm \sqrt{2\beta_8g_zz})]}{\beta_7g_zy};$$

$$\frac{ng_xy}{V_y} = \frac{-(ng_x)(-\frac{\rho g_z}{2\mu}(\beta_1x^2 + \beta_2y^2 + \beta_3z^2) + C_1x + C_3y + C_5z + \beta_5g_zt \pm \sqrt{2\beta_8g_zz})}{\beta_7g_z}; \text{ (cancel "y")}$$

Step 7

$$V_z = -\frac{L(D-J)}{LA-CE}$$

$$V_z = -\frac{(\beta_7 g_z y)(qg_x z - \beta_6 g_z x)}{(\beta_7 g_z y)(-\frac{\rho g_x}{2\mu}(ax^2+by^2+cz^2)+C_1x+C_3y+C_5z+fg_xt \pm \sqrt{2hg_x x}) - CE}$$

$$CE = -(ng_x y)(-\frac{\rho g_z}{2\mu}(\beta_1 x^2+\beta_2 y^2+\beta_3 z^2)+C_{14}x+C_{15}y+C_{16}z+\beta_5 g_z t \pm \sqrt{2\beta_8 g_z z}$$

$$\frac{qg_x z}{V_z} = \frac{-(qg_x z)\{[(\beta_7 g_z y)(-\frac{\rho g_x}{2\mu}(ax^2+by^2+cz^2)+C_1x+C_3y+C_5z+fg_xt \pm \sqrt{2hg_x x}] - [CE]\}}{(\beta_7 g_z y)(qg_x z - \beta_6 g_z x)}$$

($Note$: y can be cancelled from numerator and denominator

Summary for the fractional terms of the x–direction solution

$\frac{ng_x y}{V_y}$ and $\frac{qg_x z}{V_z}$ in terms of x, y, z and t

$$\frac{ng_x y}{V_y} = \frac{-(ng_x)(-\frac{\rho g_z}{2\mu}(\beta_1 x^2+\beta_2 y^2+\beta_3 z^2)+C_1x+C_3y+C_5z+\beta_5 g_z t \pm \sqrt{2\beta_8 g_z z})}{\beta_7 g_z}; \text{ (cancel "y")}$$

$$\frac{qg_x z}{V_z} = \frac{-(qg_x z)\{[(\beta_7 g_z y)(-\frac{\rho g_x}{2\mu}(ax^2+by^2+cz^2)+C_1x+C_3y+C_5z+fg_xt \pm \sqrt{2hg_x x}] - [CE]\}}{(\beta_7 g_z y)(qg_x z - \beta_6 g_z x)}$$

$$\left(CE = -(ng_x y)(-\frac{\rho g_z}{2\mu}(\beta_1 x^2+\beta_2 y^2+\beta_3 z^2)+C_{14}x+C_{15}y+C_{16}z+\beta_5 g_z t \pm \sqrt{2\beta_8 g_z z}\right)$$

$$d+f+h+n+q=1; \quad \lambda_4+\lambda_5+\lambda_6+\lambda_7+\lambda_8=1; \quad \beta_4+\beta_5+\beta_6+\beta_7+\beta_8=1$$

Analysis of N-S Solutions

x–direction solution

$$V_x = -\frac{\rho g_x}{2\mu}(ax^2 + by^2 + cz^2) + C_1x + C_3y + C_5z + fg_xt \pm \sqrt{2hg_xx} + \frac{ng_xy}{V_y} + \frac{qg_xz}{V_z} + \frac{\psi_y(V_y)}{V_y} + \frac{\psi_z(V_z)}{V_z} + C_9$$

$$P(x) = d\rho g_x x; \quad (a + b + c + d + h + n + q = 1) \quad V_y \neq 0,\ V_z \neq 0$$

$$\frac{ng_xy}{V_y} = \frac{-(ng_x)(-\frac{\rho g_z}{2\mu}(\beta_1x^2 + \beta_2y^2 + \beta_3z^2) + C_1x + C_3y + C_5z + \beta_5g_zt \pm \sqrt{2\beta_8g_zz})}{\beta_7g_z}$$

$$\frac{qg_xz}{V_z} = \frac{-(qg_xz)\{[(\beta_7g_zy)(-\frac{\rho g_x}{2\mu}(ax^2 + by^2 + cz^2) + C_1x + C_3y + C_5z + fg_xt \pm \sqrt{2hg_xx}] - [CE]\}}{(\beta_7g_zy)(qg_xz - \beta_6g_zx)}$$

$$(CE = -(ng_xy)(-\frac{\rho g_z}{2\mu}(\beta_1x^2 + \beta_2y^2 + \beta_3z^2) + C_{14}x + C_{15}y + C_{16}z + \beta_5g_zt \pm \sqrt{2\beta_8g_zz})$$

One observes above that the most important insight of the above solution is the indispensability of the gravity term in incompressible fluid flow. Observe that if gravity, g_x, were zero, the first three terms, the seventh, the eighth, the ninth, the tenth terms of the velocity solution and $P(x)$ would all be zero. This result can be stated emphatically that without gravity forces on earth, there would be no incompressible fluid flow on earth as is known.

More Observations Comparison of the Navier-Stokes solutions with equations of motion under gravity and liquid pressure of elementary physics

Motion equations of elementary physics:

(B): $V_f = V_0 + gt$; (C): $V_f^2 = V_0^2 + 2gx$; (D): $V = \sqrt{2gx}$; (E): $x = V_0t + \frac{1}{2}gt^2$

Liquid Pressure,

The liquud pressure, P at the bottom of a liquid of depth h units is given by $P = \rho gh$

Observe the following above:

1. Observe that the first three terms are parabolic in x, y, and z; the minus sign showing the usual inverted parabola when a projectile is fired upwards at an acute angle to the horizontal. Also note the " gt " in $V = gt$ of (B) of the motion equations and the fg_xt in the Navier-Stokes solution.

2. Observe the $P - \rho gh$ of the liquid pressure and the $P(x) = d\rho g_x x$ of the Navier-Stokes solution. Note that d is a ratio term.

3. Observe the " $\sqrt{2gx}$ " in $V = \sqrt{2gx}$ of (D) and the $\sqrt{2hg_xx}$ in the Navier-Stokes solution. There are eight main terms (ignoring the arbitrary functions) in the Navier-Stokes solution. Of these eight terms, six terms, namely, $-\frac{a\rho g_x}{2\mu}x^2$, $-\frac{b\rho g_x}{2\mu}y^2$, $-\frac{c\rho g_x}{2\mu}z^2$, fg_xt, $\sqrt{2hg_xx}$ and $d\rho g_x x$ are similar (except for the constants involved) to the terms in the equations of motion. This similarity means that the approach used in solving the Navier-Stokes equation is sound. One should also note that to obtain these six terms simultaneously on integration, only the equation with the gravity term as the subject of the equation will yield these six terms. The author suggests that this form of the equation with the gravity term as the subject of the equation be called the standard form of the Navier-Stokes equation, since in this form, one can immediately split-up the equation using ratios, and integrate.

Velocity Profile, Polynomial and Radical Parabolas, Laminar and Turbulent flow

For communication purposes, each of the terms containing the even powers x^2, y^2 and z^2 will be called a polynomial parabola, and each of the terms containing the square roots $\pm\sqrt{x}$ $\pm\sqrt{y}$ and $\pm\sqrt{z}$ will be called a radical parabola. For each polynomial parabola, the axis of symmetry is in the direction of fluid flow; but for each radical parabola, the axis of symmetry is at right angles to the direction of fluid flow.

The fluid flow in the Navier-Stokes solution may be characterized as follows. The *x*–direction solution consists of linear, parabolic, and hyperbolic terms. The first three terms characterize polynomial parabolas. The characteristic curve for the integral of the *x*–nonlinear term is a radical parabola. The integral of the *y*–nonlinear term is similar parabolically to that of the *x*–nonlinear term. The integral of the *z*–nonlinear term is a combination of two radical parabolas and a hyperbola. If the above *x*–direction flow is repeated simultaneously in the *y*– and *z*– directions, the flow is chaotic and consequently turbulent.

In the N-S solution, during fluid flow, both the polynomial and radical parabolas are present at any speed. The polynomial parabolas are prominent and dominate flow while the radical parabolas are dormant at low speeds. At a low speed, a radical parabola (or a polynomial parabola susceptible to radicalization) is not active, since the radicand of the parabola is small and consequently, the root is small. When the speed becomes large, the "*x*" in $\sqrt{2hg_x x}$ becomes large and therefore the radical parabola becomes active. One can also observe how gravity interacts with the "*x*" of the radicand. By "g" and "*x*" being factors of the radicand (instead of "g" being outside the radical), "g" is closely aligned with *x*. Note that the radical parabola will be moving at right angles to the direction of fluid flow, the direction of which is also that of the axis of symmetry of the dominating polynomial parabola. Consequently, the flow profile becomes relatively more uniform or flattened due to the radical parabola moving at right angles to the direction of fluid flow. When viscosity increases, speed decreases, and the radicand (the factor x in $\sqrt{2hg_x x}$ decreases) of the radical parabola decreases. Consequently, the disruptive behavior of the radical parabola diminishes. When the fluid flows over an obstacle, the radical parabolas temporarily become significant resulting in turbulence. For a low value of x (i.e., from low fluid velocity), the viscous term dominates and the inertial term is not significant. At high fluid velocity, the factor "x" of the radicand is large. Also when density increases, velocity increases and the radicand increases, adding to the effect of the radical parabola.

Analogy:

Imagine a crowded marathon race involving one thousand runners at various positions on the race route, all running in the same direction. Imagine also that at certain points on the route, during the race, some of the runners at various positions suddenly begin to run to the left or to the right in directions at right angles to the direction of the race route; and imagine the resulting collisions and chaos. The polynomial parabolas are those runners following the route of the race, and the radical parabolas are those runners making ninety-degree turns from various positions. Literally, the radical parabolas disrupt the laminar flow.

Uniqueness of the solution of the Navier-Stokes equation

When each term of the linearized Navier-Stokes equation was made subject of the N-S equation, only the equation with the gravity term as the subject of the equation produced a solution. Similarly, the solution of the Navier-Stokes equation solution is unique.

Option 5
Solutions of 4-D Navier-Stokes Equations

In the above method, the solution can easily be extended to any number of dimensions..

Adding $\mu \frac{\partial^2 V_x}{\partial s^2}$ and $\rho V_s \frac{\partial V_x}{\partial s}$ to the 3-D x–direction equation yields the 4-D N-S equation

$$-\mu\left(\frac{\partial^2 V_x}{\partial x^2}+\frac{\partial^2 V_x}{\partial y^2}+\frac{\partial^2 V_x}{\partial z^2}+\frac{\partial^2 V_x}{\partial s^2}\right)+\frac{\partial p}{\partial x}+\rho\frac{\partial V_x}{\partial t}+\rho V_x\frac{\partial V_x}{\partial x}+\rho V_y\frac{\partial V_x}{\partial y}+\rho V_z\frac{\partial V_x}{\partial z}+\rho V_s\frac{\partial V_x}{\partial s}=\rho g_x$$

whose solution is given by

$$V_x(x,y,z,s,t)=$$
$$-\frac{\rho g_x}{2\mu}(ax^2+by^2+cz^2+es^2)+C_1x+C_3y+C_5z+C_6s+fg_xt\pm\sqrt{2hg_xx}+\frac{ng_xy}{V_y}+\frac{qg_xz}{V_z}+\frac{rg_xs}{V_s}+$$
$$\underbrace{\frac{\psi_y(V_y)}{V_y}+\frac{\psi_z(V_z)}{V_z}+\frac{\psi_s(V_s)}{V_s}}_{\text{arbitrary functions}}+C_9$$
$$P(x)=d\rho g_xx \qquad (a+b+c+d+e+f+h+n+q+r=1) \qquad V_x\neq 0,\ V_y\neq 0,\ V_s\neq 0,$$

For n–dimensions one can repeat the above as many times as one wishes.

Option 5b
Two-term Linearized Navier-Stokes Equation (one nonlinear term)

By linearization as in Option 1, if one replaces $\rho V_y\frac{\partial V_x}{\partial y}+\rho V_z\frac{\partial V_x}{\partial z}$ by $2\rho\frac{\partial V_x}{\partial t}$ in

$$-\mu\frac{\partial^2 V_x}{\partial x^2}-\mu\frac{\partial^2 V_x}{\partial y^2}-\mu\frac{\partial^2 V_x}{\partial z^2}+\frac{\partial p}{\partial x}+\rho\frac{\partial V_x}{\partial t}+\rho V_x\frac{\partial V_x}{\partial x}+\rho V_y\frac{\partial V_x}{\partial y}+\rho V_z\frac{\partial V_x}{\partial z}=\rho g_x$$ one obtains

$$-\mu\left(\frac{\partial^2 V_x}{\partial x^2}+\frac{\partial^2 V_x}{\partial y^2}+\frac{\partial^2 V_x}{\partial z^2}\right)+\frac{\partial p}{\partial x}+3\rho\left(\frac{\partial V_x}{\partial t}\right)+\rho V_x\frac{\partial V_x}{\partial x}=\rho g_x$$, whose solution is

$$V_x(x,y,z,t)=-\frac{\rho g_x}{2\mu}(ax^2+by^2+cz^2)+C_1x+C_3y+C_5z+\frac{fg_xt}{3}\pm\sqrt{2hg_xx}+C_6$$

Conclusion (for Option 4)

One will begin from the general case and end with the special cases.

Solutions of the Navier--Stokes equations (general case)

x–direction **Navier-Stokes Equation** (also driver equation)

$$-\mu\frac{\partial^2 V_x}{\partial x^2}-\mu\frac{\partial^2 V_x}{\partial y^2}-\mu\frac{\partial^2 V_x}{\partial z^2}+\frac{\partial p}{\partial x}+\rho\frac{\partial V_x}{\partial t}+\rho V_x\frac{\partial V_x}{\partial x}+\rho V_y\frac{\partial V_x}{\partial y}+\rho V_z\frac{\partial V_x}{\partial z}=\rho g_x \quad x\text{–direction}$$

$$V_x(x,y,z,t)=$$

$$\overbrace{-\frac{\rho g_x}{2\mu}(ax^2+by^2+cz^2)+C_1x+C_3y+C_5z+fg_xt}^{\text{solution for linear terms}}\pm\overbrace{\sqrt{2hg_xx}+\frac{ng_xy}{V_y}+\frac{qg_xz}{V_z}+\underbrace{\frac{\psi_y(V_y)}{V_y}+\frac{\psi_z(V_z)}{V_z}}_{\text{arbitrary functions}}}^{\text{solution for non - linear terms}}+C_9$$

$$P(x)=d\rho g_xx;\quad (a+b+c+d+h+n+q=1)\quad V_y\neq 0,\ V_z\neq 0$$

$$\frac{ng_xy}{V_y}=\frac{-(ng_x)(-\frac{\rho g_z}{2\mu}(\beta_1x^2+\beta_2y^2+\beta_3z^2)+C_1x+C_3y+C_5z+\beta_5g_zt\pm\sqrt{2\beta_8g_zz})}{\beta_7g_z}$$

$$\frac{qg_xz}{V_z}=\frac{-(qg_xz)\{[(\beta_7g_zy)(-\frac{\rho g_x}{2\mu}(ax^2+by^2+cz^2)+C_1x+C_3y+C_5z+fg_xt\pm\sqrt{2hg_xx}]-[CE]\}}{(\beta_7g_zy)(qg_xz-\beta_6g_zx)}$$

$$CE=-(ng_xy)(-\frac{\rho g_z}{2\mu}(\beta_1x^2+\beta_2y^2+\beta_3z^2)+C_{14}x+C_{15}y+C_{16}z+\beta_5g_zt\pm\sqrt{2\beta_8g_zz}$$

One observes above that the most important insight of the above solution is the indispensability of the gravity term in incompressible fluid flow. Observe that if gravity, g_x, were zero, the first three terms, the 7th term, the 8th term, the 9th term, the 10th term and $P(x)$ would all be zero. This result can be stated emphatically that without gravity forces on earth, there would be no incompressible fluid flow on earth as is known. The author proposed and applied a new law, the law of definite ratio for incompressible fluid flow. This law states that in incompressible fluid f low, the other terms of the fluid flow equation divide the gravity term in a definite ratio, and also each term utilizes gravity to function. This law was applied in splitting-up the Navier-Stokes equations. The resulting sub-equations were readily integrable, and even the nonlinear sub-equations were readily integrated.

The x–direction Navier-Stokes equation was split-up into sub-equations using ratios. The sub-equations were solved and combined. The relation obtained from the integration of the linear part of the equation satisfied the linear part of the equation and the relation obtained from integrating the nonlinear part of the equation satisfied the nonlinear part of the equation. By solving algebraically and simultaneously for V_x, V_y and V_z, the (ng_xy/V_y) and (qg_xz/V_z) terms would be expressed explicitly in terms of x, y, z and t. The above x–direction solution is the solution everyone has been waiting for, for nearly 150 years. It was obtained in two simple steps, namely, splitting the equation using ratios and integrating.

Special Cases of the Navier-Stokes Equations

1. Linearized Navier-Stokes equations

One may note that there are six linear terms and three nonlinear terms in the Navier-Stokes equation. The linearized case was covered before the general case, and the experience gained in the linearized case guided one to solve the general case efficiently. In particular, the gravity term must be the subject of the equation for a solution. When the gravity term was the subject of the equation, the equation was called the driver equation. A splitting technique was applied to the linearized Navier-Stokes equations (Option 1). Twenty sub-equations were solved. (Four sets of equations with different equation subjects). The integration relations of one of the sets satisfied the linearized Navier-Stokes equation; and this set was from the equation with g_x as the subject of the equation.

In addition to finding a solution, the results of the integration revealed the roles of the terms of the Navier-Stokes equations in fluid flow. In particular, the gravity forces and $\partial p/\partial x$ are involved mainly in the parabolic as well as the forward motion of fluids; $\partial V_x/\partial t$ and $\partial^2 V_x/\partial x^2$ are involved in the periodic motion of fluids, and one may infer that as μ increases, the periodicity increases.

One should determine experimentally, if the ratio of the linear term $\partial V_x/\partial t$ to the nonlinear sum $V_x(\partial V_x/\partial x)+V_y(\partial V_x/\partial y)+V_z(\partial V_x/\partial z)$ is 1 to 3.

$$V_x(x,y,z,t) = \overbrace{-\frac{\rho g_x}{2\mu}(ax^2+by^2+cz^2)+C_1x+C_3y+C_5z+\frac{fg_x}{4}t+C_9}^{\textbf{Solution to the linearized Navier–Stokes equation}} \;;\; P(x) = d\rho g_x x$$

$$\boxed{\overbrace{-\mu\frac{\partial^2 V_x}{\partial x^2}-\mu\frac{\partial^2 V_x}{\partial y^2}-\mu\frac{\partial^2 V_x}{\partial z^2}+\frac{\partial p}{\partial x}+4\rho\frac{\partial V_x}{\partial t}=\rho g_x}^{\textbf{Linearized Equation}}}$$

2. Solutions of the Euler equation

Since one has solved the Navier-Stokes equation, one has also solved the Euler equation.

$$\text{Euler equation } (\mu=0): \quad \frac{\partial V_x}{\partial t}+V_x\frac{\partial V_x}{\partial x}+V_y\frac{\partial V_x}{\partial y}+V_z\frac{\partial V_x}{\partial z}+\frac{1}{\rho}\frac{\partial p}{\partial x}=g_x$$

$$\boxed{\begin{array}{l} V_x(x,y,z,t) - fg_x t \pm \sqrt{2hg_x x}+\dfrac{ng_x y}{V_y}+\dfrac{qg_x z}{V_z}+\underbrace{\dfrac{\psi_y(V_y)}{V_y}+\dfrac{\psi_z(V_z)}{V_z}}_{\text{arbitrary functions}}+C \\ P(x)=d\rho g_x x \qquad (f+h+n+q+d=1)\ V_y\neq 0,\ V_z\neq 0 \end{array}} \quad x\text{-direction}$$

$$\left.\frac{ng_x y}{V_y} = -\frac{n\beta_5 g_z t}{\beta_7} \pm \frac{(\sqrt{2\beta_8 g_z z})(ng_x)}{\beta_7 g_z}\right\} \text{B}$$

$$\left.\frac{qg_x z}{V_z} = \frac{(\beta_7 fg_z g_x^2 q - \beta_5 g_x g_z nq)tz \pm \sqrt{2hg_x x}\,\beta_7 g_x g_z qz \pm \sqrt{2\beta_8 g_z z}\; g_x^2 nqz}{\beta_7\beta_6 g_z^2 x - \beta_7 qg_z g_x z}\right\} \text{C}$$

Option 6
Solutions of 3-D Navier-Stokes Equations
(Method 2)

Here, the three equations below, will be added together; and a single equation will be integrated

$$\begin{cases} -\mu(\frac{\partial^2 V_x}{\partial x^2}+\frac{\partial^2 V_x}{\partial y^2}+\frac{\partial^2 V_x}{\partial z^2})+\frac{\partial p}{\partial x}+\rho(\frac{\partial V_x}{\partial t}+V_x\frac{\partial V_x}{\partial x}+V_y\frac{\partial V_x}{\partial y}+V_z\frac{\partial V_x}{\partial z})=\rho g_x \quad (1) \\ -\mu(\frac{\partial^2 V_y}{\partial x^2}+\frac{\partial^2 V_y}{\partial y^2}+\frac{\partial^2 V_y}{\partial z^2})+\frac{\partial p}{\partial y}+\rho(\frac{\partial V_y}{\partial t}+V_x\frac{\partial V_y}{\partial x}+V_y\frac{\partial V_y}{\partial y}+V_z\frac{\partial V_y}{\partial z})=\rho g_y \quad (2) \\ -\mu(\frac{\partial^2 V_z}{\partial x^2}+\frac{\partial^2 V_z}{\partial y^2}+\frac{\partial^2 V_z}{\partial z^2})+\frac{\partial p}{\partial z}+\rho(\frac{\partial V_z}{\partial t}+V_x\frac{\partial V_z}{\partial x}+V_y\frac{\partial V_z}{\partial y}+V_z\frac{\partial V_z}{\partial z})=\rho g_z \quad (3) \end{cases}$$

Step 1: Apply the axiom, if $a=b$ and $c=d$, then $a+c=b+d$; and therefore, add the left sides and add the right sides of the above equations . That is, $(1)+(2)+(3)=\rho g_x+\rho g_y+\rho g_z$

$$-\mu\frac{\partial^2 V_x}{\partial x^2}-\mu\frac{\partial^2 V_x}{\partial y^2}-\mu\frac{\partial^2 V_x}{\partial z^2}-\mu\frac{\partial^2 V_y}{\partial x^2}-\mu\frac{\partial^2 V_y}{\partial y^2}-\mu\frac{\partial^2 V_y}{\partial z^2}-\mu\frac{\partial^2 V_z}{\partial x^2}-\mu\frac{\partial^2 V_z}{\partial y^2}-\mu\frac{\partial^2 V_z}{\partial z^2}+\frac{\partial p}{\partial x}+\frac{\partial p}{\partial y}$$
$$+\frac{\partial p}{\partial z}+\rho\frac{\partial V_x}{\partial t}+\rho\frac{\partial V_y}{\partial t}+\rho\frac{\partial V_z}{\partial t}+\rho V_x\frac{\partial V_x}{\partial x}+\rho V_y\frac{\partial V_x}{\partial y}+\rho V_z\frac{\partial V_x}{\partial z}+\rho V_x\frac{\partial V_y}{\partial x}+\rho V_y\frac{\partial V_y}{\partial y}+\rho V_z\frac{\partial V_y}{\partial z}$$
$$+\rho V_x\frac{\partial V_z}{\partial x}+\rho V_y\frac{\partial V_z}{\partial y}+\rho V_z\frac{\partial V_z}{\partial z}=(\rho g_x+\rho g_y+\rho g_z) \qquad \text{(Three lines per equation)}$$

Let $\rho g_x+\rho g_y+\rho g_z=\rho G$, where $G=|g_x+g_y+g_z|$ to obtain

$$-\mu\frac{\partial^2 V_x}{\partial x^2}-\mu\frac{\partial^2 V_x}{\partial y^2}-\mu\frac{\partial^2 V_x}{\partial z^2}-\mu\frac{\partial^2 V_y}{\partial x^2}-\mu\frac{\partial^2 V_y}{\partial y^2}-\mu\frac{\partial^2 V_y}{\partial z^2}-\mu\frac{\partial^2 V_z}{\partial x^2}-\mu\frac{\partial^2 V_z}{\partial y^2}-\mu\frac{\partial^2 V_z}{\partial z^2}$$
$$+\frac{\partial p}{\partial x}+\frac{\partial p}{\partial y}+\frac{\partial p}{\partial z}++\rho\frac{\partial V_x}{\partial t}+\rho\frac{\partial V_y}{\partial t}+\rho\frac{\partial V_z}{\partial t}+\rho V_x\frac{\partial V_x}{\partial x}+\rho V_y\frac{\partial V_x}{\partial y}+\rho V_z\frac{\partial V_x}{\partial z}$$
$$+\rho V_x\frac{\partial V_y}{\partial x}+\rho V_y\frac{\partial V_y}{\partial y}+\rho V_z\frac{\partial V_y}{\partial z}+\rho V_x\frac{\partial V_z}{\partial x}+\rho V_y\frac{\partial V_z}{\partial y}+\rho V_z\frac{\partial V_z}{\partial z}=\rho G$$

Step 2: Solve the above 25-term equation using the ratio method. (24 ratio terms)

The ratio terms to be used are respectively the following: (Sum of the ratio terms = 1)

a ,b, c, d, f, m, n, q, r, β_1, β_2, β_3, β_4, β_5, β_6, λ_1, λ_2, λ_3, λ_4, λ_5, λ_6, λ_7, λ_8, λ_9

$-\mu\frac{\partial^2 V_x}{\partial x^2}=a\rho G$; $\quad -\mu\frac{\partial^2 V_x}{\partial y^2}=b\rho G$; $\quad -\mu\frac{\partial^2 V_x}{\partial z^2}=c\rho G$; $\quad -\mu\frac{\partial^2 V_y}{\partial x^2}=d\rho G$;

$-\mu\frac{\partial^2 V_y}{\partial y^2}=f\rho G$; $\quad -\mu\frac{\partial^2 V_y}{\partial z^2}=h\rho G$; $\quad -\mu\frac{\partial^2 V_z}{\partial x^2}=m\rho G$; $\quad -\mu\frac{\partial^2 V_z}{\partial y^2}=n\rho G$;

$-\mu\frac{\partial^2 V_z}{\partial z^2}=r\rho G$; $\quad \frac{\partial p}{\partial x}=\beta_1\rho G$; $\quad \frac{\partial p}{\partial y}=\beta_2\rho G$; $\quad \frac{\partial p}{\partial z}=\beta_3\rho G$;

$\rho\frac{\partial V_x}{\partial t}=\beta_4\rho G$; $\quad \rho\frac{\partial V_y}{\partial t}=\beta_5\rho G$; $\quad \rho\frac{\partial V_z}{\partial t}=\beta_6\rho G$; $\quad \rho V_x\frac{\partial V_x}{\partial x}=\lambda_1\rho G$;

$\rho V_y\frac{\partial V_x}{\partial y}=\lambda_2\rho G$; $\quad \rho V_z\frac{\partial V_x}{\partial z}=\lambda_3\rho G$; $\quad \rho V_x\frac{\partial V_y}{\partial x}=\lambda_4\rho G$; $\quad \rho V_y\frac{\partial V_y}{\partial y}=\lambda_5\rho G$;

$\rho V_z\frac{\partial V_y}{\partial z}=\lambda_6\rho G$; $\quad \rho V_x\frac{\partial V_z}{\partial x}=\lambda_7\rho G$; $\quad \rho V_y\frac{\partial V_z}{\partial y}=\lambda_8\rho G$; $\quad \rho V_z\frac{\partial V_z}{\partial z}=\lambda_9\rho G$

1 $\frac{\partial^2 V_x}{\partial x^2} = -\frac{a}{\mu}\rho G$ $\frac{\partial V_x}{\partial x} = -\frac{a}{\mu}\rho G x + C_1$ $V_x = -\frac{a}{\mu}\rho G\frac{x^2}{2} + C_1 x + C_2$	**2** $-\mu\frac{\partial^2 V_x}{\partial y^2} = b\rho G$ $\frac{\partial^2 V_x}{\partial y^2} = -\frac{b}{\mu}\rho G$ $\frac{\partial V_x}{\partial y} = -\frac{b}{\mu}\rho G y + C_3$ $V_x = -\frac{b}{\mu}\rho G\frac{y^2}{2} + C_3 y + C_4$	**3** $-\mu\frac{\partial^2 V_x}{\partial z^2} = c\rho G$ $-\mu\frac{\partial^2 V_x}{\partial z^2} = c\rho G$ $\frac{\partial^2 V_x}{\partial z^2} = -\frac{c}{\mu}\rho G$ $\frac{\partial V_x}{\partial z} = -\frac{c}{\mu}\rho G z + C_5$ $V_x = -\frac{c}{\mu}\rho G\frac{z^2}{2} + C_5 z + C_6$
4 $-\mu\frac{\partial^2 V_y}{\partial x^2} = d\rho G$ $-\mu\frac{\partial^2 V_y}{\partial x^2} = d\rho G$ $\frac{\partial^2 V_y}{\partial x^2} = -\frac{d}{\mu}\rho G$ $\frac{\partial V_y}{\partial x} = -\frac{d}{\mu}\rho G x + C_7$ $V_y = -\frac{d}{\mu}\rho G\frac{x^2}{2} + C_7 x + C_8$	**5** $-\mu\frac{\partial^2 V_y}{\partial y^2} = f\rho G$ $\frac{\partial^2 V_y}{\partial y^2} = -\frac{f}{\mu}\rho G$ $\frac{\partial V_y}{\partial y} = -\frac{f}{\mu}\rho G y + C_9$ $V_y = -\frac{f}{\mu}\rho G\frac{y^2}{2} + C_9 y + C_{10}$	**6** $-\mu\frac{\partial^2 V_y}{\partial z^2} = h\rho G$ $\frac{\partial^2 V_y}{\partial z^2} = -\frac{h}{\mu}\rho G$ $\frac{\partial V_y}{\partial z} = -\frac{h}{\mu}\rho G z + C_{11}$ $V_y = -\frac{h}{\mu}\rho G\frac{z^2}{2} + C_{11} z + C_{12}$
7 $-\mu\frac{\partial^2 V_z}{\partial x^2} = m\rho G$ $\frac{\partial^2 V_z}{\partial x^2} = -\frac{m}{\mu}\rho G$ $\frac{\partial V_z}{\partial x} = -\frac{m}{\mu}\rho G x + C_{13}$ $V_z = -\frac{m}{\mu}\rho G\frac{x^2}{2} + C_{13} x + C_{14}$	**8** $-\mu\frac{\partial^2 V_z}{\partial y^2} = n\rho G$ $\frac{\partial^2 V_z}{\partial y^2} = -\frac{n}{\mu}\rho G$ $\frac{\partial V_z}{\partial y} = -\frac{n}{\mu}\rho G y + C_{15}$ $V_z = -\frac{m}{\mu}\rho G\frac{y^2}{2} + C_{15} y + C_{16}$	**9** $-\mu\frac{\partial^2 V_z}{\partial z^2} = r\rho G$ $\frac{\partial^2 V_z}{\partial z^2} = -\frac{r}{\mu}\rho G$ $\frac{\partial V_z}{\partial z} = -\frac{r}{\mu}\rho G z + C_{17}$ $V_z = -\frac{r}{\mu}\rho G\frac{z^2}{2} + C_{17} z + C_{18}$

10 $\frac{\partial p}{\partial x} = \beta_1\rho G$ $\frac{dp}{dx} = \beta_1\rho G$ $P(x) = \beta_1\rho G x + C_{19}$	**11** $\frac{\partial p}{\partial y} = \beta_2\rho G$ $\frac{dp}{dy} = \beta_2\rho G$ $P(y) = \beta_2\rho G y + C_{20}$	**12** $\frac{\partial p}{\partial z} = \beta_3\rho G$ $\frac{dp}{dz} = \beta_3\rho G$ $P(z) = \beta_3\rho G z + C_{21}$

13 $\rho\dfrac{\partial V_x}{\partial t}=\beta_4\rho G$ $\dfrac{dV_x}{dt}=\beta_4 G$ $V_x=\beta_4 Gt+C_{22}$	**14** $\rho\dfrac{\partial V_y}{\partial t}=\beta_5\rho G$ $\dfrac{dV_y}{dt}=\beta_5 G$ $V_y=\beta_5 Gt+C_{23}$	**15** $\rho\dfrac{\partial V_z}{\partial t}=\beta_6\rho G$ $\dfrac{dV_z}{dt}=\beta_6 G$ $V_z=\beta_6 Gt+C_{24}$	**16** $\rho V_x\dfrac{\partial V_x}{\partial x}=\lambda_1\rho G$ $V_x\dfrac{\partial V_x}{\partial x}=\lambda_1 G$ $V_x\dfrac{dV_x}{dx}=\lambda_1 G$ $V_x dV_x=\lambda_1 G\,dx$ $\dfrac{V^2{}_x}{2}=\lambda_1 G\,x$ $V^2{}_x=2\lambda_1 G\,x$ $V_x=\pm\sqrt{2\lambda_1 G\,x}+C_{25}$
17 $\rho V_y\dfrac{\partial V_x}{\partial y}=\lambda_2\rho G$ $V_y\dfrac{dV_x}{dy}=\lambda_2 G$ $V_y dV_x=\lambda_2 G\,dy$ $V_y V_x=\lambda_2 G\,y+\psi_y(V_y)$ $V_x=\dfrac{\lambda_2 G\,y}{V_y}+\dfrac{\psi_y(V_y)}{V_y}$	**18** $\rho V_z\dfrac{\partial V_x}{\partial z}=\lambda_3\rho G$ $V_z\dfrac{dV_x}{dz}=\lambda_3 G$ $V_z dV_x=\lambda_3 G\,dz$ $V_z V_x=\lambda_3 G\,z+\psi_z(V_z)$ $V_x=\dfrac{\lambda_3 G\,z}{V_z}+\dfrac{\psi_z(V_z)}{V_z}$	**19** $\rho V_x\dfrac{\partial V_y}{\partial x}=\lambda_4\rho G$ $V_x\dfrac{dV_y}{dx}=\lambda_4 G$ $V_x dV_y=\lambda_4 G\,dx$ $V_x V_y=\lambda_4 G\,x+\psi_x(V_x)$ $V_y=\dfrac{\lambda_4 G\,x}{V_x}+\dfrac{\psi_x(V_x)}{V_x}$	**20** $\rho V_y\dfrac{\partial V_y}{\partial y}=\lambda_5\rho G$ $V_y\dfrac{dV_y}{dy}=\lambda_5 G$ $V_y dV_y=\lambda_5 G\,dy$ $\dfrac{V_y^2}{2}=\lambda_5 G\,y$ $V_y^2=2\lambda_5 G\,y$ $V_y=\pm\sqrt{2\lambda_5 G\,y}+C_{26}$
21 $\rho V_z\dfrac{\partial V_y}{\partial z}=\lambda_6\rho G$ $V_z\dfrac{dV_y}{dz}=\lambda_6 G$ $V_z dV_y=\lambda_6 G\,dz$ $V_z V_y=\lambda_6 G\,z+\psi_z(V_z)$ $V_y=\dfrac{\lambda_6 G\,z}{V_z}+\dfrac{\psi_z(V_z)}{V_z}$	**22** $\rho V_x\dfrac{\partial V_z}{\partial x}=\lambda_7\rho G$ $V_x\dfrac{dV_z}{dx}=\lambda_7 G$ $V_x dV_z=\lambda_7 G\,dx$ $V_x V_z=\lambda_7 G\,x+\psi_x(V_x)$ $V_z=\dfrac{\lambda_7 G\,x}{V_x}+\dfrac{\psi_x(V_x)}{V_x}$	**23** $\rho V_y\dfrac{\partial V_z}{\partial y}=\lambda_8\rho G$ $V_y\dfrac{dV_z}{dy}=\lambda_8 G$ $V_y dV_z=\lambda_8 G\,dy$ $V_y V_z=\lambda_8 G\,y+\psi_y(V_y)$ $V_z=\dfrac{\lambda_8 G\,y}{V_y}+\dfrac{\psi_y(V_y)}{V_y}$	**24** $\rho V_z\dfrac{\partial V_z}{\partial z}=\lambda_9\rho G$ $V_z\dfrac{dV_z}{dz}=\lambda_9 G$ $V_z dV_z=\lambda_9 G\,dz$ $\dfrac{V_z^2}{2}=\lambda_9 G\,z$ $V_z^2=2\lambda_9 G\,z$ $V_z=\pm\sqrt{2\lambda_9 G\,z}+C_{27}$

Step 3 : One Collects the integrals of the sub-equations, above, for $V_x, V_y, V_z, P(x), P(y), P(z)$

For V_x, $P(x)$	For V_y, $P(y)$	For V_z, $P(z)$
Sum of integrals from sub-equations #1, #2, #3, #13, #16, #17, #18, #10	Sum of integrals from sub-equations #4, #5, #6, #14, #19, #20, #21,#11	Sum of integrals from sub-equations #7, #8, #9, #15, #22, #23, #24, #12,
$V_x = -\frac{a}{\mu}\rho G\frac{x^2}{2} + C_1x + C_2$	$V_y = -\frac{d}{\mu}\rho G\frac{x^2}{2} + C_7x + C_8$	$V_z = -\frac{m}{\mu}\rho G\frac{x^2}{2} + C_{13}x + C_{14}$
$V_x = -\frac{b}{\mu}\rho G\frac{y^2}{2} + C_3y + C_4$	$V_y = -\frac{f}{\mu}\rho G\frac{y^2}{2} + C_9y + C_{10}$	$V_z = -\frac{m}{\mu}\rho G\frac{y^2}{2} + C_{15}y + C_{16}$
$V_x = -\frac{c}{\mu}\rho G\frac{z^2}{2} + C_5z + C_6$	$V_y = -\frac{h}{\mu}\rho G\frac{z^2}{2} + C_{11}z + C_{12}$	$V_z = -\frac{r}{\mu}\rho G\frac{z^2}{2} + C_{17}z + C_{18}$
$V_x = \beta_4Gt + C_{22}$	$V_y = \beta_5Gt + C_{21}$	$V_z = \beta_6Gt + C_{24}$
$V_x = \pm\sqrt{2\lambda_1G\,x} + C_{25}$	$V_y = \frac{\lambda_4G\,x}{V_x} + \frac{\psi_x(V_x)}{V_x}$	$V_z = \frac{\lambda_7G\,x}{V_x} + \frac{\psi_x(V_x)}{V_x}$ #
$V_x = \frac{\lambda_2G\,y}{V_y} + \frac{\psi_y(V_y)}{V_y}$	$V_y = \pm\sqrt{2\lambda_5G\,y} + C_{26}$	$V_z = \frac{\lambda_8G\,y}{V_y} + \frac{\psi_y(V_y)}{V_y}$
$V_x = \frac{\lambda_3G\,z}{V_z} + \frac{\psi_z(V_z)}{V_z}$	$V_y = \frac{\lambda_6G\,z}{V_z} + \frac{\psi_z(V_z)}{V_z}$	$V_z = \pm\sqrt{2\lambda_9G\,z} + C_{27}$
$P(x) = \beta_1\rho Gx + C_{19}$	$P(y) = \beta_2\rho Gy + C_{20}$	$P(z) = \beta_3\rho Gz + C_{21}$

From above,

For V_x, Sum of integrals from sub-equations #1, #2, #3, #13, #16, #17, #18, #10

$V_x(x,y,z,t)$

$$= -\frac{a}{\mu}\rho G\frac{x^2}{2} + C_1x - \frac{b}{\mu}\rho G\frac{y^2}{2} + C_3y - \frac{c}{\mu}\rho G\frac{z^2}{2} + C_5z + \beta_4Gt \pm \sqrt{2\lambda_1G\,x} + \frac{\lambda_2G\,y}{V_y} + \frac{\lambda_3G\,z}{V_z} + \underbrace{\frac{\psi_y(V_y)}{V_y} + \frac{\psi_z(V_z)}{V_z}}_{\text{arbitrary functions}}$$

$$P(x) = \beta_1\rho Gx + C_{19}$$

For V_y : Sum of integrals from sub-equations #4, #5, #6,#14, #19, #20, #21,#11

$V_y(x,y,z,t)$

$$= -\frac{d}{\mu}\rho G\frac{x^2}{2} + C_7x - \frac{f}{\mu}\rho G\frac{y^2}{2} + C_9y - \frac{h}{\mu}\rho G\frac{z^2}{2} + C_{11}z + \beta_5Gt + \pm\sqrt{2\lambda_5G\,y} + \frac{\lambda_4G\,x}{V_x} + \frac{\lambda_6G\,z}{V_z} + \underbrace{\frac{\psi_x(V_x)}{V_x}\frac{\psi_z(V_z)}{V_z}}_{\text{arbitrary functions}}$$

$$P(y) = \beta_2\rho Gy + C_{20}$$

For V_z: Sum of integrals from sub-equations #7, #8, #9,#15, #22, #23, #24, #12,

$$V_z = -\frac{m}{\mu}\rho G\frac{x^2}{2} + C_{13}x - \frac{n}{\mu}\rho G\frac{y^2}{2} + C_{15}y - \frac{r}{\mu}\rho G\frac{z^2}{2} + C_{17}z + \beta_6Gt \pm \sqrt{2\lambda_9G\,z} + \frac{\lambda_7G\,x}{V_x} + \frac{\lambda_8G\,y}{V_y} + \underbrace{\frac{\psi_x(V_x)}{V_x} + \frac{\psi_y(V_y)}{V_y}}_{\text{arbitrary functions}}$$

$$P(z) = \beta_3\rho Gz + C_{21}$$

Step 4: Simplify the sums of the integrals from above..(**Method 2 solutions of N-S equations**

$$V_x(x,y,z,t) = -\frac{\rho G}{2\mu}(ax^2 + by^2 + cz^2) + C_1 x + C_3 y + C_5 z + \beta_4 Gt \pm \sqrt{2\lambda_1 G\, x} + \frac{\lambda_2 G\, y}{V_y} + \frac{\lambda_3 G\, z}{V_z}$$

$$P(x) = \beta_1 \rho G x + C_{19} \qquad (V_y \neq 0,\ V_z \neq 0) \qquad + \underbrace{\frac{\psi_y(V_y)}{V_y} + \frac{\psi_z(V_z)}{V_z}}_{\text{arbitrary functions}}$$

$$V_y(x,y,z,t) = -\frac{\rho G}{2\mu}(dx^2 + fy^2 + hz^2) + C_7 x + C_9 y + C_{11} z + C_{10}\beta_5 Gt \pm \sqrt{2\lambda_5 G\, y} + \frac{\lambda_4 G\, x}{V_x} + \frac{\lambda_6 G\, z}{V_z}$$

$$P(y) = \beta_2 \rho G y + C_{20} \qquad (V_x \neq 0,\ V_z \neq 0) \qquad + \underbrace{\frac{\psi_x(V_x)}{V_x} + \frac{\psi_z(V_z)}{V_z}}_{\text{arbitrary functions}}$$

$$V_z(x,y,z,t) = -\frac{\rho G}{2\mu}(mx^2 + ny^2 + rz^2) + C_{13} x + C_{15} y + C_{17} z + \beta_6 Gt \pm \sqrt{2\lambda_9 G\, z} + \frac{\lambda_7 G\, x}{V_x} + \frac{\lambda_8 G\, y}{V_y}$$

$$P(z) = \beta_3 \rho G z + C_{21} \qquad (V_y \neq 0,\ V_y \neq 0) \qquad + \underbrace{\frac{\psi_x(V_x)}{V_x} + \frac{\psi_y(V_y)}{V_y}}_{\text{arbitrary functions}}$$

The above are solutions for $V_x\ V_y,\ V_z\ \ P(x),\ P(y),\ P(z)$ .of the Navier-Stokes Equations

Comparison of Method 1 (Option 4) and Method 2 (Option 6) of Solutions of Navier-Stokes Equations

Method 1: x–direction solution of Navier-Stokes equation

$$V_x(x,y,z,t) = -\frac{\rho g_x}{2\mu}(ax^2 + by^2 + cz^2) + C_1x + C_3y + C_5z + fg_xt \pm \sqrt{2hg_xx + \frac{ng_xy}{V_y} + \frac{qg_xz}{V_z} + \underbrace{\frac{\psi_y(V_y)}{V_y} + \frac{\psi_z(V_z)}{V_z}}_{\text{arbitrary functions}} + C_9}$$

$$P(x) = d\rho g_xx;\ (a+b+c+d+h+n+q=1)\quad (V_y \neq 0,\ V_z \neq 0) \qquad \text{(A)}$$

Method 2: x–direction solution of Navier-Stokes equation

$$V_x(x,y,z,t) = -\frac{\rho G}{2\mu}(ax^2 + by^2 + cz^2) + C_1x + C_3y + C_5z + \beta_4Gt \pm \sqrt{2\lambda_1Gx + \frac{\lambda_2Gy}{V_y} + \frac{\lambda_3Gz}{V_z} + \underbrace{\frac{\psi_y(V_y)}{V_y} + \frac{\psi_z(V_z)}{V_z}}_{\text{arbitrary functions}}}$$

$$P(x) = \beta_1\rho Gx + C_{19} \qquad (V_y \neq 0,\ V_z \neq 0) \qquad \textbf{(B)}$$

It is pleasantly surprising that the above solutions (A) and (B) are almost identical (except for the constants), even though they were obtained by different approaches as in Option 4 and Option 6. Such an agreement confirms the validity of the solution method for the system of magneto-hydrodynamic equations (see viXra:1405.0251.. For the system of magnetohydrodynamic equations, there is only a single "driver" equation. For the system of N-S equations, there are three driver equations, since each equation contains the gravity term. Therefore, one was able to solve each of the three simultaneous equations separately (as in Method 1); but in addition, one obtained an identical solution (except for the constants) in solving the simultaneous N-S system by adding the three equations in the system and integrating a single driver equation. In Method 1, the gravity term was ρg. In Method 2, the gravity term was ρG, where G is the magnitude of the vector sum of the gravity terms. Note that in Method 1, the sum of the ratio terms (8 ratio terms for each equation) equals unity, but in Method 2, the sum of the ratio t erms (24 ratio terms) for the single driver equation solved equals unity. Note that in Method 2, only a single "driver" equation was solved, but in Method 1, three "driver" equations were solved. In Method 2, one could say that the system of N-S equations was "more simultaneously" solved than in Method 1.

To summarize, solving the Navier-Stokes equations by the first method helped one to solve the magnetohydrodynamic equations (not presented in this paper.. See viXra:1405.0251) and solving the magnetohydrodynamic equations encouraged one to solve the Navier-Stokes equations by the second method.
(" Navier-Stokes equations "scratched the back" of magnetohydrodynamic equations; and in return, magnetohydrodynamic equations "scratched the back" of Navier-Stokes equations")

About integrating only a single equation
If one asked for help in solving the N-S equations, and one was told to add the three equations together and then solve them, one would think that one was being given a nonsensical advice; but now, after studying the above Option 6 method, one would appreciate such a suggestion.

Option 7
Solutions of 3-D Linearized Navier-Stokes Equations
Method 2

Here, the three equations below, will be added together; and a single equation will be integrated.

$$\begin{cases} -\mu\frac{\partial^2 V_x}{\partial x^2}-\mu\frac{\partial^2 V_x}{\partial y^2}-\mu\frac{\partial^2 V_x}{\partial z^2}+\frac{\partial p}{\partial x}+4\rho\frac{\partial V_x}{\partial t}=\rho g_x \quad (1) \\ -\mu\frac{\partial^2 V_y}{\partial x^2}-\mu\frac{\partial^2 V_y}{\partial y^2}-\mu\frac{\partial^2 V_y}{\partial z^2}+\frac{\partial p}{\partial y}+4\rho\frac{\partial V_y}{\partial t}=\rho g_y \quad (2) \\ -\mu\frac{\partial^2 V_z}{\partial x^2}-\mu\frac{\partial^2 V_z}{\partial y^2}-\mu\frac{\partial^2 V_z}{\partial z^2}+\frac{\partial p}{\partial z}+4\rho\frac{\partial V_z}{\partial t}=\rho g_z \quad (3) \end{cases}$$

Step 1: Apply the axiom, if $a = b$ and $c = d$, then $a + c = b + d$; and therefore, add the left sides and add the right sides of the above equations . That is, $(1)+(2)+(3)=\rho g_x+\rho g_y+\rho g_z$

$$-\mu\frac{\partial^2 V_x}{\partial x^2}-\mu\frac{\partial^2 V_x}{\partial y^2}-\mu\frac{\partial^2 V_x}{\partial z^2}+\frac{\partial p}{\partial x}+4\rho\frac{\partial V_x}{\partial t}-\mu\frac{\partial^2 V_y}{\partial x^2}\ \mu\frac{\partial^2 V_y}{\partial y^2}-\mu\frac{\partial^2 V_y}{\partial z^2}+\frac{\partial p}{\partial y}+4\rho\frac{\partial V_y}{\partial t}$$
$$-\mu\frac{\partial^2 V_z}{\partial x^2}-\mu\frac{\partial^2 V_z}{\partial y^2}-\mu\frac{\partial^2 V_z}{\partial z^2}+\frac{\partial p}{\partial z}4\rho\frac{\partial V_z}{\partial t}=\rho g_x+\rho g_y+\ \rho g_z$$ (Two lines per equation)

Let $\rho g_x+\rho g_y+\rho g_z=\rho G$, where $G=\left|g_x+g_y+g_z\right|$ to obtain

$$-\mu\frac{\partial^2 V_x}{\partial x^2}-\mu\frac{\partial^2 V_x}{\partial y^2}-\mu\frac{\partial^2 V_x}{\partial z^2}+\frac{\partial p}{\partial x}+4\rho\frac{\partial V_x}{\partial t}-\mu\frac{\partial^2 V_y}{\partial x^2}-\mu\frac{\partial^2 V_y}{\partial y^2}-\mu\frac{\partial^2 V_y}{\partial z^2}+\frac{\partial p}{\partial y}+4\rho\frac{\partial V_y}{\partial t}$$
$$-\mu\frac{\partial^2 V_z}{\partial x^2}-\mu\frac{\partial^2 V_z}{\partial y^2}-\mu\frac{\partial^2 V_z}{\partial z^2}+\frac{\partial p}{\partial z}+4\rho\frac{\partial V_z}{\partial t}=\rho G$$ (Two lines per equation)

Step 2: Solve the above 15-term equation using the ratio method. (14 ratio terms)
The ratio terms to be used are respectively the following: (Sum of the ratio terms = 1)
$a, b, c, d, f, h, j, m, n, q, r, s, u, v, w.$ (Sum of the ratio terms = 1)

$$-\mu\frac{\partial^2 V_x}{\partial x^2}=a\rho G;\quad -\mu\frac{\partial^2 V_x}{\partial y^2}=b\rho G;\quad -\mu\frac{\partial^2 V_x}{\partial z^2}=c\rho G;\quad \frac{\partial p}{\partial x}=d\rho G$$

$$4\rho\frac{\partial V_x}{\partial t}=f\rho G;\quad -\mu\frac{\partial^2 V_y}{\partial x^2}=h\rho G;\quad -\mu\frac{\partial^2 V_y}{\partial y^2}=j\rho G;\quad -\mu\frac{\partial^2 V_y}{\partial z^2}=m\rho G;$$

$$\frac{\partial p}{\partial y}=n\rho G;\quad 4\rho\frac{\partial V_y}{\partial t}=q\rho G;\quad -\mu\frac{\partial^2 V_z}{\partial x^2}=r\rho G;\quad -\mu\frac{\partial^2 V_z}{\partial y^2}=s\rho G;$$

$$-\mu\frac{\partial^2 V_z}{\partial z^2}=u\rho G;\quad \frac{\partial p}{\partial z}=v\rho G;\quad 4\rho\frac{\partial V_z}{\partial t}=w\rho G$$

1. $-\mu\frac{\partial^2 V_x}{\partial x^2}=a\rho G$ $\frac{\partial^2 V_x}{\partial x^2}=-\frac{a}{\mu}\rho G$ $\frac{\partial V_x}{\partial x}=-\frac{a}{\mu}\rho Gx+C_1$ $V_x=-\frac{\rho Ga}{2\mu}x^2+C_1x+C_2$	**2.** $-\mu\frac{\partial^2 V_x}{\partial y^2}=b\rho G$ $\frac{\partial^2 V_x}{\partial y^2}=-\frac{b}{\mu}\rho G$ $\frac{\partial V_x}{\partial y}=-\frac{b}{\mu}\rho Gy+C_3$ $V_x=-\frac{\rho Gb}{2\mu}y^2+C_3y+C_4$	**3** $-\mu\frac{\partial^2 V_x}{\partial z^2}=c\rho G$ $\frac{\partial^2 V_x}{\partial z^2}=-\frac{c}{\mu}\rho G$ $\frac{\partial V_x}{\partial z}=-\frac{c}{\mu}\rho Gz+C_5$ $V_x=-\frac{\rho Gc}{2\mu}z^2+C_5z+C_6$

4 $\frac{\partial p}{\partial x} = d\rho G$ $P(x) = d\rho Gx + C_7$	**5** $4\rho\frac{\partial V_x}{\partial t} = f\rho G$ $\frac{\partial V_x}{\partial t} = \frac{fG}{4}$ $V_x = \frac{fG}{4}t + C_8$	**6** $-\mu\frac{\partial^2 V_y}{\partial x^2} = h\rho G$ $\frac{\partial^2 V_y}{\partial x^2} = -\frac{h}{\mu}\rho G$ $\frac{\partial V_y}{\partial x} = -\frac{h}{\mu}\rho Gx + C_9$ $V_y = -\frac{\rho Gh}{2\mu}x^2 + C_9 x + C_{10}$
7 $-\mu\frac{\partial^2 V_y}{\partial y^2} = j\rho G$ $\frac{\partial^2 V_y}{\partial y^2} = -\frac{j}{\mu}\rho G$ $\frac{\partial V_y}{\partial y} = -\frac{j}{\mu}\rho Gy + C_{11}$ $V_y = -\frac{\rho Gj}{2\mu}y^2 + C_{11}y + C_{12}$	**8** $-\mu\frac{\partial^2 V_y}{\partial z^2} = m\rho G$ $\frac{\partial^2 V_y}{\partial z^2} = -\frac{m}{\mu}\rho G$ $\frac{\partial V_y}{\partial z} = -\frac{m}{\mu}\rho Gz + C_{13}$ $V_y = -\frac{\rho Gm}{2\mu}z^2 + C_{13}z + C_{14}$	**9** $\frac{\partial p}{\partial y} = n\rho G$ $P(y) = n\rho Gy + C_{15}$
10 $4\rho\frac{\partial V_y}{\partial t} = q\rho G$ $\frac{\partial V_y}{\partial t} = \frac{qG}{4}$ $V_y = \frac{qG}{4}t + C_{16}$	**11** $-\mu\frac{\partial^2 V_z}{\partial x^2} = r\rho G$ $\frac{\partial^2 V_z}{\partial x^2} = -\frac{r}{\mu}\rho G$ $\frac{\partial V_z}{\partial x} = -\frac{r}{\mu}\rho Gx + C_{17}$ $V_z = -\frac{\rho Gr}{2\mu}x^2 + C_{17}x + C_{18}$	**12** $-\mu\frac{\partial^2 V_z}{\partial y^2} = s\rho G$ $\frac{\partial^2 V_z}{\partial y^2} = -\frac{s}{\mu}\rho G$ $\frac{\partial V_z}{\partial y} = -\frac{s}{\mu}\rho Gy + C_{19}$ $V_z = -\frac{\rho Gs}{2\mu}y^2 + C_{19}y + C_{20}$

13 $-\mu\frac{\partial^2 V_z}{\partial z^2} = u\rho G$ $\frac{\partial^2 V_z}{\partial z^2} = -\frac{u}{\mu}\rho G$ $\frac{\partial V_z}{\partial z} = -\frac{u}{\mu}\rho Gz + C_{21}$ $V_z = -\frac{\rho Gu}{2\mu}z^2 + C_{21}z + C_{22}$	**14** $\frac{\partial p}{\partial z} = v\rho G$ $P(z) = v\rho Gz + C_{23}$	**15** $4\rho\frac{\partial V_z}{\partial t} = w\rho G$ $\frac{\partial V_z}{\partial t} = \frac{wG}{4}$ $V_z = \frac{wG}{4}t + C_{24}$

Step 3: One collect the solutions from Step 2 for (V_x, V_y, V_z, $P(x)$, $P(y)$, $P(z)$)

For V_x, Sum of integrals from sub-equations #1, #2, #3, #5, and for $P(x)$, from #4

$$V_x(x,y,z,t) = -\frac{\rho G}{2\mu}(ax^2 + by^2 + cz^2) + C_1x + C_3y + C_5z + \frac{fG}{4}t + C_8; \quad P(x) = d\rho Gx + C_7$$

For V_y Sum of integrals from sub-equations #6, #7, #8, #10, and for $P(y)$, from #9.

$$V_y(x,y,z,t) = -\frac{\rho G}{2\mu}(hx^2 + jy^2 + mz^2) + C_9x + C_{11}y + C_{13}z + \frac{qG}{4}t + C_{16}; \quad P(y) = n\rho Gy + C_{15}$$

For V_z: Sum of integrals from sub-equations #11, #12, #13, and for $P(z)$, from #14

$$V_z(x,y,z,t) = -\frac{\rho G}{2\mu}(rx^2 + sy^2 + uz^2) + C_{17}x + C_{19}y + C_{21}z + \frac{wG}{4}t + C_{24}; \quad P(z) = v\rho Gz + C_{23}$$

Comparison of the above methods for the solutions of Linearized Navier-Stokes Equations

Note below that the solutions by the two different methods are the same except for the constants involved. Now, one has two different methods for solving the system of Navier-Stokes equations. Such an agreement and consistency confirm the validity of the method used in solving the magnetohydrodynamic equations.

Solutions by Method 1

$$V_x(x,y,z,t) = -\frac{\rho g_x}{2\mu}(ax^2 + by^2 + cz^2) + C_1x + C_3y + C_5z + \frac{fg_x}{4}t + C_9; \quad P(x) = d\rho gx$$

$$V_y(x,y,z,t) = -\frac{\rho g_y}{2\mu}(hx^2 + jy^2 + mz^2) + C_1x + C_3y + C_5z + \frac{qg_y}{4}t + C; \quad P(y) = n\rho g_y y$$

$$V_z(x,y,z,t) = -\frac{\rho g_z}{2\mu}(rx^2 + sy^2 + uz^2) + C_1x + C_3y + C_5z + \frac{wg_z}{4}t + C; \quad P(z) = v\rho g_z z$$

Solutions by Method 2

$$V_x(x,y,z,t) = -\frac{\rho G}{2\mu}(ax^2 + by^2 + cz^2) + C_1x + C_3y + C_5z + \frac{fG}{4}t + C_8; \quad P(x) = d\rho Gx + C_7$$

$$V_y(x,y,z,t) = -\frac{\rho G}{2\mu}(hx^2 + jy^2 + mz^2) + C_9x + C_{11}y + C_{13}z + \frac{qG}{4}t + C_{16}; \quad P(y) = n\rho Gy + C_{15}$$

$$V_z(x,y,z,t) = -\frac{\rho G}{2\mu}(rx^2 + sy^2 + uz^2) + C_{17}x + C_{19}y + C_{21}z + \frac{wG}{4}t + C_{24}; \quad P(z) = v\rho Gz + C_{23}$$

Overall Conclusion

The Navier-Stokes (N-S) equations in 3-D and 4-D have been solved analytically for the first time by two different methods. In Method 1, the three equations were separately integrated. In Method 2, the three equations were first added together and a single equation was integrated. The solutions from these two methods were the same, except for the constants involved. The N-S solution is unique. The experience gained in solving the linearized equation helped the author to propose a new law, the law of definite ratio for incompressible fluid flow. This law states that in incompressible fluid flow, the other terms of the fluid flow equation divide the gravity term in a definite ratio, and each term utilizes gravity to function. The sum of the terms of the ratio is always unity. The application of this law helped speed-up the solutions of the non-linearized N-S equations, since there was no more experimentation as to the subject of the equation. It was also shown that without gravity forces on earth, there would be no incompressible fluid flow on earth as is known.
The solutions and relations revealed the role of each term of the Navier-Stokes equations in fluid flow. Most importantly, the gravity term is the indispensable term in fluid flow, and it is involved in the parabolic as well as the forward motion. The pressure gradient term is also involved in the parabolic motion. The viscosity terms are involved in parabolic, periodic and decreasingly exponential motion. As the viscosity increases, periodicity increases. The variable acceleration term is also involved in the periodic and decreasingly exponential motion. The convective acceleration terms produce square root function behavior and behavior of fractional terms containing square root functions with variables in the denominator. In terms of the velocity profile, he x–direction solution consists of linear, parabolic, and hyperbolic terms. The firs t three terms characterize polynomial parabolas. The characteristic curve for the integral of the x–nonlinear term is a radical parabola. The integral of the y–nonlinear term is similar parabolically to that of the x–nonlinear term. The integral of the z–nonlinear term is a combination of two radical parabolas and a hyperbola. If the above x–direction flow is repeated simultaneously in the y– and z– directions, the flow is chaotic and consequently turbulent.
The following statements can be made:
(a) The N-S equations have unique solutions; (b) The N-S equations have parabolic solutions; **3.** The N-S equations have square root function solutions. 4. The N-S equations do not have periodic solutions but have periodic relations. **5.**. The N-S equations do not have decreasingly exponential solutions but have decreasingly exponential relations.
In applications, the ratio terms a, b, c, d, f, h, n, q and others may perhaps be determined using information such as initial and boundary conditions or may have to be determined experimentally. The author came to the experimental determination conclusion after referring to preliminaries.. The question is how did the grandmother determine the terms of the ratio for her grandchildren? Note that so far as the general solutions of the N-S equations are concerned, one needs not find the specific values of the ratio terms.
Finally, for any fluid flow design, one should always maximize the role of gravity for cost-effectiveness, durability, and dependability. Perhaps, Newton's law for fluid flow should read "Sum of everything else equals ρg" ; and this would imply the proposed new law that the other terms divide the gravity term in a definite ratio, and each term utilizes gravity to function.

P.S.

Maples software was used to help express the implicit terms in terms of x, y, z, and t, by solving System P (p.261, 262). None of the academic programs could solve the system of solutions M. The author would like to find a software that can solve the original system , System M, for comparison purposes.

Option 8

Spin-off: CMI Millennium Prize Problem Requirements

Proof 1

For the linearized Navier-Stokes equations

Proof of the existence of solutions of the Navier-Stokes equations

Since from page 11, it has been shown that the smooth equations given by $V_x(x,y,z,t) = -\frac{\rho g_x}{2\mu}(ax^2 + by^2 + cz^2) + C_1x + C_3y + C_5z + \frac{fg_x}{4}t + C_9$; $P(x) = d\rho g_x x$ are solutions of the linearized equation, $-\mu(\frac{\partial^2 V_x}{\partial x^2} + \frac{\partial^2 V_x}{\partial y^2} + \frac{\partial^2 V_x}{\partial z^2}) + \frac{\partial p}{\partial x} + 4\rho\frac{\partial V_x}{\partial t} = \rho g_x$, it has been shown that smooth solutions to the above differential equation exist. and the proof is complete.

From, above, if $y = 0,\ z = 0,$ $V_x(x,t) = -\frac{\rho g_x}{2\mu}ax^2 + C_1x + \frac{fg_x}{4}t + C_9$; $P(x) = d\rho gx + C_{10}$

Therefore, $V_x(x,0) = V_x^0(x) = -\frac{\rho g_x}{2\mu}ax^2 + C_{10}x + C_9$

Finding $P(x,t)$

1. $V_x(x,t) = -\frac{\rho g_x}{2\mu}(ax^2) + C_1x + \frac{fg_x}{4}t + C_9$; $P(x) = d\rho g_x x$ **2**. $\frac{\partial p}{\partial x} = d\rho g$;

Required: To find $P(x,t)$ (that is, find a formula for P in terms of x and t)

$$\frac{dp}{dt} = \frac{dp}{dx}\frac{dx}{dt}$$

$$\frac{dp}{dt} = \frac{dp}{dx}V_x \qquad (\frac{dx}{dt} = V_x)$$

$$\frac{dp}{dt} = d\rho g_x\left(-\frac{\rho g_x}{2\mu}(ax^2) + C_1x + \frac{fg_x}{4}t + C_9\right) \qquad (\frac{dp}{dx} = d\rho g_x)$$

$$P(x,t) = \int d\rho g_x\left(-\frac{\rho g_x}{2\mu}(ax^2) + C_1x + \frac{fg_x}{4}t + C_9\right)dt$$

$$P(x,t) = d\rho g_x\left(-\frac{a\rho g_x}{2\mu}x^2t + C_1xt + \frac{fg_x}{8}t^2 + C_9\text{t}\right) + C_{10}$$

For the corresponding coverage for the original Navier-Stokes equation, see the next page

Proof 2

For the Non-linearized Navier-Stokes equations (Original Equations)

Proof of the existence of solutions of the Navier-Stokes equations

From page 22, if $y=0,\ z=0$ in

Solution to Linear part

$$V_x(x,y,z,t) = -\frac{\rho g_x}{2\mu}(ax^2+by^2+cz^2)+C_1x+C_3y+C_5z+\underbrace{fg_xt}_{\text{continued}} \pm\sqrt{2hg_xx}+\frac{ng_xy}{V_y}+\frac{qg_xz}{V_z}+\frac{\psi_y(V_y)}{V_y}+\frac{\psi_z(V_z)}{V_z}$$

solution of Euler equation

$$P(x)=d\rho g_x x$$

one obtains

$$\boxed{V_x(x,t)=-\frac{\rho g_x}{2\mu}ax^2+C_1x+fg_xt\pm\sqrt{2hg_xx}+C_9;\quad P(x)=d\rho g_xx;}$$

$$V_x(x,0)=V_x^0(x)=-\frac{\rho g_x}{2\mu}ax^2+C_1x\pm\sqrt{2hg_xx}+C_9;\quad P(x)=d\rho g_xx;$$

Since previously, from p.21, it has been shown that the smooth equations given by

$V_x(x,t)=-\frac{\rho g_x}{2\mu}ax^2+C_1x+fg_xt\pm\sqrt{2hg_xx}+C_9;\ P(x)=d\rho g_xx;$ are solutions of

$-\mu\frac{\partial^2 V_x}{\partial x^2}+\frac{\partial p}{\partial x}+\rho\frac{\partial V_x}{\partial t}+\rho V_x\frac{\partial V_x}{\partial x}=\rho g_x$ (deleting the $y-$ and $z-$ terms of (A)), p.20, one has shown that smooth solutions to the above differential equation exist, and the proof is complete.

Finding $P(x,t)$:

1. $V_x(x,t)=-\frac{\rho g_x}{2\mu}ax^2+C_1x+fg_xt\pm\sqrt{2hg_xx}+C_9;\ P(x)=d\rho g_xx;$ **2.** $\frac{\partial p}{\partial x}=d\rho g;$

$$\frac{dp}{dt}=\frac{dp}{dx}\frac{dx}{dt}$$

$$\frac{dp}{dt}=\frac{dp}{dx}V_x \qquad (\frac{dx}{dt}=V_x)$$

$$\frac{dp}{dt}=d\rho g_x\left(-\frac{\rho g_x}{2\mu}(ax^2)+C_1x\pm\sqrt{2hg_xx}+fg_xt+C_9\right) \qquad (\frac{dp}{dx}=d\rho g_x)$$

$$P(x,t)=\int d\rho g_x\left(-\frac{\rho g_x}{2\mu}(ax^2)+C_1x\pm\sqrt{2hg_xx}+fg_xt+C_9\right)dt$$

$$P(x,t)=d\rho g_x\left(-\frac{a\rho g_x}{2\mu}x^2t+C_1xt\pm t\sqrt{2hg_xx}+\frac{fg_xt^2}{2}+C_9t\right)+C_{10}$$

Appendix 8

Magnetohydrodynamic Equations Solutions

Abstract

The system of magnetohydrodynamic (MHD) equations have been solved analytically in this paper. The author applied the technique used in solving the Navier-Stokes equations and applied a new law, the law of definite ratio for MHD. This law states that in MHD, the other terms of the system of equations divide the gravity term in a definite ratio, and each term utilizes gravity to function. The sum of the terms of the ratio is always unity. It is shown that without gravity forces on earth, there would be no magnetohydrodynamics on earth as is known. The equations in the system of equations were added to produce a single equation which was then integrated. Ratios were used to split-up this single equation into sub-equations which were readily integrated, and even, the non-linear sub-equations were readily integrated. Twenty-seven sub-equations were integrated. The linear part of the relation obtained from the integration of the linear part of the equation satisfied the linear part of the equation; and the relation from the integration of the non-linear part satisfied the non-linear part of the equation.The solutions revealed the role of each term in magnetohydrodynamics. In particular, the gravity term is the indispensable term in magnetohydrodynamics.
The solutions of the MHD equations were compared with the solutions of the N-S equations, and there were similarities and dissimilarities.

Solutions of the Magnetohydrodynamic Equations

This system consists of four equations and one is to solve for $V_x, V_y, V_z, B_x, B_y, B, P(x)$

Magnetohydrodynamic Equations

1. $$\frac{\partial V_x}{\partial x}+\frac{\partial V_y}{\partial y}+\frac{\partial V_z}{\partial z}=0 \text{ <-- continuity equation}$$

2. $$\overbrace{\rho\frac{\partial V_x}{\partial t}+\rho V_x\frac{\partial V_x}{\partial x}+\rho V_y\frac{\partial V_x}{\partial y}+\rho V_z\frac{\partial V_x}{\partial z}}^{\text{Navier–Stokes}}=\overbrace{-\frac{\partial p}{\partial x}+\frac{1}{\mu}(\nabla\times B)\times B+\rho g_x}^{\text{Lorentz force}}$$

3. $$\rho\frac{\partial B}{\partial t}=\nabla\times(V\times B)+\eta\nabla^2 B$$
$$\rho\frac{\partial B}{\partial t}=\nabla\times(V\times B)+\eta\left(\frac{\partial^2 B}{\partial x^2}+\frac{\partial^2 B}{\partial y^2}+\frac{\partial^2 B}{\partial z^2}\right)$$
(η = magnetic diffusivity)

4. $$\nabla\bullet B=0$$
$$\frac{\partial B_x}{\partial x}+\frac{\partial B_y}{\partial y}+\frac{\partial B_z}{\partial z}=0$$

Step 1:

1. If ρ is constant: (for incompressible fluid)
$$\frac{\partial V_x}{\partial x}+\frac{\partial V_y}{\partial y}+\frac{\partial V_z}{\partial z}=0 \text{ <-- continuity equation}$$

2. $$\overbrace{\rho\frac{\partial V_x}{\partial t}+\rho V_x\frac{\partial V_x}{\partial x}+\rho V_y\frac{\partial V_x}{\partial y}+\rho V_z\frac{\partial V_x}{\partial z}}^{\text{Navier - Stokes}}=\overbrace{-\frac{\partial p}{\partial x}+\frac{1}{\mu}(\nabla\times B)\times B+\rho g_x}^{\text{Lorentz force}}$$
$$\rho\frac{\partial V_x}{\partial t}+\rho V_x\frac{\partial V_x}{\partial x}+\rho V_y\frac{\partial V_x}{\partial y}+\rho V_z\frac{\partial V_x}{\partial z}=-\frac{\partial p}{\partial x}+\frac{1}{\mu}\left(B_z\left(\frac{\partial B_x}{\partial z}-\frac{\partial B_z}{\partial x}\right)-B_y\left(\frac{\partial B_y}{\partial x}-\frac{\partial B_x}{\partial y}\right)\right)+\rho g_x$$
$$\boxed{\rho\frac{\partial V_x}{\partial t}+\rho V_x\frac{\partial V_x}{\partial x}+\rho V_y\frac{\partial V_x}{\partial y}+\rho V_z\frac{\partial V_x}{\partial z}=-\frac{\partial p}{\partial x}+\frac{1}{\mu}\left(B_z\frac{\partial B_x}{\partial z}-B_z\frac{\partial B_z}{\partial x}-B_y\frac{\partial B_y}{\partial x}+B_y\frac{\partial B_x}{\partial y}\right)+\rho g_x}$$

3. $$\rho\frac{\partial B}{\partial t}=\nabla\times(V\times B)+\eta\nabla^2 B$$
$$\rho\frac{\partial B}{\partial t}=\frac{\partial}{\partial y}(V_xB_y-V_yB_x)-\frac{\partial}{\partial z}(V_zB_x-V_xB_z)+\eta\left(\frac{\partial^2 B}{\partial x^2}+\frac{\partial^2 B}{\partial y^2}+\frac{\partial^2 B}{\partial z^2}\right)$$
$$\boxed{\rho\frac{\partial B}{\partial t}=\frac{\partial}{\partial y}V_xB_y-\frac{\partial}{\partial y}V_yB_x-\frac{\partial}{\partial z}V_zB_x+\frac{\partial}{\partial z}V_xB_z+\eta\frac{\partial^2 B_x}{\partial x^2}+\eta\frac{\partial^2 B_x}{\partial y^2}+\eta\frac{\partial^2 B_x}{\partial z^2}}$$

4. $$\boxed{\begin{array}{c}\nabla\bullet B=0\\ \dfrac{\partial B_x}{\partial x}+\dfrac{\partial B_y}{\partial y}+\dfrac{\partial B_z}{\partial z}=0\end{array}}$$

Step 2:
After the "vector juggling" one obtains the following system of equations which one will solve.

$$\begin{cases} 1.\ \dfrac{\partial V_x}{\partial x}+\dfrac{\partial V_y}{\partial y}+\dfrac{\partial V_z}{\partial z}=0 \\ 2.\ \rho\dfrac{\partial V_x}{\partial t}+\rho V_x\dfrac{\partial V_x}{\partial x}+\rho V_y\dfrac{\partial V_x}{\partial y}+\rho V_z\dfrac{\partial V_x}{\partial z}+\dfrac{\partial p}{\partial x}-\dfrac{1}{\mu}B_z\dfrac{\partial B_x}{\partial z}+\dfrac{1}{\mu}B_z\dfrac{\partial B_z}{\partial x}+\dfrac{1}{\mu}B_y\dfrac{\partial B_y}{\partial x}-\dfrac{1}{\mu}B_y\dfrac{\partial B_x}{\partial y}=\rho g_x \\ 3. \\ \dfrac{\rho\partial B_x}{\partial t}-V_x\dfrac{\partial B_y}{\partial y}-B_y\dfrac{\partial V_x}{\partial y}+V_y\dfrac{\partial B_x}{\partial y}+B_x\dfrac{\partial V_y}{\partial y}+V_z\dfrac{\partial B_x}{\partial z}+B_x\dfrac{\partial V_z}{\partial z}-V_x\dfrac{\partial B_z}{\partial z}-B_z\dfrac{\partial V_x}{\partial z}-\dfrac{\eta\partial^2 B_x}{\partial x^2}-\dfrac{\eta\partial^2 B_x}{\eta\partial y^2}-\dfrac{\eta\partial^2 B_x}{\eta\partial z^2}=0 \\ 4.\ \dfrac{\partial B_x}{\partial x}+\dfrac{\partial B_y}{\partial y}+\dfrac{\partial B_z}{\partial z}=0 \end{cases}$$

At a glance, and from the experience gained in solving the Navier-Stokes equations, one can identify equation (2) as the driver equation, since it contains the gravity term, and the gravity term is the subject of the equation. However, since the system of equations is to be solved simultaneously and there is only a single "driver", the gravity term, all the terms in the system of equations will be placed in the driver equation, Equation 2. As suggested by Albert Einstein, Friedrich Nietzsche , and Pablo Picasso, one will think like a child at the next step.

Step 3: Thinking like a ninth grader, one will apply the following axiom:

If $a=b$ and $c=d$, then $a+c=b+d$; and therefore, add the left sides and add the right sides of the above equations . That is, $(1)+(2)+(3)+(4)=\rho g_x$

$$\begin{cases} \dfrac{\partial V_x}{\partial x}+\dfrac{\partial V_y}{\partial y}+\dfrac{\partial V_z}{\partial z}+\rho\dfrac{\partial V_x}{\partial t}+\rho V_x\dfrac{\partial V_x}{\partial x}+\rho V_y\dfrac{\partial V_x}{\partial y}+\rho V_z\dfrac{\partial V_x}{\partial z}+\dfrac{\partial p}{\partial x}-\dfrac{1}{\mu}B_z\dfrac{\partial B_x}{\partial z}+\dfrac{1}{\mu}B_z\dfrac{\partial B_z}{\partial x}+\dfrac{1}{\mu}B_y\dfrac{\partial B_y}{\partial x}- \\ \dfrac{1}{\mu}B_y\dfrac{\partial B_x}{\partial y}+\dfrac{\rho\partial B_x}{\partial t}-V_x\dfrac{\partial B_y}{\partial y}-B_y\dfrac{\partial V_x}{\partial y}+V_y\dfrac{\partial B_x}{\partial y}+B_x\dfrac{\partial V_y}{\partial y}+V_z\dfrac{\partial B_x}{\partial z}+B_x\dfrac{\partial V_z}{\partial z}-V_x\dfrac{\partial B_z}{\partial z}-B_z\dfrac{\partial V_x}{\partial z}- \\ \dfrac{\eta\partial^2 B_x}{\partial x^2}-\dfrac{\eta\partial^2 B_x}{\eta\partial y^2}-\dfrac{\eta\partial^2 B_x}{\eta\partial z^2}+\dfrac{\partial B_x}{\partial x}+\dfrac{\partial B_y}{\partial y}+\dfrac{\partial B_z}{\partial z}=\rho g_x \qquad \text{(Three lines per equation)} \end{cases}$$

Step 4: Writing all the linear terms first

$$\begin{cases} \dfrac{\partial V_x}{\partial x}+\dfrac{\partial V_y}{\partial y}+\dfrac{\partial V_z}{\partial z}+\rho\dfrac{\partial V_x}{\partial t}+\dfrac{\partial p}{\partial x}+\dfrac{\rho\partial B_x}{\partial t}-\dfrac{\eta\partial^2 B_x}{\partial x^2}-\dfrac{\eta\partial^2 B_x}{\eta\partial y^2}-\dfrac{\eta\partial^2 B_x}{\eta\partial z^2}+\dfrac{\partial B_x}{\partial x}+\dfrac{\partial B_y}{\partial y}+\dfrac{\partial B_z}{\partial z} \\ +\rho V_x\dfrac{\partial V_x}{\partial x}+\rho V_y\dfrac{\partial V_x}{\partial y}+\rho V_z\dfrac{\partial V_x}{\partial z}-\dfrac{1}{\mu}B_z\dfrac{\partial B_x}{\partial z}+\dfrac{1}{\mu}B_z\dfrac{\partial B_z}{\partial x}+\dfrac{1}{\mu}B_y\dfrac{\partial B_y}{\partial x}-\dfrac{1}{\mu}B_y\dfrac{\partial B_x}{\partial y}-V_x\dfrac{\partial B_y}{\partial y}-B_y\dfrac{\partial V_x}{\partial y} \\ +V_y\dfrac{\partial B_x}{\partial y}+B_x\dfrac{\partial V_y}{\partial y}+V_z\dfrac{\partial B_x}{\partial z}+B_x\dfrac{\partial V_z}{\partial z}-V_x\dfrac{\partial B_z}{\partial z}-B_z\dfrac{\partial V_x}{\partial z}=\rho g_x \qquad \text{(Three lines per equation)} \end{cases}$$

(Since all the terms are now in the same driver equation, let ρg_x "drive them" simultaneously.)

Step 5: Solve the above 28-term equation using the ratio method. (27 ratio terms)

The ratio terms to be used are respectively the following: (Sum of the ratio terms = 1)

$\beta_1, \beta_2, \beta_3, a, b, c, d, f, m, q, r, s, \omega_1, \omega_2, \omega_3, \omega_4, \omega_5, \omega_6, \lambda_1, \lambda_2, \lambda_3, \lambda_4, \lambda_5, \lambda_6, \lambda_7, \lambda_8, \lambda_9$

1. $\dfrac{\partial V_x}{\partial x}=\beta_1\rho g_x$ $\dfrac{dV_x}{dx}=\beta_1\rho g_x$ $V_x=\beta_1\rho g_x x+C_{16}$	2. $\dfrac{\partial V_y}{\partial y}=\beta_2\rho g_x$ $\dfrac{dV_y}{dy}=\beta_2\rho g_x$ $V_y=\beta_2\rho g_x y+C_{17}$	3. $\dfrac{\partial V_z}{\partial z}=\beta_3\rho g_x$ $\dfrac{dV_z}{dz}=\beta_3\rho g_x$ $V_z=\beta_3\rho g_x z+C_{18}$	4. $\rho\dfrac{\partial V_x}{\partial t}=a\rho g_x$ $\dfrac{\partial V_x}{\partial t}=ag_x$ $V_x=ag_x t+C_1$

5.	6.	7.
$\frac{\partial p}{\partial x} = b\rho g_x$ $\frac{dp}{dx} = b\rho g_x$ $P(x) = b\rho g_x x + C$	$\rho \frac{\partial B_x}{\partial t} = c\rho g_x$ $\frac{\partial B_x}{\partial t} = cg_x$ $\frac{dB_x}{dt} = cg_x$ $B_x = cg_x t + C_{1b}$	$-\eta \frac{\partial^2 B_x}{\partial x^2} = d\rho g_x$ $\frac{d^2 B_x}{dx^2} = -\frac{d\rho g_x}{\eta}$ $\frac{dB_x}{dx} = -\frac{d\rho g_x x}{\eta} + C_2$ $B_x = -\frac{d\rho g_x x^2}{2\eta} + C_2 x + C_3$

8.	9.	10.
$-\eta \frac{\partial^2 B_x}{\partial y^2} = f\rho g_x$ $\frac{d^2 B_x}{dy^2} = -\frac{f\rho g_x}{\eta}$ $\frac{dB_x}{dy} = -\frac{f\rho g_x y}{\eta} + C_4$ $B_x = -\frac{f\rho g_x y^2}{2\eta} + C_4 y + C_5$	$-\eta \frac{\partial^2 B_x}{\partial z^2} = m\rho g_x$ $\frac{d^2 B_x}{dz^2} = -\frac{m\rho g_x}{\eta}$ $\frac{dB_x}{dz} = -\frac{m\rho g_x z}{\eta} + C_6$ $B_x = -\frac{m\rho g_x z^2}{2\eta} + C_6 x + C_7$	$\frac{\partial B_x}{\partial x} = q\rho g_x$ $\frac{dB_x}{dx} = q\rho g_x$ $B_x = q\rho g_x x + C_{19}$

11.	12.	13.	14.
$\frac{\partial B_y}{\partial y} = r\rho g_x$ $\frac{dB_y}{dy} = r\rho g_x$ $B_y = r\rho g_x y + C_{20}$	$\frac{\partial B_z}{\partial z} = s\rho g_x$ $\frac{dB_z}{dz} = s\rho g_x$ $B_z = s\rho g_x z + C_{21}$	$\rho V_x \frac{\partial V_x}{\partial x} = \omega_1 \rho g_x$ $V_x \frac{dV_x}{dx} = \omega_1 g_x$ $V_x dV_x = \omega_1 g_x dx$ $\frac{V_x^{\ 2}}{2} = \omega_1 g_x x$ $V_x^{\ 2} = 2\omega_1 g_x x$ $V_x = \pm\sqrt{2\omega_1 g_x x} + C_2$	$\rho V_y \frac{\partial V_x}{\partial y} = \omega_2 \rho g_x$ $V_y dV_x = \omega_2 g_x dy$ $V_y V_x = \omega_2 g_x y + \psi_y(V_y)$ $V_x = \frac{\omega_2 g_x y}{V_y} + \frac{\psi_y(V_y)}{V_y}$ $V_y \neq 0$

15.	16	17.
$\rho V_z \frac{\partial V_x}{\partial z} = \omega_3 \rho g_x$ $V_z \frac{dV_x}{dz} = \omega_3 g_x$ $V_z dV_x = \omega_3 g_x dz$ $V_z V_x = \omega_3 g_x z + \psi_z(V_z)$ $V_x = \frac{\omega_3 g_x z}{V_z} + \frac{\psi_z(V_z)}{V_z}$ $V_z \neq 0$	$B_z \frac{\partial B_x}{\partial z} = -\omega_4 \mu \rho g_x$ $B_z dB_x = -\omega_4 \mu \rho g_x dz$ $B_z B_x = -\omega_4 \mu \rho g_x z + \psi_z(B_z)$ $B_x = -\frac{\omega_4 \mu \rho g_x z}{B_z} + \frac{\psi_z(B_z)}{B_z}$ $B_z \neq 0$	$B_z \frac{\partial B_z}{\partial x} = \omega_5 \mu \rho g_x$ $B_z \frac{dB_z}{dx} = \omega_5 \mu \rho g_x$ $B_z dB_z = \omega_5 \mu \rho g_x dx$ $\frac{B_z^{\ 2}}{2} = \omega_5 \mu \rho g_x x$ $B_z^{\ 2} = 2\omega_5 \mu \rho g_x x$ $B_z = \pm\sqrt{2\omega_5 \mu \rho g_x x} + C$

18.

$$B_y \frac{\partial B_y}{\partial x} = \omega_6 \mu \rho g_x$$
$$B_y \frac{dB_y}{dx} = \omega_6 \mu \rho g_x$$
$$B_y dB_y = \omega_6 \mu \rho g_x dx$$
$$\frac{B_y{}^2}{2} = \omega_6 \mu \rho g_x x$$
$$B_y{}^2 = 2\omega_6 \mu \rho g_x x$$
$$B_y = \pm\sqrt{2\omega_6 \mu \rho g_x x} + C$$

19.

$$-\frac{1}{\mu} B_y \frac{\partial B_x}{\partial y} = \lambda_1 \rho g_x$$
$$B_y \frac{dB_x}{dy} = -\lambda_1 \mu \rho g_x$$
$$B_y dB_x = -\lambda_1 \mu \rho g_x dy$$
$$B_y B_x = -\lambda_1 \mu \rho g_x y + \psi_y(B_y)$$
$$B_x = -\frac{\lambda_1 \mu \rho g_x y}{B_y} + \frac{\psi_y(B_y)}{B_y}$$
$$B_y \neq 0$$

20

$$-V_x \frac{\partial B_y}{\partial y} = \lambda_2 \rho g_x$$
$$V_x \frac{dB_y}{dy} = -\lambda_2 \rho g_x$$
$$V_x dB_y = -\lambda_2 \rho g_x dy$$
$$V_x B_y = -\lambda_2 \rho g_x y + \psi_x(V_x)$$
$$B_y = \frac{-\lambda_2 \rho g_x y}{V_x} + \frac{\psi_x(V_x)}{V_x}$$
$$V_x \neq 0$$

21.

$$-B_y \frac{\partial V_x}{\partial y} = \lambda_3 \rho g_x$$
$$B_y \frac{dV_x}{dy} = -\lambda_3 \rho g_x$$
$$B_y dV_x = -\lambda_3 \rho g_x dy$$
$$B_y V_x = -\lambda_3 \rho g_x y + \psi_y(B_y)$$
$$V_x = -\frac{\lambda_3 \rho g_x y}{B_y} + \frac{\psi_y(B_y)}{B_y}$$
$$B_y \neq 0$$

22.

$$V_y \frac{\partial B_x}{\partial y} = \lambda_4 \rho g_x$$
$$V_y \frac{dB_x}{dy} = \lambda_4 \rho g_x$$
$$V_y dB_x = \lambda_4 \rho g_x dy$$
$$V_y B_x = \lambda_4 \rho g_x y + \psi_y(V_y)$$
$$B_x = \frac{\lambda_4 \rho g_x y}{V_y} + \frac{\psi_y(V_y)}{V_y}$$
$$V_y \neq 0$$

23.

$$B_x \frac{\partial V_y}{\partial y} = \lambda_5 \rho g_x$$
$$B_x \frac{dV_y}{dy} = \lambda_5 \rho g_x$$
$$B_x dV_y = \lambda_5 \rho g_x dy$$
$$B_x V_y = \lambda_5 \rho g_x y + \psi_x(B_x)$$
$$V_y = \frac{\lambda_5 \rho g_x y}{B_x} + \frac{\psi_x(B_x)}{B_x}$$
$$B_x \neq 0$$

24.

$$V_z \frac{\partial B_x}{\partial z} = \lambda_6 \rho g_x$$
$$V_z \frac{dB_x}{dz} = \lambda_6 \rho g_x$$
$$V_z dB_x = \lambda_6 \rho g_x dz$$
$$V_z B_x = \lambda_6 \rho g_x z + \psi_z(V_z)$$
$$B_x = \frac{\lambda_6 \rho g_x z}{V_z} + \frac{\psi_z(V_z)}{V_z}$$
$$V_z \neq 0$$

25.

$$B_x \frac{\partial V_z}{\partial z} = \lambda_7 \rho g_x$$
$$B_x \frac{dV_z}{dz} = \lambda_7 \rho g_x$$
$$B_x dV_z = \lambda_7 \rho g_x dz$$
$$B_x V_z = \lambda_7 \rho g_x z + \psi_x(B_x)$$
$$V_z = \frac{\lambda_7 \rho g_x z}{B_x} + \frac{\psi_x(B_x)}{B_x}$$
$$B_x \neq 0$$

26

$$-V_x \frac{\partial B_z}{\partial z} = \lambda_8 \rho g_x$$
$$V_x \frac{dB_z}{dz} = -\lambda_8 \rho g_x$$
$$V_x dB_z = -\lambda_8 \rho g_x dz$$
$$V_x B_z = -\lambda_8 \rho g_x z + \psi_x(V_x)$$
$$B_z = -\frac{\lambda_8 \rho g_x z}{V_x} + \frac{\psi_x(V_x)}{V_x}$$
$$V_x \neq 0$$

27.

$$-B_z \frac{\partial V_x}{\partial z} = \lambda_9 \rho g_x$$
$$B_z \frac{dV_x}{dz} = -\lambda_9 \rho g_x$$
$$B_z dV_x = -\lambda_9 \rho g_x dz$$
$$B_z V_x = -\lambda_9 \rho g_x z + \psi_z(B_z)$$
$$V_x = -\frac{\lambda_9 \rho g_x z}{B_z} + \frac{\psi_z(B_z)}{B_z}$$
$$B_z \neq 0$$

Step 6: One collects the integrals of the sub-equations, above, for V_x, V_y, V_z, B_x, B_y, B_z, $P(x)$

$V_x(x,y,z,t) =$ (sum of integrals from sub-equations #1, #4, #13, #14, #15, #21, #27)

$$\beta_1\rho g_x x + a g_x t \pm \sqrt{2\omega_1 g_x x} + \frac{\omega_2 g_x y}{V_y} - \frac{\lambda_3 \rho g_x y}{B_y} + \frac{\omega_3 g_x z}{V_z} - \frac{\lambda_9 \rho g_x z}{B_z} + \underbrace{\frac{\psi_z(V_z)}{V_z} + \frac{\psi_y(B_y)}{B_y} + \frac{\psi_y(V_y)}{V_y} + \frac{\psi_z(B_z)}{B_z}}_{\text{arbitrary functions}} + C_1;$$

(integral from sub–equation #5)

$$P(x) = b\rho g_x x + C_2$$

(sum of integrals from sub–equations #2, #23)

$$V_y(y) = \beta_2 \rho g_x y + \frac{\lambda_5 \rho g_x y}{B_x} + \underbrace{\frac{\psi_x(B_x)}{B_x}}_{\text{arbitrary function}} + C_3$$

(sum of integrals from sub–equations #3, #25)

$$V_z(z) = \beta_3 \rho g_x z + \frac{\lambda_7 \rho g_x z}{B_x} + \underbrace{\frac{\psi_x(B_x)}{B_x}}_{\text{arbitrary function}} + C_4$$

(sum of integrals from sub-equations #6, #7, #8, #9, #10, #16, #19, #22, #24)

$B_x(x,y,z,t) =$

$$B_x = -\frac{\rho g_x}{2\eta}(dx^2 + fy^2 + mz^2) + q\rho g_x x + C_2 x + C_4 y + C_6 z + c g_x t - \frac{\lambda_1 \mu \rho g_x y}{B_y} + \frac{\lambda_4 \rho g_x y}{V_y} - \frac{\omega_4 \mu \rho g_x z}{B_z} +$$

$$\frac{\lambda_6 \rho g_x z}{V_z} + \underbrace{\frac{\psi_z(B_z)}{B_z} + \frac{\psi_y(B_y)}{B_y} + \frac{\psi_y(V_y)}{V_y} + \frac{\psi_z(V_z)}{V_z}}_{\text{arbitrary functions}} + C_7$$

(sum of integrals from sub–equations #11, #18, #20)

$$B_y = r\rho g_x y \pm \sqrt{2\omega_6 \mu \rho g_x x} - \frac{\lambda_2 \rho g_x y}{V_x} + \underbrace{\frac{\psi_x(V_x)}{V_x}}_{\text{arbitrary function}} + C_8$$

(sum of integrals from sub–equations #12, #17, #26)

$$B_z = s\rho g_x z \pm \sqrt{2\omega_5 \mu \rho g_x x} - \frac{\lambda_8 \rho g_x z}{V_x} + \underbrace{\frac{\psi_x(V_x)}{V_x}}_{\text{arbitrary function}} + C_{21}$$

Step 7: Find the test derivatives for the linear part

1.	2.	3.	4.	5.	6.
$\frac{\partial V_x}{\partial x}=(\beta_1\rho g_x)$	$\frac{\partial V_y}{\partial y}=(\beta_2\rho g_x)$	$\frac{\partial V_z}{\partial z}=(\beta_3\rho g_x)$	$\frac{\partial V_x}{\partial t}=(ag_x)$	$\frac{\partial p}{\partial x}=(b\rho g_x)$	$\frac{dB_x}{dt}=(cg_x)$

7.	8.	9.	10.	11.	12.
$\frac{\partial^2 B_x}{\partial x^2}=-\frac{d\rho g_x}{\eta}$	$\frac{\partial^2 B_x}{\partial y^2}=-\frac{f\rho g_x}{\eta}$	$\frac{\partial^2 B_x}{\partial z^2}=-\frac{m\rho g_x}{\eta}$	$\frac{\partial B_x}{\partial x}=q\rho g_x$	$\frac{\partial B_y}{\partial y}=r\rho g_x$	$\frac{\partial B_z}{\partial z}=s\rho g_x$

Test derivatives for the nonlinear part

13.	14.	15.	16	17.
$\frac{\partial V_x}{\partial x}=\frac{\omega_1 g_x}{V_x}$	$\frac{\partial V_x}{\partial y}=\frac{\omega_2 g_x}{V_y}$	$\frac{\partial V_x}{\partial z}=\frac{\omega_3 g_x}{V_z}$	$\frac{\partial B_x}{\partial z}=-\frac{\omega_4\mu\rho g_x}{B_z}$	$\frac{\partial B_z}{\partial x}=\frac{\omega_5\mu\rho g_x}{B_z}$

18.	19.	20.	21.	22.
$\frac{\partial B_y}{\partial x}=\frac{\omega_6\mu\rho g_x}{B_y}$	$\frac{\partial B_x}{\partial y}=-\frac{\lambda_1\mu\rho g_x}{B_y}$	$\frac{\partial B_y}{\partial y}=-\frac{\lambda_2\rho g_x}{V_x}$	$\frac{\partial V_x}{\partial y}=-\frac{\lambda_3\rho g_x}{B_y}$	$\frac{\partial B_x}{\partial y}=\frac{\lambda_4\rho g_x}{V_y}$

23.	24.	25.	26.	27.
$\frac{\partial V_y}{\partial y}=\frac{\lambda_5\rho g_x}{B_x}$	$\frac{\partial B_x}{\partial z}=\frac{\lambda_6\rho g_x}{V_z}$	$\frac{\partial V_z}{\partial z}=\frac{\lambda_7\rho g_x}{B_x}$	$\frac{\partial B_z}{\partial z}=-\frac{\lambda_8\rho g_x}{V_x}$	$\frac{\partial V_x}{\partial z}=-\frac{\lambda_9\rho g_x}{B_z}$

Step 8: Substitute the above test derivatives respectively in the following 28-term equation

$$\begin{cases}\frac{\partial V_x}{\partial x}+\frac{\partial V_y}{\partial y}+\frac{\partial V_z}{\partial z}+\rho\frac{\partial V_x}{\partial t}+\frac{\partial p}{\partial x}+\frac{\rho\partial B_x}{\partial t}-\frac{\eta\partial^2 B_x}{\partial x^2}-\frac{\eta\partial^2 B_x}{\eta\partial y^2}-\frac{\eta\partial^2 B_x}{\eta\partial z^2}+\frac{\partial B_x}{\partial x}+\frac{\partial B_y}{\partial y}+\frac{\partial B_z}{\partial z}\\ +\rho V_x\frac{\partial V_x}{\partial x}+\rho V_y\frac{\partial V_x}{\partial y}+\rho V_z\frac{\partial V_x}{\partial z}-\frac{1}{\mu}B_z\frac{\partial B_x}{\partial z}+\frac{1}{\mu}B_z\frac{\partial B_z}{\partial x}+\frac{1}{\mu}B_y\frac{\partial B_y}{\partial x}-\frac{1}{\mu}B_y\frac{\partial B_x}{\partial y}-V_x\frac{\partial B_y}{\partial y}-B_y\frac{\partial V_x}{\partial y}\\ +V_y\frac{\partial B_x}{\partial y}+B_x\frac{\partial V_y}{\partial y}+V_z\frac{\partial B_x}{\partial z}+B_x\frac{\partial V_z}{\partial z}-V_x\frac{\partial B_z}{\partial z}-B_z\frac{\partial V_x}{\partial z}=\rho g_x\end{cases}$$

(Three lines per equation)

$$\begin{cases}(\beta_1\rho g_x)+(\beta_2\rho g_x)+(\beta_3\rho g_x)+\rho(ag_x)+(b\rho g_x)+\rho(cg_x)-\eta(-\frac{d\rho g_x}{\eta})-\eta(-\frac{f\rho g_x}{\eta})-\eta(-\frac{m\rho g_x}{\eta})+\\ (q\rho g_x)+(r\rho g_x)+(s\rho g_x)+\rho V_x(\frac{\omega_1 g_x}{V_x})+\rho V_y(\frac{\omega_2 g_x}{V_y})+\rho V_z(\frac{\omega_3 g_x}{V_z})-\frac{1}{\mu}B_z(-\frac{\omega_4\mu\rho g_x}{B_z})+\\ \frac{1}{\mu}B_z(\frac{\omega_5\mu\rho g_x}{B_z})+\frac{1}{\mu}B_y(\frac{\omega_6\mu\rho g_x}{B_y})-\frac{1}{\mu}B_y(-\frac{\lambda_1\mu\rho g_x}{B_y})-V_x(-\frac{\lambda_2\rho g_x}{V_x})-B_y(-\frac{\lambda_3\rho g_x}{B_y})+V_y(\frac{\lambda_4\rho g_x}{V_y})+\\ B_x(\frac{\lambda_5\rho g_x}{B_x})+V_z(\frac{\lambda_6\rho g_x}{V_z})+B_x(\frac{\lambda_7\rho g_x}{B_x})-V_x(-\frac{\lambda_8\rho g_x}{V_x})-B_z(-\frac{\lambda_9\rho g_x}{B_z})\overset{?}{=}\rho g_x\end{cases}$$

(Four lines per equation)

$$\begin{cases}\beta_1\rho g_x+\beta_2\rho g_x+\beta_3\rho g_x+a\rho g_x+b\rho g_x+c\rho g_x+d\rho g_x+f\rho g_x+m\rho g_x q\rho g_x+r\rho g_x+s\rho g_x+\omega_1\rho g_x\\ +\omega_3\rho g_x+\omega_5\rho g_x+\omega_6\rho g_x+\lambda_1\mu\rho g_x+\lambda_2\rho g_x+\lambda_3\rho g_x+\lambda_4\rho g_x+\lambda_5\rho g_x+\omega_2\rho g_x+\omega_3\rho g_x\\ +\lambda_6\rho g_x+\lambda_7\rho g_x+\lambda_8\rho g_x+\lambda_9\rho g_x\overset{?}{=}\rho g_x\end{cases}$$

(Three lines per equation)

$$\begin{cases}\beta_1 g_x+\beta_2 g_x+\beta_3 g_x+ag_x+bg_x+cg_x+dg_x+fg_x+mg_xqg_x+rg_x+sg_x+\omega_1 g_x+\omega_3 g_x+\omega_5 g_x \\ +\omega_6 g_x+\lambda_1 g_x+\lambda_2 g_x+\lambda_3 g_x+\lambda_4 g_x+\lambda_5 g_x+\omega_2 g_x+\omega_3 g_x+\lambda_6 g_x+\lambda_7 g_x+\lambda_8 g_x+\lambda_9 g_x \overset{?}{=} g_x\end{cases} \text{ (2 lines)}$$

$$\begin{cases}g_x(\beta_1+\beta_2+\beta_3+a+b+c+d+f+m+q+r+s+\omega_1+\omega_3+\omega_5+\lambda_3+\lambda_4+\lambda_5+\omega_2+\omega_3+\lambda_6+\lambda_7 \\ +\omega_6+\lambda_1+\lambda_2+\lambda_8+\lambda_9)\overset{?}{=} g_x\end{cases} \qquad \text{(Two lines per equation)}$$

$$g_x(1)\overset{?}{=} g_x \qquad \text{(Sum of the ratio terms = 1)}$$

$$g_x\overset{?}{=} g_x \quad \text{Yes}$$

Since an identity is obtained, the solutions to the 28-term equation are as follows

$V_x(x,y,z,t) =$ (sum of integrals from sub - equations #1, #4,#13,#14,#15,#21,#27)

$$\beta_1\rho g_x x + ag_x t \pm \sqrt{2\omega_1 g_x x} + \frac{\omega_2 g_x y}{V_y} - \frac{\lambda_3\rho g_x y}{B_y} + \frac{\omega_3 g_x z}{V_z} - \frac{\lambda_9\rho g_x z}{B_z} + \underbrace{\frac{\psi_z(V_z)}{V_z} + \frac{\psi_y(B_y)}{B_y} + \frac{\psi_y(V_y)}{V_y} + \frac{\psi_z(B_z)}{B_z}}_{\text{arbitrary functions}} + C_1;$$

(integral from sub–equation #5)

$$P(x) = b\rho g_x x + C_2$$

(sum of integrals from sub–equations #2,#23)

$$V_y = \beta_2\rho g_x y + \frac{\lambda_5\rho g_x y}{B_x} + \underbrace{\frac{\psi_x(B_x)}{B_x}}_{\text{arbitrary function}} + C_3$$

(sum of integrals from sub–equations #3, #25)

$$V_z = \beta_3\rho g_x z + \frac{\lambda_7\rho g_x z}{B_x} + \underbrace{\frac{\psi_x(B_x)}{B_x}}_{\text{arbitrary function}} + C_4$$

(sum of integrals from sub - equations #6, #7, #8, #9, #10, #16,#19, #22, #24)

$B_x(x,y,z,t) =$

$$B_x = -\frac{\rho g_x}{2\eta}(dx^2+fy^2+mz^2) + q\rho g_x x + C_2 x + C_4 y + C_6 z + + cg_x t - \frac{\lambda_1\mu\rho g_x y}{B_y} + \frac{\lambda_4\rho g_x y}{V_y} - \frac{\omega_4\mu\rho g_x z}{B_z} +$$

$$\frac{\lambda_6\rho g_x z}{V_z} + \underbrace{\frac{\psi_z(B_z)}{B_z} + \frac{\psi_y(B_y)}{B_y} + \frac{\psi_y(V_y)}{V_y} + \frac{\psi_z(V_z)}{V_z}}_{\text{arbitrary functions}} + C_7$$

(sum of integrals from sub–equations #11,#18,#20)

$$B_y = r\rho g_x y \pm \sqrt{2\omega_6\mu\rho g_x x} - \frac{\lambda_2\rho g_x y}{V_x} + \underbrace{\frac{\psi_x(V_x)}{V_x}}_{\text{arbitrary function}} + C_8$$

(sum of integrals from sub–equations #12,#17,#26)

$$B_z = s\rho g_x z \pm \sqrt{2\omega_5\mu\rho g_x x} - \frac{\lambda_8\rho g_x z}{V_x} + \underbrace{\frac{\psi_x(V_x)}{V_x}}_{\text{arbitrary function}} + C_{21}$$

Step 9: The linear part of the relation satisfies the linear part of the equation (in Step 8; and the non-linear part of the relation satisfies the non-linear part of the equati. The solutions are above.

Analogy for the Identity Checking Method: If one goes shopping with American dollars and Japanese yens (without any currency conversion) and after shopping, if one wants to check the cost of the items purchased, one would check the cost of the items purchased with dollars against the receipts for the dollars; and one would also check the cost of the items purchased with yens against the receipts for the yens purchase. However, if one converts one currency to the other, one would only have to check the receipts for only a single currency, dollars or yens. This conversion case is similar to the linearized N-S equations, where there was no partitioning in identity checking.

Important insight

One observes above that the most important insight of the above solutions is the indispensability of the gravity term in MHD.. Observe that if gravity, g_x, were zero, all the non-constant terms in each solution would be zero. These results can be stated emphatically that without gravity forces on earth, there would be no magnetohydrodynamics on earth as is known. It would not therefore be meaningful to write a system of MHD equations without the gravity term, since there would be no magnetohydrodynmics.

Supporter Equation Contributions (see also viXra:1405.0251)

Note above that there are 28 terms in the driver equation, and 27 supporter equations, Each supporter equation provides useful information about the driver equation. The more of these supporter equations that are integrated, the more the information one obtains about the driver equation. However, without solving a supporter equation, one can sometimes write down some characteristics of the integration relation of the supporter equation by referring to the subjects of the supporter equations of the Navier-Stokes equations. For example, if one uses $(\eta\partial^2 B_x/\partial x^2)$ as the subject of a supporter equation here, the curve for the integration relation obtained would be parabolic, periodic, and decreasingly exponential. Using $\rho(\partial V/\partial t)$ as the subject of the supporter equation, the curve would be periodic and decreasingly exponential. Using $(\partial p/\partial x)$, the curve would be parabolic.

Comparison of Solutions of Navier-Stokes Equations and Solutions of Magnetohydrodynamic Equations

Navier-Stokes x–direction solution

$$V_x(x,y,z,t) = -\frac{\rho g_x}{2\mu}(ax^2+by^2+cz^2)+C_1x+C_3y+C_5z+fg\pm\sqrt{2hgx}+\frac{ngy}{V_y}+\frac{qgz}{V_z}+\underbrace{\frac{\psi_y(V_y)}{V_y}+\frac{\psi_z(V_z)}{V_z}}_{\text{arbitrary functions}}$$

$$P(x) = d\rho g_x x \qquad (V_y \neq 0,\ V_z \neq 0)$$

For magnetohydrodynamic solutions, see previous page

1. V_x for MHD system resembles the V_x for the Euler solution part of N-S solution.
2. $P(x)$) for N-S and MHD equations are the same.
3. V_y and V_z for MHD are different from those of N-S solution.
4. B_x is parabolic and resembles V_x for N-S, except for the absence of the square root function.
5. B_y and B_z resemble the Euler solution part of the N-S solution.

Conclusion

The author proposed and applied a new law to solve the system of magnetohydrodynamic equations. This law states that in magnetohydrodynamics, all the other terms in the system of equations divide the gravity term in a definite ratio, and each term utilizes gravity to function. The experience gained in solving the Navier-Stokes equations guided the author to solve the MHD equations. It was shown that without gravity forces on earth, there would be no magnetohydrodynamics on earth as is known. The equations in the system of equations were added to produce a single equation which was then integrated. Ratios were used to split-up the single equation, and the resulting sub-equations were readily integrated; and even, the nonlinear sub-equations were readily integrated. Twenty-seven sub-equations were integrated. The linear part of the relation obtained from the integration of the linear part of the equation satisfied the linear part of the equation; and the relation from the integration of the non-linear part satisfied the non-linear part of the equation. Comparison of the solutions of MHD equations with the solutions of the N-S equations revealed the following: (a) V_x for MHD system resembles the V_x for the non-linear part of the N-S solution; (b) $P(x)$) for N-S and MHD equations are the same; (c) V_y and V_z for MHD are different from those of N-S solutions; (d) B_x is parabolic and resembles V_x for N-S solution, except for the absence of the square root function; and (e) B_y and B_z resemble the non-linear part of N-S solution.

By solving algebraically and simultaneously for V_x, V_y, V_z, B_x, B_y, B,, the solutions could be expressed in term of x, y, z and t.

In applications, the ratio terms may perhaps be determined using information such as i nitial and boundary conditions or may have to be determined experimentally. Finally, for any magnetohydrodynamic design, one should always maximize the role of gravity for cost -effectiveness, durability, and dependability. Perhaps, a law for magnetohydrodynamics should read "Sum of everything else equals ρg" ; and this would imply the proposed new law that the other terms in the system of equations divide the gravity term in a definite ratio, and each term utilizes gravity to function.

Note: The liquid pressure, P at the bottom of a liquid of depth h units is given by $\boxed{P = \rho g h}$. From the MHD solutions in this paper, $\boxed{P(x) = b\rho g x}$ from integrating $\frac{dp}{dx} = b\rho g$ where b is a ratio term. Each of the other terms in the MHD equation must also be set equal to the product of a ratio term and ρg. This result implies that the approach used in solving the MHD equations is sound.

P.S.

The author spent more time on " vector juggling" than on the integration of the equations, since no complete system without vector notation was available either in textbooks or on-line. The integration took less time because of the experience with the N-S equations. Any error in the vector juggling part, if any, can be integrated within minutes.

Appendix 9

"Lesson 9: Navier-Stokes Equations Solved Simply" (see also p.189)

"5% of the people think; 10% of the people think that they think; and the other 85% would rather die than think."----Thomas Edison

"The simplest solution is usually the best solution"---Albert Einstein

Abstract

Coincidences. The US Supreme Court consists of nine members, one of whom is the Chief Justice of the Court. So also, a one-direction Navier-Stokes equation consists of nine members, one of which is the indispensable gravity term, without which there would be no incompressible fluid flow as shown by the solutions of the N-S equations (viXra:1512.0334). Another coincidence is that numerologically, the number, 9, is equivalent to the 1800's (1 + 8 + 0 + 0 = 9) time period during which the number of the members of the Supreme Court became fixed at 9, while the formulation of the nine-term N-S equations was completed. Another coincidence is that the solutions of the N-S equations were completed (viXra:1512.0334) by the author in the year, 2016 (2 + 0 +1+ 6 = 9). Using a new introductory approach, this paper covers the author's previous solutions of the N-S equations (viXra:1512.0334). In particular, the N-S solutions have been compared to the equations of motion and liquid pressure of elementary physics. The N-S solutions are (except for the constants involved) very similar or identical to the equations of motion and liquid pressure of elementary physics. The results of the comparative analysis show that the N--S equations have been properly solved. It could be stated that the solutions of the N-S equations have existed since the time the equations of motion and liquid pressure of elementary physics were derived. A one-direction Navier-Stokes equation has also been derived from the equations of motion and liquid pressure of elementary physics. Insights into the solutions include how the polynomial parabolas, the radical parabolas, and the hyperbolas interact to produce turbulent flow. It is argued that the solutions and methods of solving the N-S equations are unique, and that only the approach by the author will ever produce solutions to the N-S equations. By a solution, the equation must be properly integrated and the integration results must be tested in the original equation for identity before the integration results are claimed as solutions.

Options

Introduction

The following examples are to convince the reader that the approach used in splitting the N-S equation and pairing the terms is valid. This approach works for partial differential equations if the single term on the right side of the equation is a constant. One can view this splitting and paring as dividing the constant term on the right side of the equation (using ratios) by the terms on the left side of the equation. It is to be understood that, to claim a solution to a partial differential equation, the integration results must be checked for identity in the original partial differential equation.

As suggested by Albert Einstein, one will think like a child at the beginning of the solution of the N-S equation. Actually, one will think like a ninth grader.

Suppose one performs the following operation:

Example 1: Addition of only two numbers

$$20 + 25 = 45 \qquad (1)$$

$$20 = 45 \times \tfrac{20}{45} = 45 \times \tfrac{4}{9} \qquad (2)$$

$$25 = 45 \times \tfrac{25}{45} = 45 \times \tfrac{5}{9} \qquad (3)$$

Equations (2), and (3), can be written as follows:

$$20 = 45a \qquad (4)$$

$$25 = 45b \qquad (5)$$

Above, a and b are called ratio terms

$$a = \tfrac{4}{9},\ b = \tfrac{5}{9} \qquad (a + b = 1\ ,\ \tfrac{4}{9} + \tfrac{5}{9} = \tfrac{9}{9} = 1)$$

Adding equations (4) and (5),

$$20 + 25 = 45(a + b)$$

One can conclude that the sum of the ratio terms is always 1.

The next example is a preparation towards the "main dish".

Example 2: There is only a single Navier-Stokes term on the left side of the equation.

If $\frac{dp}{dx} = \rho g_x \qquad$ (A)

find $P(x)$

Solution

$$\frac{dp}{dx} = \rho g_x \qquad (1)$$

$$dp = \rho g_x dx$$

$$P(x) = \int \rho g_x dx$$

$P(x) = \rho g_x x + C \quad (2) \qquad$ (Integrating)

Check solution for identity in (A),

From (2), $\frac{dp}{dx} = \rho g_x$

Substituting for the left side of (A),

$\rho g_x \overset{?}{=} \rho g_x \qquad$ Yes

Extra 1 : $\frac{dp}{dy} = \rho g_y \qquad (1)$

$$dp = \rho g_y dy$$

$$P(y) = \int \rho g_y dy$$

$$P(y) = \rho g_y y + C \qquad (2)$$

Checking:

$\frac{dp}{dy} = \rho g_y$ Substituting for the left side of (1)

$\rho g_y \overset{?}{=} \rho g_y \quad$ Yes

Extra 2

$$\frac{dp}{dh} = \rho g \qquad (1)$$

$$dp = \rho g dh$$

$$P(h) = \rho g h + C \qquad (2)$$

Note::

In elementary physics, the liquid pressure at the bottom of a liquid of depth h units is given by $P = \rho g h$

Example 2 and its results hint at how to solve the Navier-Stokes equations.

What was done for the $\frac{dp}{dh}$ term to produce a correct equation for the liquid pressure must be repeated for each of the terms on the left side of the N-S equation, with the gravity term as the subject of the equation.

Example 3: There are two Navier-Stokes terms on the left side of the equation.

If $\frac{\partial p}{\partial x} + \rho \frac{\partial V_x}{\partial t} = \rho g_x$, (A)

find $P(x)$ and $V_x(t)$

Solution

Step 1: Assume that the terms on the left side of the equation are dividing the term on the right side of the equation in the ratio $d : f$. That is, the ratio terms for $\frac{\partial p}{\partial x}$ and $\rho \frac{\partial V_x}{\partial t}$ are d and f respectively.

Step 2: For $\frac{\partial p}{\partial x}$

$\frac{\partial p}{\partial x} = d\rho g_x$ (d is a ratio term)

$\frac{dp}{dx} = d\rho g_x$

(One drops the partials symbol, since a single independent variable is involved)

$dp = d\rho g_x dx$

$P(x) = \int d\rho g_x dx$

$P(x) = d\rho g_x x + C$ (integrating)

Step 3: For $\rho \frac{\partial V_x}{\partial t}$

$\rho \frac{\partial V_x}{\partial t} = f\rho g_x$ (f is a ratio term)

$\frac{\partial V_x}{\partial t} = f g_x$ (Dividing out the ρ)

$\frac{dV_x}{dt} = f g_x$

(One drops the partials symbol, since a single independent variable is involved)

$dV_x = f g_x dt$

$V_x(t) = \int f g_x dt$;

$V_x(t) = f g_x t + C$ (integrating)

Therefore, $P(x) = d\rho g_x x + C$ (B); and $V_x(t) = f g_x t + C$ (D)

To check for identity:

From (B), $\frac{dp}{dx} = d\rho g_x$ and from (D), $\frac{dV_x}{dt} = f g_x$

Substituting for $\frac{\partial p}{\partial x}$ and $\frac{\partial V_x}{\partial t}$ from (B) and (D) respectively on the left side of (A), above.

$$d\rho g_x + f\rho g_x \overset{?}{=} \rho g_x$$

$$\rho g_x (d + f) \overset{?}{=} \rho g_x$$

$$\rho g_x (1) \overset{?}{=} \rho g_x \quad \text{Yes} \quad (d + f = 1)$$

Since an identity is obtained, $P(x) = d\rho g_x x + C$, $V_x(t) = f g_x t + C$

Again, the results are similar to the equations of motion and liquid pressure of elementary physics.

Example 4: There are now three Navier-Stokes terms on the left side of the equation.

If $-\mu\frac{\partial^2 V_x}{\partial x^2} + \frac{\partial p}{\partial x} + \rho\frac{\partial V_x}{\partial t} = \rho g_x$, (A)

find V_x and $P(x)$.

Solution: Let the ratio terms for $-\mu\frac{\partial^2 V_x}{\partial x^2}, \frac{\partial p}{\partial x}, \rho\frac{\partial V_x}{\partial t}$ be a, d and f respectively.

(That is, the terms on the left side of the equation divide the term on the right in the ratio $a : d : f$)

Step 1: For $-\mu\frac{\partial^2 V_x}{\partial x^2}$

$-\mu\frac{\partial^2 V_x}{\partial x^2} = a\rho g_x$ (a is a ratio term)

$-\mu\frac{d^2 V_x}{dx^2} = a\rho g_x$

(One drops the partials symbol)

$\frac{d^2 V_x}{dx^2} = -\frac{a\rho g_x}{\mu}$

$\frac{dV_x}{dx} = -\frac{a\rho g_x x}{\mu} + C$ (integrating)

$V_x = -\frac{a\rho g_x x^2}{2\mu} + C_1 x + C_2$

(integrating again)

Step 2: For $\frac{\partial p}{\partial x}$

From Step 2 of Example 3 ,

For $\frac{\partial p}{\partial x}$

$\frac{\partial p}{\partial x} = d\rho g_x$ (d is a ratio term)

$\frac{dp}{dx} = d\rho g_x$

(One drops the partials symbol)

$dp = d\rho g_x dx$

$P(x) = \int d\rho g_x dx$

$P(x) = d\rho g_x x + C$

(integrating.)

Step 3: For $\rho\frac{\partial V_x}{\partial t}$

From Step 3 of Example 3

$\rho\frac{\partial V_x}{\partial t} = f\rho g_x$ (f is a ratio term)

$\frac{\partial V_x}{\partial t} = g_x$ (Dividing out the ρ)

$\frac{dV_x}{dt} = fg_x$

(One drops the partials symbol since a single independent variable is involved)

$dV_x = fg_x dt$

$V_x = \int fg_x dt$

$V_x = fg_x t + C$ (integrating)

Since there are two V_x's, one adds the results from Step 1, and Step 3.

$V_x = -\frac{a\rho g_x x^2}{2\mu} + C_1 x + fg_x t + C_3$ (B); and $P(x) = d\rho g_x x + C$ (D)

To check for identity:

From (B) an (D) respectively, one obtains the following:

$$\boxed{\begin{array}{c}\frac{\partial V_x}{\partial x} = -\frac{a\rho g_x x}{\mu} \\ \frac{\partial^2 V_x}{\partial x^2} = -\frac{a\rho g_x}{\mu}\end{array}}, \quad \boxed{\frac{\partial V_x}{\partial t} = fg_x}, \quad \boxed{\frac{dp}{dx} = d\rho g_x}$$

Substituting for $\frac{\partial^2 V_x}{\partial x^2}$, $\frac{\partial V_x}{\partial t}$ and $\frac{\partial p}{\partial x}$ from (B) and (D) respectively on the left side of (A) above,

$$-\mu(-\frac{a\rho g_x}{\mu}) + d\rho g_x + \rho f g_x \stackrel{?}{=} \rho g_x$$

$$a\rho g_x + d\rho g_x + \rho f g_x \stackrel{?}{=} \rho g_x$$

$$\rho g_x(a + d + f) \stackrel{?}{=} \rho g_x$$

$$\rho g_x(1) \stackrel{?}{=} \rho g_x \quad \text{Yes} \quad (a + d + f = 1)$$

Since an identity is obtained,, $V_x = -\frac{a\rho g_x x^2}{2\mu} + C_1 x + fg_x t + C_3$; $P(x) = d\rho g_x x + C$

Again, the results are similar to the equations of motion and liquid pressure of elementary physics.

Example 5: There are now five Navier-Stokes terms, including two non-linear terms on the left side of the equation.

If $\rho V_x \frac{\partial V_x}{\partial x} + \rho V_y \frac{\partial V_x}{\partial y} - \mu \frac{\partial^2 V_x}{\partial x^2} + \frac{\partial p}{\partial x} + \rho \frac{\partial V_x}{\partial t} = \rho g_x$ (A)

find V_x and $P(x)$.

Solution

Let the ratio terms for $V_y \frac{\partial V_x}{\partial y}$, $-\mu \frac{\partial^2 V_x}{\partial x^2}$, $\rho \frac{\partial V_x}{\partial t}$, and $\frac{\partial p}{\partial x}$ be h, n, a, f and d respectively.
(That is, the terms on the left side of the equation divide the term on the right in the ratio $h:n:a:d:f$)

Step 1: For $\rho V_x \frac{\partial V_x}{\partial x}$

$\rho V_x \frac{dV_x}{dx} = h\rho g_x$

$V_x \frac{dV_x}{dx} = hg_x$ (divide out ρ)

$V_x dV_x = hg_x dx$

$\frac{V_x^2}{2} = hg_x x$

$\boxed{V_x = \pm\sqrt{2hg_x x + C_7}}$

(integrating)

Step 2: For $\rho V_y \frac{\partial V_x}{\partial y}$

$\rho V_y \frac{\partial V_x}{\partial y} = n\rho g_x$

$V_y \frac{dV_x}{dy} = ng_x$

$V_y dV_x = ng_x dy$

$V_y V_x = ng_x y + \psi_y(V_y)$

$\boxed{V_x = \frac{ng_x y}{V_y} + \frac{\psi_y(V_y)}{V_y}}$ (integrate)

where $\frac{\psi_y(V_y)}{V_y}$ is an arbitrary function

Note that this is an implicit solution, since the solution contains V_y

Step 3: For $-\mu \frac{\partial^2 V_x}{\partial x^2}$

$-\mu \frac{\partial^2 V_x}{\partial x^2} = a\rho g_x$ (a is a ratio term)

$-\mu \frac{d^2 V_x}{dx^2} = a\rho g_x$

(One drops the partials symbol, since a single independent variable is involved)

$-\frac{d^2 V_x}{dx^2} = -\frac{a\rho g_x}{\mu}$

$\frac{dV_x}{dx} = -\frac{a\rho g_x x}{\mu} + C$ (integrating)

$\boxed{V_x = -\frac{a\rho g_x x^2}{2\mu} + C_1 x + C_2}$

(integrating again)

Step 4: For $\rho \frac{\partial V_x}{\partial t}$

From Step 3 of Example 3

$\rho \frac{\partial V_x}{\partial t} = f\rho g_x$ (f is a ratio term)

$\frac{\partial V_x}{\partial t} = g_x$ (Dividing out the ρ)

$\frac{dV_x}{dt} = fg_x$

((One drops the partials symbol since a single independent variable is involved)

$dV_x = fg_x dt$

$V_x(t) = \int fg_x dt$

$\boxed{V_x(t) = fg_x t + C}$ (integrating)

Step 5: For $\frac{\partial p}{\partial x}$

From Step 2 of Example 3,

$\frac{\partial p}{\partial x} = d\rho g_x$ (d is a ratio term)

$\frac{dp}{dx} = d\rho g_x$ (One drops the partials symbol)

$dp = d\rho g_x dx$

$P(x) = \int d\rho g_x dx$

$\boxed{P(x) = d\rho g_x x + C}$ (integrating)

Since there are four V_x's, one adds the results from Steps 1, 2, 3 and 4.

$V_x = \pm\sqrt{2hg_x x} - \frac{a\rho g_x x^2}{2\mu} + C_1 x + fg_x t + \frac{ng_x y}{V_y} + \frac{\psi_y(V_y)}{V_y} + C_3$;   (B)

and $P(x) = d\rho g_x x + C$ (D)

To check for identity

From (B) an (D) respectively,

$V_x = -\frac{a\rho g_x x^2}{2\mu} + C_1 x + fg_x t + \pm\sqrt{2hg_x x} + \frac{ng_x y}{V_y} + \frac{\psi_y(V_y)}{V_y} + C_3$, $P(x) = d\rho g_x x + C$, one obtains the following derivatives. However, one takes the derivative for the linear terms separately and for the non-linear terms separately.

Test derivatives for the linear part (From Example 4))			Test derivatives for the non-linear part	
$\frac{\partial^2 V_x}{\partial x^2} = -\frac{a\rho g_x}{\mu}$	$\frac{\partial V_x}{\partial t} = fg_x$	$\frac{\partial p}{\partial x} = d\rho g_x$	$V_x^2 = 2hg_x x$ $2V_x \frac{\partial V_x}{\partial x} = 2hg_x$ $\frac{\partial V_x}{\partial x} = \frac{hg_x}{V_x}, V_x \neq 0$	$V_x = \frac{ng_x y}{V_y}$ $\frac{\partial V_x}{\partial y} = \frac{ng_x}{V_y}$ $\frac{\partial V_x}{\partial y} = \frac{ng_x}{V_y}$

Substituting for $\frac{\partial^2 V_x}{\partial x^2}, \frac{\partial Vx}{\partial t}, \frac{\partial p}{\partial x}$ $\frac{\partial V_x}{\partial x} \frac{\partial V_x}{\partial y}$ from above table, respectively in the left side of (A),

$$V_x \frac{\partial V_x}{\partial x} + V_y \frac{\partial V_x}{\partial y} - \mu \frac{\partial^2 V_x}{\partial x^2} + \frac{\partial p}{\partial x} + \rho \frac{\partial Vx}{\partial t} = \rho g_x$$

$$\rho V_x \frac{hg_x}{V_x} + \rho V_y \frac{ng_x}{V_y} - \mu(-\frac{a\rho g_x}{\mu}) + d\rho g_x + \rho fg_x \overset{?}{=} \rho g_x$$

$$\rho hg_x + \rho ng_x + a\rho g_x + d\rho g_x + \rho fg_x \overset{?}{=} \rho g_x$$

$$\rho g_x (h + n + a + d + f) \overset{?}{=} \rho g_x$$

$$\rho g_x (1) \overset{?}{=} \rho g_x \quad \text{Yes } (h + n + a + d + f = 1)$$

Since an identity is obtained, $V_x = -\frac{a\rho g_x x^2}{2\mu} + C_1 x + fg_x t \pm \sqrt{2hg_x x} + \frac{ng_x y}{V_y} + \frac{\psi_y(V_y)}{V_y} + C_3$;

$P(x) = d\rho g_x x + C$

Again,, except for the fifth term, the results are similar to the equations of motion and liquid pressure of elementary physics.. Note that after solving the N-S equations for the y and z-directions, the system of equations will be solved to express $(ng_x y/V_y)$ in terms of motion equations of elementary physics .
From above, if there were 100 terms on the left side of the equation, one could repeat the above procedure for each term. Therefore, using the ratio method, the N-S equations can be solved in any number of dimensions.
After having solved for a viscosity term, two convective acceleration terms, a variable acceleration term, and the pressure gradient term, one is ready to solve a complete Navier-Stokes equation.
In fact , one is ready to solve a 100-term Navier-Stokes equation.
If the reader likes the approach used so far, see epsilon-delta proofs in Calculus 1 & 2 by A. A. Frempong, published by Yellowtextbooks.com

Solutions of 3-D Navier-Stokes Equations

By imitating the previous examples, one will now solve the x–direction Navier-Stokes equation.

$$-\mu\frac{\partial^2 V_x}{\partial x^2}-\mu\frac{\partial^2 V_x}{\partial y^2}-\mu\frac{\partial^2 V_x}{\partial z^2}+\frac{\partial p}{\partial x}+\rho\frac{\partial V_x}{\partial t}+\rho V_x\frac{\partial V_x}{\partial x}+\rho V_y\frac{\partial V_x}{\partial y}+\rho V_z\frac{\partial V_x}{\partial z}=\rho g_x \qquad \text{(A)}$$

The subject of the above equation is the gravity term, because when each term of the linearized N-S equation was used as the subject of the equation, only the equation with the gravity term as the subject produced a solution (see viXra:1512.0334).

Step 1: The first step here, is to split-up the equation into eight sub-equations using the ratio method. The ratio terms for the terms on the left side of the equation are a, b, c, d, f, h, n, q, respectively. $(a+b+c+d+f+h+n+q=1)$

$$1.\ -\mu\frac{\partial^2 V_x}{\partial x^2}=a\rho g_x;\quad 2.\ -\mu\frac{\partial^2 V_x}{\partial y^2}=b\rho g_x;\quad 3.\ -\mu\frac{\partial^2 V_x}{\partial z^2}=c\rho g_x;\quad 4.\ \frac{\partial p}{\partial x}=d\rho g_x;$$

$$5.\ \frac{\partial V_x}{\partial t}=fg_x\quad 6.\ V_x\frac{\partial V_x}{\partial x}=hg_x;\quad 7.\ V_y\frac{\partial V_x}{\partial y}=ng_x;\quad 8.\ V_z\frac{\partial V_x}{\partial z}=qg_x$$

Step 2: Solve the differential equations in Step 1.

Note that after splitting the equations, the equations can be solved using techniques of ordinary differential equations, since there would be a single independent variable in each equation. One can also view each of the ratio terms a, b, c, d, f, h, n, q as a fraction (a real number) of $\boxed{g_x}$ contributed by each expression on the left-hand side of equation (A) above.

Solutions of the eight sub-equations

1. $-\mu\frac{\partial^2 V_x}{\partial x^2}=a\rho g_x;$

$$\frac{d^2V_x}{dx^2}=-\frac{a\rho g_x}{\mu}$$

$$\frac{dV_x}{dx}=-\frac{a\rho g_x}{\mu}x+C_1 \text{ (integr.)}$$

$$V_{x1}=-\frac{a\rho g_x}{2\mu}x^2+C_1x+C_2$$

2. $-\mu\frac{\partial^2 V_x}{\partial y^2}=b\rho g_x$

$$\frac{d^2V_x}{dy^2}=-\frac{b\rho g_x}{\mu}$$

$$\frac{dV_x}{dy}=-\frac{b\rho g_x}{\mu}x+C_1$$

$$V_{x2}=-\frac{b\rho g_x}{2\mu}y^2+C_1y+C_2$$

3. $-\mu\frac{\partial^2 V_x}{\partial z^2}=c\rho g_x;$

$$\frac{d^2V_x}{dz^2}=-\frac{c\rho g_x;}{\mu}$$

$$\frac{dV_x}{dz}=-\frac{c\rho g_x;}{\mu}z+C_5$$

$$V_{x3}=-\frac{c\rho g_x;}{2\mu}z^2+C_5z+C_6$$

4. $\frac{\partial p}{\partial x}=d\rho g_x;$

$$\frac{1}{\rho}\frac{\partial p}{\partial x}=dg_x$$

$$\frac{\partial p}{\partial x}=d\rho g_x$$

$$p=d\rho g_x x+C_7$$

5. $\boxed{\frac{\partial V_x}{\partial t}=fg_x}$

$$V_{x4}=fgt$$

6. $V_x\frac{\partial V_x}{\partial x}=hg_x;$

$$V_x\frac{dV_x}{dx}=hg_x$$

$$V_xdV_x=hg_xdx$$

$$\frac{V_x^2}{2}=hg_xx$$

$$V_{x5}=\pm\sqrt{2hg_xx+C_7}$$

7. $V_y\frac{\partial V_x}{\partial y}=ng_x;$

$$V_y\frac{dV_x}{dy}=ng_x$$

$$V_ydV_x=ng_x\mathrm{d}y$$

$$V_yV_x=ng_x\mathrm{y}+\psi_y(V_y)$$

$$V_{x6}=\frac{ng_x\mathrm{y}}{V_y}+\frac{\psi_y(V_y)}{V_y}$$

8. $V_z\frac{\partial V_x}{\partial z}=qg_x$

$$V_z\frac{dV_x}{dz}=qg_x$$

$$V_zdV_x=qg_x\mathrm{d}z;$$

$$V_zV_x=qg_x\mathrm{z}+\psi_z(V_z)$$

$$V_{x7}=\frac{qg_x\mathrm{z}}{V_z}+\frac{\psi_z(V_z)}{V_z}$$

Note:
$\psi_y(V_y)$, $\psi_z(V_z)$ are arbitrary functions, (integration constants)
$V_y\neq 0$
$V_z\neq 0$

Step 3: One combines the above solutions

$$V_x(x,y,z,t) = V_{x1} + V_{x2} + V_{x3} + V_{x4} + V_{x5} + V_{x6} + V_{x7}$$

$$= -\frac{ag_x}{2k}x^2 + C_1x - \frac{bg_x}{2k}y^2 + C_3y - \frac{cg_x}{2k}z^2 + C_5z + fg_xt \pm \sqrt{2hg_xx} + \frac{ng_xy}{V_y} + \frac{qg_xz}{V_z} + \frac{\psi_y(V_y)}{V_y} + \frac{\psi_z(V_z)}{V_z}$$

relation for linear terms — relation for non-linear terms

$$-\frac{\rho g_x}{2\mu}(ax^2 + by^2 + cz^2) + C_1x + C_3y + C_5z + fg_xt \pm \sqrt{2hg_xx} + \frac{ng_xy}{V_y} + \frac{qg_xz}{V_z} + \underbrace{\frac{\psi_y(V_y)}{V_y} + \frac{\psi_z(V_z)}{V_z}}_{\text{arbitrary functions}} + C_9$$

$$P(x) = d\rho g_xx; \quad (a+b+c+d+f+h+n+q=1) \quad V_y \neq 0,\ V_z \neq 0$$

Step 4: Find the test derivatives

Test derivatives for the linear part					Test derivatives for the non-linear part		
$\frac{\partial^2 V_x}{\partial x^2} = -\frac{a\rho g_x}{\mu}$	$\frac{\partial^2 V_x}{\partial y^2} = -\frac{b\rho g_x}{\mu}$	$\frac{\partial^2 V_x}{\partial z^2} = -\frac{c\rho g_x}{\mu}$	$\frac{\partial p}{\partial x} = d\rho g_x$	$\frac{\partial V_x}{\partial t} = fg_x$	$V_x^2 = 2hg_xx$ $2V_x\frac{\partial V_x}{\partial x} = 2hg_x$ $\frac{\partial V_x}{\partial x} = \frac{hg_x}{V_x}, V_x \neq 0$	$\frac{\partial V_x}{\partial y} = \frac{ng_x}{V_y}$	$\frac{\partial V_x}{\partial z} = \frac{qg_x}{V_z}$

Step 5: Substitute the derivatives from Step 4 in equation (A) for the checking.

$$-\mu\frac{\partial^2 V_x}{\partial x^2} - \mu\frac{\partial^2 V_x}{\partial y^2} - \mu\frac{\partial^2 V_x}{\partial z^2} + \frac{\partial p}{\partial x} + \rho\frac{\partial V_x}{\partial t} + \rho V_x\frac{\partial V_x}{\partial x} + \rho V_y\frac{\partial V_x}{\partial y} + \rho V_z\frac{\partial V_x}{\partial z} = \rho g_x \quad \textbf{(A)}$$

$$-\mu\left(-\frac{a\rho g_x}{\mu} - \frac{b\rho g_x}{\mu} - \frac{c\rho g_x}{\mu}\right) + d\rho g_x + f\rho g_x + \rho\left(V_x\frac{hg_x}{V_x}\right) + \rho V_y\left(\frac{ng_x}{V_y}\right) + \rho V_z\left(\frac{qg_x}{V_z}\right) \stackrel{?}{=} \rho g_x$$

$$a\rho g_x + b\rho g_x + c\rho g_x + d\rho g_x + f\rho g_x + h\rho g_x + n\rho g_x + q\rho g_x \stackrel{?}{=} \rho g_x$$

$$a\rho g_x + b\rho g_x + c\rho g_x + d\rho g_x + f\rho g_x + h\rho g_x + n\rho g_x + q\rho g_x \stackrel{?}{=} \rho g_x$$

$$\rho g_x(a+b+c+d+f+h+n+q) \stackrel{?}{=} \rho g_x$$

$$\rho g_x(1) \stackrel{?}{=} \rho g_x \quad \text{Yes} \quad (a+b+c+d+f+h+n+q=1)$$

Step 6: The linear part of the relation satisfies the linear part of the equation; and the non-linear part of the relation satisfies the non-linear part of the equation.(

Analogy for the Identity Checking Method: If one goes shopping with American dollars and Japanese yens (without any currency conversion) and after shopping, if one wants to check the cost of the items purchased, one would check the cost of the items purchased with dollars against the receipts for the dollars; and one would also check the cost of the items purchased with yens against the receipts for the yens purchase. However, if one converts one currency to the other, one would only have to check the receipts for only a single currency, dollars or yens. This conversion case is similar to the linearized equations (see viXra:1512.0334), where there was no partitioning in identity checking. Note that for the Euler equations (viXra:1512.0332), there was no partitioning in taking derivatives for identity checking.

Note: After expressing $\frac{ng_xy}{V_y}$ and $\frac{q_eg_xz}{V_z}$ in terms of x, y, z, and t, there would be no partitioning in identity checking.

Summary of solutions for V_x , V_y, V_z ($P(x) = d\rho g_x x$; $P(y) = \lambda_4 \rho g_y y$, $P(z) = \beta_4 \rho g_z z$)

$$V_x = -\frac{\rho g_x}{2\mu}(ax^2 + by^2 + cz^2) + C_1x + C_3y + C_5z + fg_xt \pm \sqrt{2hg_xx} + \frac{ng_xy}{V_y} + \frac{qg_xz}{V_z} + \frac{\psi_y(V_y)}{V_y} + \frac{\psi_z(V_z)}{V_z} + C_9$$

$P(x) = d\rho g_x x$; $(a + b + c + d + h + n + q = 1)$ $V_y \neq 0$, $V_z \neq 0$

$$V_y = -\frac{\rho g_y}{2\mu}(\lambda_1 x^2 + \lambda_2 y^2 + \lambda_3 z^2) + C_{10}x + C_{11}y + C_{12}z + \lambda_5 g_y t \pm \sqrt{2\lambda_7 g_y y} + \frac{\lambda_6 g_y x}{V_x} + \frac{\lambda_8 g_y z}{V_z} + \frac{\psi_x(V_x)}{V_x} + \frac{\psi_z(V_z)}{V_z}$$

$$V_z = -\frac{\rho g_z}{2\mu}(\beta_1 x^2 + \beta_2 y^2 + \beta_3 z^2) + C_{14}x + C_{15}y + C_{16}z + \beta_5 g_z t \pm \sqrt{2\beta_8 g_z z} + \frac{\beta_6 g_z x}{V_x} + \frac{\beta_7 g_z y}{V_y} + \frac{\psi_x(V_x)}{V_x} + \frac{\psi_y(V_y)}{V_y}$$

The above solutions are unique, because from the experience in Option 1 (viXra:1512.0334)., only the equations with the gravity terms as the subjects of the equations produced the solutions.

$$\begin{cases} V_x = -\frac{\rho g_x}{2\mu}(ax^2 + by^2 + cz^2) + C_1x + C_3y + C_5z + fg_xt \pm \sqrt{2hg_xx} + \frac{ng_xy}{V_y} + \frac{qg_xz}{V_z} + \frac{\psi_y(V_y)}{V_y} + \frac{\psi_z(V_z)}{V_z} \\ V_y = -\frac{\rho g_y}{2\mu}(\lambda_1 x^2 + \lambda_2 y^2 + \lambda_3 z^2) + C_{10}x + C_{11}y + C_{12}z + \lambda_5 g_y t \pm \sqrt{2\lambda_7 g_y y} + \frac{\lambda_6 g_y x}{V_x} + \frac{\lambda_8 g_y z}{V_z} + \frac{\psi_x(V_x)}{V_x} + \frac{\psi_z(V_z)}{V_z} \\ V_z = -\frac{\rho g_z}{2\mu}(\beta_1 x^2 + \beta_2 y^2 + \beta_3 z^2) + C_{14}x + C_{15}y + C_{16}z + \beta_5 g_z t \pm \sqrt{2\beta_8 g_z z} + \frac{\beta_6 g_z x}{V_x} + \frac{\beta_7 g_z y}{V_y} + \frac{\psi_x(V_x)}{V_x} + \frac{\psi_y(V_y)}{V_y} \end{cases}$$

One will next solve the above system of solutions for V_x , V_y, V_z in order to express $\frac{ng_xy}{V_y}$ and $\frac{q_eg_xz}{V_z}$ in terms of x, y, z, and t The author used the help of the Maples software for the simultaneous algebraic solutions for V_x , V_y, V_z . The basic expressions are of the forms $-\frac{\rho g_x}{2\mu}ax^2$, $-\frac{\rho g_x}{2\mu}by^2$, $-\frac{\rho g_x}{2\mu}cz^2$, fg_xt, $\sqrt{2hg_xx}$, and $d\rho g_x x$; These expressions are similar to the terms of the equations of motion under gravity and liquid pressure of elementary physics. Note that the explicit solutions will be the results of the basic operations (addition, subtraction, multiplication, division, power finding and root extraction) on the expressions in Step 1 below.

Solving for V_x , V_y, V_z $\frac{ng_xy}{V_y}$, and $\frac{q_eg_xz}{V_z}$

Let $V_x = x$, $V_y = y$ and $V_z = z$. (x, y and z are being used for simplicity. They will be changed back to V_x , V_y, and V_z later, and they do not represent the variables x, y and z in the solutions.

Step 1 From the above system of solutions, let

$$A = -\frac{\rho g_x}{2\mu}(ax^2 + by^2 + cz^2) + C_1x + C_3y + C_5z + fg_xt \pm \sqrt{2hg_xx}$$

$$B = -\frac{\rho g_y}{2\mu}(\lambda_1 x^2 + \lambda_2 y^2 + \lambda_3 z^2) + C_{10}x + C_{11}y + C_{12}z + \lambda_5 g_y t \pm \sqrt{2\lambda_7 g_y y}$$

$$C = -\frac{\rho g_z}{2\mu}(\beta_1 x^2 + \beta_2 y^2 + \beta_3 z^2) + C_{14}x + C_{15}y + C_{16}z + \beta_5 g_z t \pm \sqrt{2\beta_8 g_z z}$$

$D = qg_xz$; $E = ng_xy$; $F = \lambda_6 g_y x$

$G = \lambda_8 g_y z$; $J = \beta_6 g_z x$; $L = \beta_7 g_z y$

Step 2 Then the solutions to the N-S system of equations become (ignoring the arbitrary functions)

$$\left.\begin{aligned} x &= A + \frac{D}{z} + \frac{E}{y} \\ y &= B + \frac{F}{x} + \frac{G}{z} \\ z &= C + \frac{J}{x} + \frac{L}{y} \end{aligned}\right\} \quad M$$

Step 3	**Step 4**
$xyz = Ayz + Dy + Ez$ (1) $xyz = Bxz + Fz + Gx$ (2) $xyz = Cxy + Jy + Lx$ (3)) } N	$0 = Ayz + Dy + Ez - Bxz - Fz - Gx$ (4) $0 = Ayz + Dy + Ez - Cxy - Jy - Lx$ (5) $0 = Bxz + Fz + Gx - Cxy - Jy + -Lx$ (6) } P

Maples software was used to solve system P to obtain

Step 5a

$$x = \frac{L(FCD - FCJ - JLA + JCE)}{C(-BLD + BLJ + GLA - GCE)}$$

$$V_x = \frac{L(FCD - FCJ - JLA + JCE)}{C(-BLD + BLJ + GLA - GCE)} \quad \text{(back to } V_x\text{)}$$

$$y = -\frac{L}{C};$$

$$V_y = -\frac{L}{C} \quad \text{(changing back to } V_y \text{ as agreed to)}$$

$$z = -\frac{L(D-J)}{LA - CE};$$

$$V_z = -\frac{L(D-J)}{LA - CE} \quad \text{(changing back to } V_z \text{ as agrred to)}$$

Note:
None of the popular academic programs could solve the system in M.
Maples solved system P (step 4 above) for x, y,and z in terms of A, B, C, D. E. F, G. J. and L.
Note also that x, y and z are not the same as the x, y and z in the system of equations..
They were used for convenience and simplicity .

Step 5b : Apply the following and substitute for A, B, C, D. E. F, G., J. and L in steps 6-7 below

$$A = -\frac{\rho g_x}{2\mu}(ax^2 + by^2 + cz^2) + C_1x + C_3y + C_5z + fg_xt \pm \sqrt{2hg_xx}$$

$$B = -\frac{\rho g_y}{2\mu}(\lambda_1x^2 + \lambda_2y^2 + \lambda_3z^2) + C_{10}x + C_{11}y + C_{12}z + \lambda_5g_yt \pm \sqrt{2\lambda_7g_yy}$$

$$C = -\frac{\rho g_z}{2\mu}(\beta_1x^2 + \beta_2y^2 + \beta_3z^2) + C_{14}x + C_{15}y + C_{16}z + \beta_5g_zt \pm \sqrt{2\beta_8g_zz}$$

$$D = qg_xz;\ E = ng_xy;\ F = \lambda_6g_yx$$

$$G = \lambda_8g_yz;\ \ J = \beta_6g_zx;\ \ L = \beta_7g_zy$$

Step 6

$$V_y = -\frac{L}{C} = -\frac{(\beta_7g_zy)}{(-\frac{\rho g_z}{2\mu}(\beta_1x^2 + \beta_2y^2 + \beta_3z^2) + C_1x + C_3y + C_5z + \beta_5g_zt \pm \sqrt{2\beta_8g_zz}}$$

$$\frac{ng_xy}{V_y} = ng_xy \div -(\frac{(\beta_7g_zy)}{(-\frac{\rho g_z}{2\mu}(\beta_1x^2 + \beta_2y^2 + \beta_3z^2) + C_1x + C_3y + C_5z + \beta_5g_zt \pm \sqrt{2\beta_8g_zz}})$$

$$\frac{ng_xy}{V_y} = -\frac{(ng_xy)[(-\frac{\rho g_z}{2\mu}(\beta_1x^2 + \beta_2y^2 + \beta_3z^2) + C_1x + C_3y + C_5z + \beta_5g_zt \pm \sqrt{2\beta_8g_zz})]}{\beta_7g_zy};$$

$$\frac{ng_xy}{V_y} = \frac{-(ng_x)(-\frac{\rho g_z}{2\mu}(\beta_1x^2 + \beta_2y^2 + \beta_3z^2) + C_1x + C_3y + C_5z + \beta_5g_zt \pm \sqrt{2\beta_8g_zz})}{\beta_7g_z}; \text{ (cancel "y")}$$

Step 7: $V_z = -\frac{L(D-J)}{LA-CE} = \frac{JL-DL)}{LA-CE}$

$$\frac{qg_x z}{V_z} = (qg_x z) \bullet \frac{(\beta_7 g_z y)[-\frac{\rho g_x}{2\mu}(ax^2+by^2+cz^2)+C_1x+C_3y+C_5z+fg_xt \pm \sqrt{2hg_xx}] -}{(\beta_7 g_z y)[\beta_6 g_z x - qg_x z]}$$

$$\frac{(ng_x y)[-\frac{\rho g_z}{2\mu}(\beta_1 x^2+\beta_2 y^2+\beta_3 z^2)+C_{14}x+C_{15}y+C_{16}z+\beta_5 g_z t \pm\sqrt{2\beta_8 g_z z}]}{(\beta_7 g_z y)[\beta_6 g_z x - qg_x z]}$$

$$\frac{qg_x z}{V_z} = (qg_x z) \bullet \frac{(\beta_7 g_z)[-\frac{\rho g_x}{2\mu}(ax^2+by^2+cz^2)+C_1x+C_3y+C_5z+fg_xt \pm \sqrt{2hg_xx}] -}{(\beta_7 g_z)[\beta_6 g_z x - qg_x z]}$$

$$\frac{(ng_x)[-\frac{\rho g_z}{2\mu}(\beta_1 x^2+\beta_2 y^2+\beta_3 z^2)+C_{14}x+C_{15}y+C_{16}z+\beta_5 g_z t \pm\sqrt{2\beta_8 g_z z}]}{(\beta_7 g_z)[\beta_6 g_z x - qg_x z]}$$

Summary for the fractional terms of the x–direction solution

$\frac{ng_x y}{V_y}$ and $\frac{qg_x z}{V_z}$ in terms of x, y, z and t

$$\frac{ng_x y}{V_y} = \frac{-(ng_x)(-\frac{\rho g_z}{2\mu}(\beta_1 x^2+\beta_2 y^2+\beta_3 z^2)+C_1x+C_3y+C_5z+\beta_5 g_z t \pm\sqrt{2\beta_8 g_z z})}{\beta_7 g_z}; \text{ (cancel "y")}$$

$$\frac{qg_x z}{V_z} = (qg_x z) \bullet \frac{(\beta_7 g_z)[-\frac{\rho g_x}{2\mu}(ax^2+by^2+cz^2)+C_1x+C_3y+C_5z+fg_xt \pm \sqrt{2hg_xx}] -}{(\beta_7 g_z y)[\beta_6 g_z x - qg_x z]}$$

$$\frac{(ng_x)[-\frac{\rho g_z}{2\mu}(\beta_1 x^2+\beta_2 y^2+\beta_3 z^2)+C_{14}x+C_{15}y+C_{16}z+\beta_5 g_z t \pm\sqrt{2\beta_8 g_z z}]}{(\beta_7 g_z)[\beta_6 g_z x - qg_x z]}$$

$$\left(CE = -(ng_x y)(-\frac{\rho g_z}{2\mu}(\beta_1 x^2+\beta_2 y^2+\beta_3 z^2)+C_{14}x+C_{15}y+C_{16}z+\beta_5 g_z t \pm\sqrt{2\beta_8 g_z z} \right)$$

$$d+f+h+n+q=1; \quad \lambda_4+\lambda_5+\lambda_6+\lambda_7+\lambda_8=1; \quad \beta_4+\beta_5+\beta_6+\beta_7+\beta_8=1$$

OR Compactly, x-direction solution of N-S Equation (in explicit solutions)

$\mathbf{V}_x(x,y,z,t) = A+B+F$ **Expanded** V_x

$$\mathbf{A} = -\frac{\rho g_x}{2\mu}(ax^2+by^2+cz^2)+C_1x+C_3y+C_5z+fg_xt \pm \sqrt{2hg_xx}+C_9$$

$$\mathbf{B} = \frac{-(ng_x)(-\frac{\rho g_z}{2\mu}(\beta_1 x^2+\beta_2 y^2+\beta_3 z^2)+C_1x+C_3y+C_5z+\beta_5 g_z t \pm\sqrt{2\beta_8 g_z z})}{\beta_7 g_z} + \frac{\psi_y(V_y)}{V_y}$$

$$\mathbf{F} = (qg_x z) \bullet \frac{(\beta_7 g_z)[-\frac{\rho g_x}{2\mu}(ax^2+by^2+cz^2)+C_1x+C_3y+C_5z+fg_xt \pm \sqrt{2hg_xx}] -}{(\beta_7 g_z y)[\beta_6 g_z x - qg_x z]}$$

$$\frac{(ng_x)[-\frac{\rho g_z}{2\mu}(\beta_1 x^2+\beta_2 y^2+\beta_3 z^2)+C_{14}x+C_{15}y+C_{16}z+\beta_5 g_z t \pm\sqrt{2\beta_8 g_z z}]}{(\beta_7 g_z)[\beta_6 g_z x - qg_x z]} + \frac{\psi_z(V_z)}{V_z}$$

$P(x) = d\rho g_x x; \quad (a+b+c+d+h+n+q=1) \quad \beta_1+\beta_2+\beta_3+\beta_5+\beta_6+\beta_7+\beta_8=1$

Comparison of Navier-Stokes Solutions and Equations of Motion and Liquid Pressure of Elementary Physics;

Solutions of the Navier-Stokes Equations (Original) : x–direction

A $$V_x = -\frac{\rho g_x}{2\mu}(ax^2+by^2+cz^2)+C_1x+C_3y+C_5z+fg_xt \pm \sqrt{2hg_xx}+\frac{ng_xy}{V_y}+\frac{qg_xz}{V_z}+\frac{\psi_y(V_y)}{V_y}+\frac{\psi_z(V_z)}{V_z}+C_9$$

$$P(x) = d\rho g_xx; \quad (a+b+c+d+h+n+q=1) \quad V_y \neq 0,\ V_z \neq 0$$

Summary for the fractional terms of the x–direction in terms of x, y, z and t.

$$\frac{ng_xy}{V_y} = \frac{-(ng_x)(-\frac{\rho g_z}{2\mu}(\beta_1x^2+\beta_2y^2+\beta_3z^2)+C_1x+C_3y+C_5z+\beta_5g_zt \pm\sqrt{2\beta_8g_zz})}{\beta_7g_z}$$

$$\frac{qg_xz}{V_z} = \frac{-(qg_xz)\{[(\beta_7g_zy)(-\frac{\rho g_x}{2\mu}(ax^2+by^2+cz^2)+C_1x+C_3y+C_5z+fg_xt \pm \sqrt{2hg_xx}]-[CE]\}}{(\beta_7g_zy)(qg_xz-\beta_6g_zx)}$$

$$(CE = -(ng_xy)(-\frac{\rho g_z}{2\mu}(\beta_1x^2+\beta_2y^2+\beta_3z^2)+C_{14}x+C_{15}y+C_{16}z+\beta_5g_zt \pm\sqrt{2\beta_8g_zz})$$

Motion equations of elementary physics:

B (B): $V_f = V_0 + gt$; (C): $V_f^2 = V_0^2 + 2gx$; (D): $V = \sqrt{2gx}$; (E): $x = V_0t + \frac{1}{2}gt^2$

The **liquid pressure**, P at the bottom of a liquid of depth h units is given by $P = \rho gh$

Similarity 1. The solutions are (except for the constants involved) very similar to the equations of motion and liquid pressure of elementary physics.

Observe the following about the Navier-Stokes Solutions Box A

1. The first three terms are parabolic in x, y, and z; the minus sign shows the usual inverted parabola when a projectile is fired upwards at an acute angle to the horizontal; also note the " gt " in $V = gt$ of (B) of the motion equations and the fg_xt in the Navier-Stokes solution.

2. The pressure, $P = \rho gh$ of the liquid pressure and the $P(x) = d\rho g_xx$ of the Navier-Stokes solution.

Note that, only the approach in this paper could yield $P(x) = d\rho g_xx$ by integrating $dp/dx = d\rho g_x$

3. Observe the " $\sqrt{2gx}$ " in $V = \sqrt{2gx}$ of (D) and the $\sqrt{2hg_xx}$ in the Navier-Stokes solution.

In fact, the N-S solution term $\sqrt{2hg_xx}$ could have been obtained from $V_f^2 = V_0^2 + 2gx$,(C) , of the equation of motion by letting $V_0 = 0$ (for the convective term) ignoring the ratio term "h" of the N-S radicand. There are eight main terms (ignoring the arbitrary functions) in the N-S solution. Of these eight terms, six terms, namely, $-\frac{a\rho g_x}{2\mu}x^2$, $-\frac{b\rho g_x}{2\mu}y^2$, $-\frac{c\rho g_x}{2\mu}z^2$, fg_xt, $\sqrt{2hg_xx}$ and $d\rho g_xx$ are similar (except for the constants involved) to the terms in the equations of motion and liquid pressure. This similarity means that the approach used in solving the Navier-Stokes equation is sound. One should also note that to obtain these six terms simultaneously on integration, only the equation with the gravity term as the subject of the equation will yield these six terms. The author suggests that this form of the equation with the gravity term as the subject of the equation be called the standard form of the Navier-Stokes equation, since in this form, one can immediately split-up the equations using ratios, and integrate.

4. With regards to the variables x, y, and z, the parabolicity of the first three terms and the parabolicity of the eighth, ninth and tenth terms hint at inverse relations.. For examples, $V_x = x^2$ and $V_x = \pm\sqrt{x}$ are inverse relations of each other, $V_x = y^2$ and $V_x = \pm\sqrt{y}$ are inverse relations of each other, $V_x = z^2$ and $V_x = \pm\sqrt{z}$ are inverse relations of each other.
The implications of knowing these relationships is that if one knows the steps, rules or formulas for designing for laminar flow, one can deduce the steps, rules or formulas for designing for turbulent flow by reversing the steps and using opposite operations in each step of the corresponding laminar flow design. Thus for every method, or formula for laminar flow, there is a corresponding method, formula for turbulent flow design (see also

Motion equations of elementary physics	N-S Terms
$V_f = V_0 + gt$	fg_xt,
$V_f^2 = V_0^2 + 2gx$	$\sqrt{2hg_xx}$
$x = V_0t + \frac{1}{2}gt^2$	$-\frac{a\rho g_x}{2\mu}x^2$,
$V = \sqrt{2gx}$	$\sqrt{2hg_xx}$
$P = \rho gh$ (pressure equation)	$P(x) = d\rho g_xx$

Similarity 2. Sound approach used in pairing the terms of the equation

The approach used in splitting and pairing the terms of the equation is sound, especially by the results for the pressure $\boxed{P(x) = d\rho g_xx}$. To obtain $P(x) = d\rho g_xx$ by integrating $dp/dx = d\rho g_x$, only the equation with the gravity term as the subject of the equation will produce a solution. Therefore, if dp/dx is set to a constant multiplied by the gravity term (which is the subject of the
N-S equation)) then each of the other terms must also be set equal to a constant multiplied by ρg_x

$$\boxed{-\mu\frac{\partial^2 V_x}{\partial x^2} - \mu\frac{\partial^2 V_x}{\partial y^2} - \mu\frac{\partial^2 V_x}{\partial z^2} + \overbrace{\frac{\partial p}{\partial x}}^{\text{This}} + \rho\frac{\partial V_x}{\partial t} + \rho V_x\frac{\partial V_x}{\partial x} + \rho V_y\frac{\partial V_x}{\partial y} + \rho V_z\frac{\partial V_x}{\partial z} = \rho g_x}$$

From $\frac{\partial p}{\partial x} = d\rho g_x$ $P(x) = d\rho g_xx$, where d is a ratio term. Since ratio terms (constants)) were used to mathematically split-up the nine-term equation into eight sub-equations, all the terms of the equation remain "uncontaminated" and any results of the integration should differ from any other results only by constants.
When each term of the linearized Navier-Stokes equation was made subject of the N-S equation, only the equation with the gravity term as the subject of the equation produced a solution. (viXra:1512.0334).

Similarity 3a: Insight of the N-S solutions and equations of motion and fluid pressure of elementary physics

$$V_x = -\frac{\rho g_x}{2\mu}(ax^2 + by^2 + cz^2) + C_1x + C_3y + C_5z + fg_xt \pm \sqrt{2hg_xx} + \frac{ng_xy}{V_y} + \frac{qg_xz}{V_z} + \frac{\psi_y(V_y)}{V_y} + \frac{\psi_z(V_z)}{V_z} + C_9$$

$P(x) = d\rho g_xx$; $(a + b + c + d + h + n + q = 1)$ $V_y \neq 0$, $V_z \neq 0$

One observes above that the most important insight of the above solution is the **indispensability** of the gravity term in incompressible fluid flow. Observe that if gravity, g_x, were zero, the first three terms, the seventh, the eighth, the ninth, the tenth terms of the velocity solution and $P(x)$ would all be zero. For equations of motion and liquid pressure of elementary physics, each equation contains an indispensable gravity term.See above table.

Velocity profile; Laminar flow and Turbulent Flow

For communication purposes, each of the terms containing the even powers x^2, y^2 and z^2 will be called a polynomial parabola, and each of the terms containing the square roots $\pm\sqrt{x}$, $\pm\sqrt{y}$ and $\pm\sqrt{z}$ will be called a radical parabola. Also, each of the terms containing variables in the denominator will be called a hyperbola. The terms, polynomial parabola, radical parabola and hyperbola will be used interchangeably with what produces these profiles.

Turbulence Occurrence

We cover two approaches, namely Approach A and Approach B.
In Approach A, the fluid flow direction as well as that of the polynomial parabolas is horizontally to the right; while the direction of the radical parabolas is upwards. In Approach B, the fluid flow direction as well as that of the polynomial parabolas is vertically downwards in the page, while the direction of the radical polynomials is to the left.

Approach A

In the Navier-Stokes solutions, during fluid flow, the polynomial parabolas, the radical parabolas, and the hyperbolas are present at any speed. The polynomial parabolas are prominent and dominate flow while the radical parabolas are dormant at low speeds, and consequently, the flow is laminar. At a low speed, a radical parabola (or a polynomial parabola susceptible to radicalization) is not very active, since the radicand of the parabola is small and consequently, the square root is small. When the speed becomes large, and certain Reynolds numbers are reached, the "x" in $\sqrt{2hg_x x}$ becomes large and therefore the radical parabola becomes active. Note that the radical parabola will be moving at right angles to the direction of fluid flow, the direction of which is also that of the axis of symmetry of the dominating polynomial parabola. In the figure below, assume that flow is to the right. Then while the axis of symmetry of the polynomial parabola (**P**) is in the horizontal direction, the axis of symmetry of the radical parabola (**R**) would be in the vertical direction (that is, at right angles to fluid flow direction). Also, note the branches of the hyperbola (**H**) in the second and third quadrants. For each branch of the hyperbola, one end becomes asymptotic to the axis of symmetry of the polynomial parabola, while the other end becomes asymptotic to the axis of symmetry of the radical parabola, and thereby "interlocking" the polynomial parabola and the radical parabola together.

Velocity Profiles

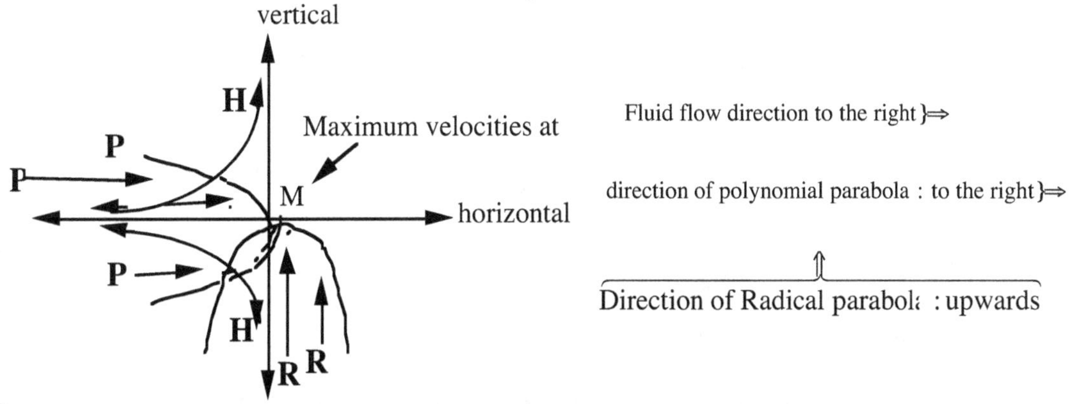

P--Polynomial parabola; R--Radical parabola; H--Hyperbola;

Therefore, the polynomial parabola, the radical parabola and the branches of the hyperbola become connected together to form a system such that any changes in one of them affect the behavior of the others.

Any action that increases the flow velocity such that certain Reynolds numbers are reached, increases the " x " in $\sqrt{2hg_x x}$ and consequently, increases the effect of the radical parabola which is in direction at right angles to the fluid flow direction. The radical parabolas would be moving, from various positions, to the left or to the right, at right angles to the direction of fluid flow, noting that the direction of fluid flow is the direction of the polynomial parabolas, and at the same time, the hyperbolas will be moving asymptotically to fluid flow direction and asymptotically to direction of the radical parabola as in the figure. Thus, while the dominating polynomial parabolas are moving to the right, and the radical parabolas are moving at right angles to direction of fluid flow, the hyperbolas would be moving asymptotically to the axes of symmetry of the polynomial and radical parabolas, resulting in deviation from laminar flow and producing flows such as vortex flow, swirling flow, and turbulent flow. Imagine the polynomial parabolas pulling to the right, while the radical parabolas are pushing upwards, with the hyperbola halves pressing against the axes of the parabolas and the resulting deviation from laminar flow to turbulence and chaos.

Approach B

In the Navier-Stokes solutions, during fluid flow, the polynomial parabolas, the radical parabolas, and the hyperbolas are present at any speed. The polynomial parabolas are prominent and dominate flow while the radical parabolas are dormant at low speeds, and consequently, the flow is laminar. At a low speed, a radical parabola (or a polynomial parabola susceptible to radicalization) is not very active, since the radicand of the parabola is small and consequently, the square root is small. When the speed becomes large, and certain Reynolds numbers are reached, the "x" in $\sqrt{2hg_x x}$ becomes large and therefore the radical parabola becomes active. Note that the radical parabola will be moving at right angles to the direction of fluid flow, the direction of which is also that of the axis of symmetry of the dominating polynomial parabola. In the figure below, assume that low is downwards. Then while the axis of symmetry of the polynomial parabola (**P**) is in the vertical direction, the axis of symmetry of the radical parabola (**R**) would be in the horizontal direction (that is, at right angles to fluid flow direction). Also, note the branches of the hyperbola (**H**) in the first and fourth quadrants.

For each branch of the hyperbola, one end becomes asymptotic to the axis of symmetry of the polynomial parabola, while the other end becomes asymptotic to the axis of symmetry of the radical parabola, and thereby "interlocking" the polynomial parabola and the radical parabola together.

Velocity Profiles

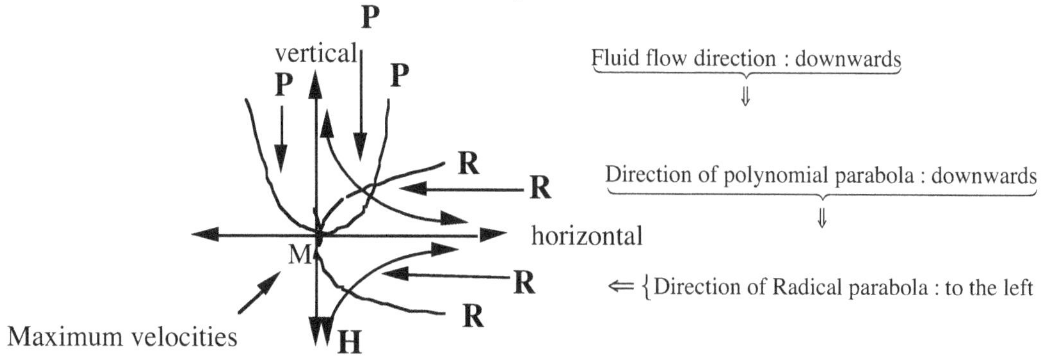

P--Polynomial parabola; R--Radical parabola; H--Hyperbola;

Therefore, the polynomial parabola, the radical parabola and the branches of the hyperbola become connected together to form a system such that any changes in one of them affect the behavior of the others.

Any action that increases the flow velocity such that certain Reynolds numbers are reached , increases the " x" in $\sqrt{2hg_x x}$ and consequently, increases the effect of the radical parabola which is in direction at right angles to the fluid flow direction. The radical parabolas would be moving, from various positions, to the left or to the right, at right angles to the direction of fluid flow, noting that
the direction of fluid flow is the direction of the polynomial parabolas, and at the same time, the hyperbolas will be moving asymptotically to fluid flow direction and asymptotically to direction of the radical parabola as in the figure. Thus, while the dominating polynomial parabolas are moving downwards, and the radical parabolas are moving at right angles to direction of fluid flow, the hyperbolas would be moving asymptotically to the axes of symmetry of the polynomial and radical parabolas, resulting in deviation from laminar flow and producing flows such as vortex flow, swirling flow, and turbulent flow. Imagine the polynomial parabolas pushing downwards, whilethe radical parabolas are pushing to the left, with the hyperbola halves pressing against the axes of the parabolas and the resulting deviation from laminar flow to turbulence and chaos.

Derivation of the Navier-Stokes equations from equations of motion under gravity and liquid pressure of elementary physics

Motion equations of elementary physics	N-S Terms
$V_f = V_0 + gt$	$fg_x t,$
$V_f^2 = V_0^2 + 2gx$	$\sqrt{2hg_x x}$
$x = V_0 t + \frac{1}{2}gt^2$	$-\frac{a\rho g_x}{2\mu}x^2,$
$V = \sqrt{2gx}$	$\sqrt{2hg_x x}$
$p = \rho gh$ (liquid pressure equation)	$p(x) = d\rho g_x x$

1-2. From $P = \rho gh,$ one obtains the N–S terms $\frac{dp}{dx}$ and ρg_x

3. From $V = gt$, one obtains the N–S term $\frac{dV}{dt}$ (note : $\frac{dV}{dt} = g$)

4. From $V = \sqrt{2gx}$, one obtains the N–S term $Vx\frac{dVx}{dx}$

$$(Vx^2 = 2gx; \leftrightarrow 2Vx\frac{dVx}{dx} = 2g \leftrightarrow Vx\frac{dVx}{dx} = g)$$

5. Adding the derivative with respect to y, one obtains the N–S term $Vy\frac{dV_x}{dy}$

6. Adding the derivative with respect to z, one obtains the N–S term $Vz\frac{dV_x}{dz}$

From 1 to 6, one obtains the terms

$$\frac{dp}{dx}, \rho g_x, \frac{dV}{dt}, Vx\frac{dV}{dx}, Vy\frac{dV_x}{dy}, Vz\frac{dV_x}{dz},$$

Adding the viscous terms $-\mu\frac{\partial^2 V_x}{\partial x^2}$, $-\mu\frac{\partial^2 V_x}{\partial y^2}$, $-\mu\frac{\partial^2 V_x}{\partial z^2}$ one obtains

$$\frac{dp}{dx}, \rho g_x, \frac{dV}{dt}, Vx\frac{dV}{dx}, Vy\frac{dV_x}{dy}, Vz\frac{dV_x}{dz}, -\mu\frac{\partial^2 V_x}{\partial x^2}, -\mu\frac{\partial^2 V_x}{\partial y^2}, -\mu\frac{\partial^2 V_x}{\partial z^2}$$

Introducing ρ and putting the terms into an equation, one obtains the x-direction Navier-Stokes Equation $-\mu\frac{\partial^2 V_x}{\partial x^2} - \mu\frac{\partial^2 V_x}{\partial y^2} - \mu\frac{\partial^2 V_x}{\partial z^2} + \frac{\partial p}{\partial x} + \rho\frac{\partial V_x}{\partial t} + \rho V_x\frac{\partial V_x}{\partial x} + \rho V_y\frac{\partial V_x}{\partial y} + \rho V_z\frac{\partial V_x}{\partial z} = \rho g_x$

The above equation is the nine-member supreme equation of incompressible fluid flow.

Overall Conclusion

The Navier-Stokes (N-S) equations in 3-D have been solved analytically. It was also shown that without gravity forces on earth, there would be no incompressible fluid flow on earth as is known. Note that so far as the general solutions of the N-S equations are concerned, one needs not find the specific values of the ratio terms involved. The N-S solutions were compared to the equations of motion under gravity and liquid pressure of elementary physics, and it was found out that, except for the constants involved, the N-S solutions are very similar or identical to the equations of motion and liquid pressure of elementary physics. Such agreement shows that the N--S equations were properly solved. It could be stated that the solutions of the N-S equations have existed since the time the equations of motion and liquid pressure of elementary physics were derived. Insights into the solutions include how the polynomial parabolas, the radical parabolas, and the hyperbolas interact to produce turbulent flow. Also, the x-direction N-S equation was derived from the equations of motion under gravity and liquid pressure of elementary physics. Finally, for any fluid flow design, one should always maximize the role of gravity for cost-effectiveness, durability, and dependability. Perhaps, Newton's law for fluid flow should read "Sum of everything else equals ρg" ; and this would imply that the other terms of the N-S equation divide the gravity term in a definite ratio, and each term utilizes gravity to function.

Uniqueness of the solution of the Navier-Stokes equation on

When each term of the linearized Navier-Stokes equation (see viXra:1512.0334) was made subject of the N-S equation, only the equation with the gravity term as the subject of the equation produced a solution. Similarly, the solution of the Navier-Stokes equation solution is unique.

About the solutions of the N-S Equations

1. In the CMI requirements paper, it is suggested that one can assume that gravity is zero. From the author's solutions of the N-S papers (See viXra:1512.0334), gravity cannot be zero, otherwise, there would be no fluid flow.. If one assumes that gravity, g, is zero, then one should also assume that $\frac{\partial p}{\partial x} = 0$ in the N-S equation, since in elementary physics $P = \rho gh$. If one is designing oil or water pipelines or water channels, one cannot assume zero gravity, since there would be no fluid flow without gravity.

2. A number of papers sometimes mention periodic solutions of N-S. equations. There are no periodic solutions, but periodic relations, because the integration results do not satisfy the N-S equations completely. Note that on integrating the N-S equations, and obtaining sines and cosines in the integration results, one should not mention "periodic" solutions until one has successfully checked the results for identity in the original equation. In any case, N-S equations have **no periodic** solutions, (perhaps quasi-periodic solutions) . See viXra:1512.0334

After comparison with the equations of motion and liquid pressure of elementary physics, the author believes that tthe Navier-Stokes equations have finally been solved analytically.

Spin-off: CMI Millennium Prize Problem Requirements Proof

For the Navier-Stokes equations (Original Equations)

Proof of the existence of solutions of the Navier-Stokes equations

From page 250, if $y = 0$, $z = 0$ in

Solution to Linear part

$$V_x(x,y,z,t) = \overbrace{-\frac{\rho g_x}{2\mu}(ax^2 + by^2 + cz^2) + C_1 x + C_3 y + C_5 z +}^{\text{Solution to Linear part}} \underbrace{fg_x t}_{\text{continued}} \underbrace{\pm\sqrt{2hg_x x} + \frac{ng_x y}{V_y} + \frac{qg_x z}{V_z} + \frac{\psi_y(V_y)}{V_y} + \frac{\psi_z(V_z)}{V_z}}_{\text{solution of Euler equation}}$$

solution of Euler equation

$P(x) = d\rho g_x x$

one obtains

$$\boxed{V_x(x,t) = -\frac{\rho g_x}{2\mu} ax^2 + C_1 x + fg_x t \pm \sqrt{2hg_x x} + C_9; \quad P(x) = d\rho g_x x;}$$

$$V_x(x,0) = V_x^0(x) = -\frac{\rho g_x}{2\mu} ax^2 + C_1 x \pm \sqrt{2hg_x x} + C_9; \quad P(x) = d\rho g_x x;$$

$V_x(x,t) = -\frac{\rho g_x}{2\mu} ax^2 + C_1 x + fg_x t \pm \sqrt{2hg_x x} + C_9$; $P(x) = d\rho g_x x$; are solutions of

$-\mu \frac{\partial^2 V_x}{\partial x^2} + \frac{\partial p}{\partial x} + \rho \frac{\partial V_x}{\partial t} + \rho V_x \frac{\partial V_x}{\partial x} = \rho g_x$ (deleting the $y-$ and $z-$ terms of (A)), p.2488,

Therefore, smooth solutions to the above differential equation exist, and the proof is complete.

Finding $P(x,t)$:

1. $V_x(x,t) = -\frac{\rho g_x}{2\mu} ax^2 + C_1 x + fg_x t \pm \sqrt{2hg_x x} + C_9$; $P(x) = d\rho g_x x$; **2.** $\frac{\partial p}{\partial x} = d\rho g$;

$$\frac{dp}{dt} = \frac{dp}{dx}\frac{dx}{dt}$$

$$\frac{dp}{dt} = \frac{dp}{dx} V_x \qquad \left(\frac{dx}{dt} = V_x\right)$$

$$\frac{dp}{dt} = d\rho g_x\left(-\frac{\rho g_x}{2\mu}(ax^2) + C_1 x \pm \sqrt{2hg_x x} + fg_x t + C_9\right) \qquad \left(\frac{dp}{dx} = d\rho g_x\right)$$

$$P(x,t) = \int d\rho g_x\left(-\frac{\rho g_x}{2\mu}(ax^2) + C_1 x \pm \sqrt{2hg_x x} + fg_x t + C_9\right) dt$$

$$P(x,t) = d\rho g_x\left(-\frac{a\rho g_x}{2\mu} x^2 t + C_1 xt \pm t\sqrt{2hg_x x} + \frac{fg_x t^2}{2} + C_9 t\right) + C_{10}$$

Appendix 9

Beal Conjecture Proved & Specialized to Prove Fermat's Last Theorem

"5% of the people think; 10% of the people think that they think; and the other 85% would rather die than think."----Thomas Edison

"The simplest solution is usually the best solution"---Albert Einstein

Abstract

Beal conjecture has been proved on a single page; and the proof has been specialized to prove Fermat's last theorem, on half of a page. The approach used in the proof is exemplified by the following system. If a system functions properly and one wants to determine if the same system will function properly with changes in the system, one would first determine the necessary conditions which allow the system to function properly, and then guided by the necessary conditions, one will determine if the changes will allow the system to function properly. So also, if one wants to prove that there are no solutions for the equation $c^z = a^x + b^y$ when $x,y,z > 2$, one should first determine why there are solutions when $x, y, z = 2$, and note the necessary condition in the solutions for $x, y, x = 2$. The necessary condition in the solutions for $x, y, x = 2$ will guide one to determine if there are solutions when $x,y,z > 2$. The proof in this paper is based on the identity $(a^2 + b^2)/c^2 = 1$ for a primitive Pythagorean triple (a, b, c). It is shown by contradiction that the uniqueness of the $x, y, x = 2$ identity excludes all other x, y, z–values, $x,y,z > 2$ from satisfying the equation $c^z = a^x + b^y$. One will first show that if $x, y, z = 2$, $c^z = a^x + b^y$ holds, noting the necessary condition in the solution; followed by showing that if $x,y,z > 2$ (x,y,z integers), $c^z = a^x + b^y$ has no solutions. Two proof versions are covered. The first version begins with only the terms in the given equation, but the second version begins with the introduction of ratio terms which are ubsequently and "miraculously" eliminated to allow the introduction of a much needed term for the necessary condition for $c^z = a^x + b^y$ to have solutions or to be true. Each proof is very simple, and even high school students can learn it. The approach used in the proof has applications in science, engineering, medicine, research, business, and any properly working system when desired changes are to be made in the system.

Beal Conjecture Proved (Version 1 Proof)

Introduction

This paper proves the equivalent Beal conjecture that the equation $c^z = a^x + b^y$ has no solutions in positive integers a,b,c,x,y,z, where a,b,c, are relatively prime and $x,y,z > 2$. Two simple proof versions are covered. The first version begins with only the terms in the given equation, but the second version begins with the introduction of ratio terms which are subsequently eliminated to allow the introduction of a much needed term for the necessary condition for $c^z = a^x + b^y$ to have solutions or to be true. For an application, the proof of Beal conjecture is specialized to prove Fermat's last theorem. A step-by-step procedure is used in the proof to facilitate easy reading.

Beal Conjecture Proved: Version 1

Given: $c^z = a^x + b^y$ (x,y,z integers; $a, b,$and c are relatively prime positive integers)

Required: To prove that $c^z = a^x + b^y$ does not have solutions if $x,y,z > 2$

Plan: One will first show that if $x,\ y,\ x = 2$, $c^z = a^x + b^y$ has solutions, followed by showing that if $x,y,z > 2$ (x,y,z, integers) , $c^z = a^x + b^y$ has no solutions.

Proof:

Step 1: $c^z = a^x + b^y$;

$$\frac{a^x + b^y}{c^z} = \frac{c^z}{c^z};$$

$$\boxed{\frac{a^x + b^y}{c^z} = 1} \quad \text{(A)}$$

(A) is the necessary condition for $c^z = a^x + b^y$ to be true. or to have solutions.

(The ratio $(a^x + b^y)$ to $c^z = 1$)

Step 2: If $x,\ y,\ x = 2$, $\frac{a^x + b^y}{c^z} = \frac{a^2 + b^2}{c^2} = 1$ is true for a primitive Pythagorean triple $(a,\ b, c)$. (Example: For the integers 3, 4, 5, $\frac{a^2 + b^2}{c^2} = 1 \quad (3^2 + 4^2)/5^2 = 25/25 = 1)$ Thus, if $x,\ y,\ x = 2$, the necessary condition $(a^n + b^n)/c^n = 1$ is satisfied and $c^z = a^x + b^y$ is true or has solutions..

Step 3: Proof for $x,y,z > 2$ **by contradiction**

If $x,y,z > 2$, and one assumes that $\frac{a^x + b^y}{c^z} = 1$, then $\frac{a^x + b^y}{c^z} = \frac{a^2 + b^2}{c^2}$ (B) (By the transitive equality property, since $\frac{a^2 + b^2}{c^2} = 1$.)

Step 4: From (B) and equating the exponents, $x > 2 = 2$ is false, since an integer greater than 2 cannot be equal to 2; similarly, $y > 2 = 2$ is false; and $z > 2 = 2$ is false. The above falsities imply contradiction, and $\frac{a^x + b^y}{c^z} \neq \frac{a^2 + b^2}{c^2}$ not as in (B); and hence, the assumption that $\frac{a^x + b^y}{c^z} = 1$, if $x,y,z > 2$, is false.

Step 5: Therefore, $\frac{a^x + b^y}{c^z}$ is not equal to 1. if $x,y,z > 2$. Since the necessary condition $\frac{a^x + b^y}{c^z} = 1$, is not satisfied if $x,y,z > 2$, the equation, $c^z = a^x + b^y$ has no solutions if $x,y,z > 2$. Therefore $c^z = a^x + b^y$ has solutions only if $x,\ y,\ x = 2$, and does not have solutions if $x,y,z > 2$. The proof is complete.

Beal Conjecture Proved: Version 2 (Using ratios)

Confirmation of Version 1 Proof

Given: $c^z = a^x + b^y$ (x,y,z integers; $a, b,$and c are relatively prime positive integers)

Required: To prove that $c^z = a^x + b^y$ does not have solutions if $x,y,z > 2$

Plan: One will first show that if $x,\ y,\ x = 2$, $c^z = a^x + b^y$ has solutions, followed by showing that if $x,y,z > 2$ (x,y,z, integers), $c^z = a^x + b^y$ has no solutions. One begins by applying ratio terms.

$c^z = a^x + b^y$ (1) (Given)

$a^x + b^y = c^z$ (2) (rewriting)

$a^x = rc^z$ (3) (r is a ratio term)

$b^y = sc^z$ (4) (s is a ratio term) ($r + s = 1$)

$rc^z + sc^z = c^z$ (5) (substitute for a^x and b^y from (3) and (4))

$c^z(r + s) = c^z$ (6)

Now, by the substitution axiom, since $r + s = 1$, $r + s$ can be replaced by any quantity = 1. One can therefore replace $r + s$ by $\frac{a^2 + b^2}{c^2}$,

since $\frac{a^2 + b^2}{c^2} = 1$ for a primitive Pythagorean triple $(a,\ b,\ c)$.

Then equation (6) becomes $c^z(\frac{a^2 + b^2}{c^2}) = c^z$ (7)

If $z = 2$, (7) becomes $c^2(\frac{a^2 + b^2}{c^2}) = c^2$ (8)

$c^2 = c^2(\frac{a^2 + b^2}{c^2})$ (8) (rewriting)

Equation (8) is true since $\frac{a^2 + b^2}{c^2} = 1$. Consequently, equations (8) and (1) hold. Therefore, if $x,\ y,\ x = 2$, $c^z = a^x + b^y$ has solutions.

> **Example on ratio terms**
> If $4 + 8 = 12$, and the ratio terms are $\frac{1}{3}$ and $\frac{2}{3}$, then $4 = \frac{1}{3} \bullet 12$, $8 = \frac{2}{3} \bullet 12$; and the sum of the ratio terms is $\frac{1}{3} + \frac{2}{3} = 1$
>
> **Elimination of the ratio terms r and s**
> The author was impressed and gratified by the substitution axiom which permitted the introduction of the much needed necessary condition $(a^x + b^y)/c^z = 1$.

Generalizing equation (7), one obtains $c^z(\frac{a^x + b^y}{c^z}) = c^z$ (9) in which the necessary condition for (9) to hold is $\frac{a^x + b^y}{c^z} = 1$. One will next show that if $x,y,z > 2$, the condition $(a^x + b^y)/c^z = 1$ is never satisfied and consequently $c^z = a^x + b^y$ has no solutions.

Proof for $x,y,z > 2$ by contradiction

If $x,y,z > 2$, and one assumes that $\frac{a^x + b^y}{c^z} = 1$, then $\frac{a^x + b^y}{c^z} = \frac{a^2 + b^2}{c^2}$ (B)

(By the transitive equality property, since $\frac{a^2 + b^2}{c^2} = 1$, for a primitive Pythagorean triple $(a,\ b,\ c)$. From (B), $x > 2 = 2$ is false, since an integer greater than 2 cannot be equal 2.. Similarly, $y > 2 = 2$ is false; and $z > 2 = 2$ is false. The above falsities imply contradiction.

Hence, the assumption that $\frac{a^x + b^y}{c^z} = 1$, when $x,y,z > 2$, is false. Therefore, $\frac{a^x + b^y}{c^z}$ is not equal to 1. ($(a^x + b^y)/c^z \neq 1$) if $x,y,z > 2$. Since the necessary condition, $\frac{a^x + b^y}{c^z} = 1$, is not satisfied if $x,y,z > 2$, the equation $c^z = a^x + b^y$ has no solutions if $x,y,z > 2$. Therefore, $c^z = a^x + b^y$ has solutions only if $x,\ y,\ x = 2$ and does not have solutions if $x,y,z > 2$.
The proof is complete.

Conclusion for Beal Conjecture

Beal conjecture has been proved on a single page. One first determined why there are solutions when $x, y, x = 2$. The necessary condition in the solutions for $x, y, x = 2$ guided one to determine if there are solutions when $x,y,z > 2$. The necessary condition is $(a^x + b^y)/c^z = 1$, where a, b, and c are relatively prime positive integers. This necessary condition is satisfied only if $x, y, z = 2$, to produce $(a^2 + b^2)/c^2 = 1$, where (a, b, c) is a primitive Pythagorean triple. If $x,y,z > 2$, the necessary $(a^x + b^y)/c^z = 1$ is never satisfied. It was shown by contradiction that the uniqueness of the $x, y, z = 2$ identity excludes all other x, y, z–values-values, $x,y,z > 2$, from satisfying the equation $c^z = a^x + b^y$. The proof is very simple, and even high school students can learn it.
The proof is very simple, and even high school students can learn it.

One will next specialize the above proof to prove Fermat's last theorem. However, note that one can also generalize the proof of Fermat's Last Theorem to obtain the proof of Beal Conjecture.

Fermat's Last Theorem Proved

To obtain the proof of Fermat's Last Theorem from the proof of Beal Conjecture, let $x = n$, $y = n$. $z = n$ in Beal Conjecture proof and delete redundant repetitions.

Thus, $x, y, z > 2$ becomes $n > 2$; $x, y, z = 2$ becomes $n = 2$; $a^x + b^y$ becomes $a^n + b^n$; c^z becomes c^n., Then one obtains the following proof

Given: $c^n = a^n + b^n$ (n an integer; a, b, and c are relatively prime positive integers)

Required: To prove that $c^n = a^n + b^n$ does not hold if $n > 2$

Plan: One will first show that if $n = 2$, $c^n = a^n + b^n$ holds, followed by showing that if $n > 2$ (n an integer), $c^n = a^n + b^n$ does not hold.

Proof

Step 1: $c^n = a^n + b^n$;

$$\frac{a^n + b^n}{c^n} = \frac{c^n}{c^n};$$

$$\boxed{\frac{a^n + b^n}{c^n} = 1} \quad \text{(A)}$$

(A) is the necessary condition for $c^n = a^n + b^n$ to be true. or to have solutions.
(The ratio $(a^n + b^n)$ to $c^n = 1$)

Step 2: If $n = 2$, $\frac{a^n + b^n}{c^n} = \frac{a^2 + b^2}{c^2} = 1$ is true for a Pythagorean triple a, b, c, (Example: For the integers 3, 4, 5, $\frac{a^2 + b^2}{c^2} = 1$ $(3^2 + 4^2)/5^2 = 25/25 = 1)$

Thus, if $n = 2$,the necessary condition $(a^n + b^n)/c^n = 1$ is satisfied and $c^n = a^n + b^n$ is true or has solutions.

Step 3: One will next show that if $n > 2$, the necessary condition, $\frac{a^n + b^n}{c^n} = 1$, is never satisfied.

Step 4: Proof for $n > 2$ by contradiction

If $n > 2$, and one assumes that $\frac{a^n + b^n}{c^n} = 1$,

then $\frac{a^n + b^n}{c^n} = \frac{a^2 + b^2}{c^2}$ (B)

(By the transitive equality property, since $\frac{a^2 + b^2}{c^2} = 1$). From (B), and equating the exponents, $n > 2 = 2$ is false, since an integer greater than 2 cannot be equal to 2. Hence, the assumption that $\frac{a^n + b^n}{c^n} = 1$, if $n > 2$, is false.

Step 5: Therefore, $\frac{a^n + b^n}{c^n}$ is not equal to 1. $\left(\frac{a^n + b^n}{c^n} \neq 1\right)$ if $n > 2$. Since the necessary condition $\frac{a^n + b^n}{c^n} = 1$, is not satisfied if $n > 2$, the equation $c^n = a^n + b^n$ has no solutions if $n > 2$.
Therefore $c^n = a^n + b^n$ has solutions only if $n = 2$ and does not have solutions if $n > 2$.
The proof is complete.

Note: Fermat's Last Theorem can also be proved using the identity $\boxed{\sin^2 x + \cos^2 x = 1}$ or its equivalents, instead of $\frac{a^2 + b^2}{c^2} = 1$ (see viXra:1605.0195

Overall Conclusion

Two versions of the proof of Beal Conjecture have been presented in this paper. Thefirst version began with only the terms of the given equation; but the second version began with the introduction of ratio terms which were later on "miraculously" eliminated to permit the introduction of a much needed necessary condition term for the equations to have solutions. The Beal proof was specialized to prove Fermat's last theorem. The necessary condition for the relevant equations involved to be true is that $(a^x + b^y)/c^z = 1$ (for Beal proof); and $(a^n + b^n)/c^n = 1$ for Fermat's proof. It was determined that the Beal equation, $c^z = a^x + b^y$ is true only if $x, y, x = 2$; and the Fermat's equation $c^n = a^n + b^n$ is true only if $n = 2$. Therefore, $c^z = a^x + b^y$ has solutions only if $x, y, z = 2$, and does not have solutions if $x,y,z > 2$. Similarly, the Fermat equation $c^n = a^n + b^n$ has solutions only if $n = 2$, and does not have solutions if $n > 2$.
One should note above that Version 2 of Beal proof confirmed Version 1 of Beal proof.

PS

Application of the approach used in proving Beal Conjecture

If a 3-ton non-portable machine functions properly in environment number 2 and one wants to determine if the same machine will function properly in environments 3, 4, and, 5 up to 1000 different environments, one option is to dismantle the machine in environment number 2, and reassemble it in each of the new environments, up 1000 environments and test the machine. Another option, the better option, is to determine the necessary conditions which allow the machine to function properly in environment 2. If the necessary conditions are not available in environments 3, 4, 5, etc, the machine will not function properly in the new environments, and no efforts should be wasted in carrying the machine to the environments and be tested.

Adonten

INDEX-Power of Ratios

D

E

F

Some useful conversion factors

Units of length

1 cm = .3937 in = .0328ft = .01094 yd
1 m = 100 cm = 39.3701 in = 3.2808 ft.= 1.0936 yd
1 in. = 2.54 cm = .0833 ft = .0254 m
1 ft = 12 in. = 30.48 cm = .3048 m = .3333 yd
1 mile = 1760 yd = 5280 ft = 1.6093 km
1 yd = 3 ft = 36 in.= .9144 m = 91.44 cm
1 km = .62137 mile = 1000 m = 100,000 cm

Units of mass

1 kg = 1000 g = 2.2046 lb =35.274 oz
1 lb (avdp) = 453.592 g = 16 oz
1 metric ton = 1000 kg = 10^6 g = 1.1023 ton
1 ton = 907.1847 kg = 2000 lb = .9072 metric ton
1 gm = .03527 oz = 1000 mg =.0022046 lb
1 oz = 28.3495 g = .0625 lb = 16 drams
1 long ton (British) = 2240 lb

Units of volume

1 liter = 1000 cm^3 = 61.0237 $in.^3$= .26417 gal = 1.0567 qt =.03531 ft^3= 2.113 pt
1 gal (U.S.) = 4 qt = 3.7854 liter = 8 pt = 231 $in.^3$ = .13368 ft^3
1 qt = 2 pt = .946353 liter =946.353 cm^3 = 57.75 $in.^3$ = .25 gal =.034201 ft^3
1 cord = 128 ft^3
1 pt = .473 liter = .5 qt
1 ft^3 = 1728 $in.^3$
1 yd^3 = 27 ft^3 = 46656 $in.^3$

Units of area

1 ft^2 (sq. ft) = 144 $in.^2$ (sq. in.)
1 yd^2 (sq. yd) = 9 ft^2 (sq. ft) = 1296 sq. in.
1 $mile^2$ (sq. mile) = 640 acres
1 m^2 = $10^4 cm^2$
1 acre = 4840 yd^2

Symbols for units

cm = centimeter
m = meter
in. = inch
ft = foot
yd = yard
km = kilometer
mi = mile

gal = gallon
qt = quart
pt = pint
oz = ounce

g = gram
kg = kilogram
lb = pound
mg = milligram

Some prefixes (International System)

Prefix	**Power**
tera	10^{12}
giga	10^9
mega	10^6
kilo	10^3
hecto	10^2
deka	10^1
deci	10^{-1}
centi	10^{-2}
milli	10^{-3}
micro	10^{-6}
nano	10^{-9}
pico	10^{-12}

Conversion Factors for Measurements

American System (British System) **Inter**conversion (Factors) Metric System

Some " **bridge**s" for converting from one system to the other

	American System	Interconversion	Metric System
Length	12 inches (in) = 1 foot (ft.) 3 feet (ft.) = 1 yard (yd) 5280 feet = 1 mile (mi) 1760 yards = 1 mile	1 in. = 2.54 cm 1 yd = 0.9144 m 1 mi = 1.61 km 1 km = 0.62 mi	1 kilometer (km) = 10^3 m = 1000 m 1 hectometer (hm) = 10^2 m = 100 m 1 dekameter (dam) = 10^1 m = 10 m **1 meter (m)** = 10^0 m = 1m 1 decimeter (dm) = 10^{-1} m = 0.1m 1 centimeter (cm) = 10^{-2} m = 0.01m 1 millimeter (mm) = 10^{-3} m = 0.001m

Some " **bridges**" for converting from one system to the other

	American System	Interconversion	Metric System
Mass	1 *l*b = 16 oz 1 ton = 2000 *l*b 1 long ton = 2240 *l*b	1 kg = 2.2 *l*b 1 *l*b = 454 g 1 oz = 28.4 g = 16 drams 1 ton = 0.9072 metric ton	1 kilogram (kg) = 10^3 g = 1000 g 1 hectogram (hg) = 10^2 g = 100 g 1 dekagram (dag) = 10^1 g = 10 g **1 gram (g)** = 10^0 g = 1 g 1 decigram (dg) = 10^{-1} g = 0.1g 1 centigram (cg)) = 10^{-2} g = 0.01g 1 milligram (mg) = 10^{-3} g = 0.001g

Some " **bridge**s" for converting from one system to the other

	American System	Interconversion	Metric System
Volume	16 fluid oz (fl-oz)= 1 pint (pt) 2 pints (pt) = 1 quart (qt) 4 quarts = 1 gallon (gal)	1 liter (*l*) = 1.057 qt 1 gal = 3.785 *l* 1 liter = 2.1 pt 1 pt = .473 *l*	1 kiloliter (k*l*) = 10^3 *l* = 1000 *l* 1 hectoliter (h*l*) = 10^2 *l* = 100 *l* 1 dekaliter (da*l*) = 10^1 *l* = 10 *l* **1 liter (*l*)** = 10^0 *l* = 1 *l* 1 deciliter (d*l*) = 10^{-1} *l* = 0.1*l* 1 centiliter (c*l*) = 10^{-2} *l* = 0.01*l* 1 milliliter (m*l*) = 10^{-3} *l* = 0.001 *l*

Must remember the following (metric system:)

- 100 cm = 1 m
- 1000 m = 1 km

- 1000 mg = 1 g
- 1000 g = 1 kg

- 1000 m*l* = 1 *l*
- 1 m*l* = 1 cc = 1 cm^3
- 1000 cc = 1 *l*

Mnemonic device (metric system)

- k – ilo – 10^3
- h – ecto – 10^2
- d – eka – 10^1
- d – eci – 10^{-1}
- c – enti – 10^{-2}
- m – illi – 10^{-3}

Say the following aloud:

Step 1: First go down vertically as kei-eitch-dii-dii-see-em, then Step 2

Step 2: Kilo-hecto-deka-deci-centi-milli, and then note how the powers decrease vertically downwards.

Examples: 1 Kilometer = 10^3 meter; 1 milligram = 10^{-3} gram;

1 centimeter = 10^{-2} meter = $\frac{1}{100}$ meter --->100 centimeters = 1 meter.

Mathematical Modeling
Some Reciprocal Relationships

1. Arithmetic If A working alone can do a piece of work in time t_A; B working alone can do the same work in time t_B; C working alone can do the same work in time t_C, and if A, B, and C working together, can do the same work in time t_{ABC}, then

$$\boxed{\frac{1}{t_{ABC}} = \frac{1}{t_A} + \frac{1}{t_B} + \frac{1}{t_C}}$$

That is, the reciprocal of the working-together time equals the sum of the reciprocals of working-alone times (individual times).

2. Geometry: For any triangle, the reciprocal of the inradius (R) equals the sum of the reciprocals of the exradii (r_1, r_2, and r_3).

Thus $$\boxed{\frac{1}{R} = \frac{1}{r_1} + \frac{1}{r_2} + \frac{1}{r_3}}$$

3. Physics (Electricity) For electrical resistances in parallel (in an electric circuit), the reciprocal of the combined resistance, R, equals the sum of the reciprocals of the separate resistances, r_1, r_2, and r_3.

Thus $$\boxed{\frac{1}{R} = \frac{1}{r_1} + \frac{1}{r_2} + \frac{1}{r_3}}$$

4. Physics (Optics)

For two thin lenses in contact, the reciprocal of the combined focal length, F, equals the sum of the reciprocals of the separate focal lengths, f_1 and f_2, .

Thus $$\boxed{\frac{1}{F} = \frac{1}{f_1} + \frac{1}{f_2}}$$

5. Physics (Optics) For spherical mirrors and thin lenses, the reciprocal of the focal length F equals the sum of the reciprocals of the object distance, d_o and the image distance d_i.

Thus $$\boxed{\frac{1}{F} = \frac{1}{d_o} + \frac{1}{d_i}}$$

6. Physics (Mechanics). If two bubbles of radii r_1, r_2, coalesce into a double bubble, the radius, R, of the partition is given by

$$\boxed{\frac{1}{R} = \frac{1}{r_1} - \frac{1}{r_2}}$$

www.ingramcontent.com/pod-product-compliance
Lightning Source LLC
Chambersburg PA
CBHW080517220326
41599CB00032B/6117
* 9 7 8 1 9 4 6 4 8 5 1 2 0 *